AF553932

BIOLOGY OF CYCLOSTOMES

BIOLOGY
OF
CYCLOSTOMES

By

Dr. D.R. Khanna
Reader in Zoology
Gurukul Kangri University
Haridwar (Uttaranchal)
&
Dr. P.R. Yadav
Lecturer
Department of Zoology
D.A.V. College
Muzaffarnagar (U.P.)

DISCOVERY PUBLISHING HOUSE
NEW DELHI-110002

First Published-2005

ISBN 81-7141-949-6

Published by
DISCOVERY PUBLISHING HOUSE
4831/24, Ansari Road, Prahlad Street,
Darya Ganj, New Delhi-110002 (India)
Phone: 23279245 • Fax: 91-11-23253475
E-mail:dphtemp@indiatimes.com

Printed at
Arora Offset Press
Laxmi Nagar, Delhi–92

Preface

Cyclostomes are the unique jawless vertebrates holding a significant position in the history of vertebrate evolution. They are the sole survivors of once flourishing group of the Paleozoic Agnathans. This title provides an integrated account of lampreys and hagfishes including their paleontology, physiology, biochemistry, endocrinology, ecology and behaviour. This comprehensive assessment will be a valuable reference for all those working on the vertebrate evolutionary and comparative studies in the related fields of biology. The text covers all aspects of the natural history of these jawless creatures.

In the preparation of this book large number of books and research papers have been consulted. So no authenticity is claimed.

Text book can not be written without the support and professional contributions of many people. This book is no exception. The authors are grateful to those teachers and colleagues whose stimulating discussions have clarified certain vague points for him, but all errors, omissions, and corrections needed are solely their responsibility.

The authors tried hard to be accurate and upto date in statement and realises the impossibility of completely avoiding errors therefore, the authors will greatly appreciate having his attention called to any questionable statement.

The authors expresses their gratitute to Mr. Wasan and staff of M/s Discovery Publishing House for their whole hearted co-operation in the publication of this book.

Authors

Contents

1

Introduction

Vertebrate Story

Mention "animal" and most people will think of a vertebrate. Vertebrates are often abundant and conspicuous parts of people's experience of the natural world. Vertebrates are also very diverse: The approximately 50,000 extant (=currently living) species of vertebrates range in size from fishes that weight as little as 0.1 gram when they are fully mature to whales that weigh nearly 100,000 kilograms. Vertebrates live in virtually all the habitats on Earth: Bizarre fishes, some with mouths so large they can swallow prey larger than their own bodies, cruise through the depths of the sea, sometimes luring prey to them with glowing lights. Some 15 kilometers above he fishes, migrating birds fly over the crest of the Himalayas, the highest mountains on Earth.

The behaviour of vertebrates are as diverse and complex as their body forms. Vertebrate life is energetically expensive, and vertebrates get the energy they need from food they eat. Carnivores eat the flesh of other animals and show a wide range of methods of capturing prey: Some predators search the environment to find prey, whereas others wait in one place for prey to come to them. Some carnivores pursue their prey at high speeds, others pull prey into their mouths by suction. In some cases the foraging behaviours that vertebrates use appear to be exactly the ones that maximize the amount of energy they obtain for the time they spend hunting; in other cases vertebrates can appear to be remarkably inept predators. Many vertebrates swallow their prey intact, sometimes while it is alive and struggling, but other vertebrates have very specific methods of dispatching prey: Venomous snakes inject

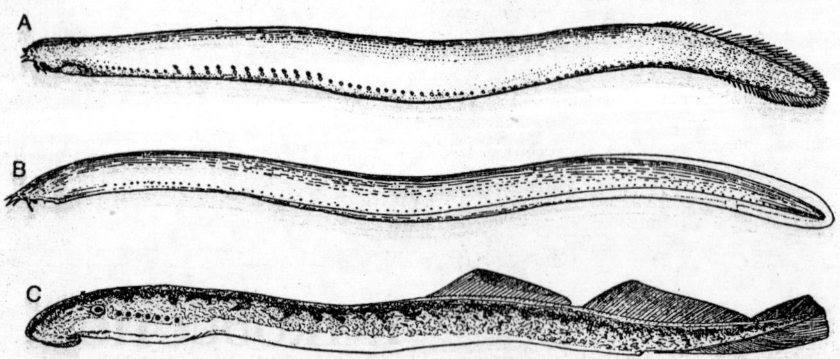

Fig. 1.1. Three types of cyclostomes. A—The slime hag, Bdellostoma; B—the hagfish, Myxine; C—the lamprey, Petromyzon.

complex mixtures of toxins, and cats (of all sizes from house cats to tigers) kill their prey with a distinctive bite on the neck. Herbivores eat plants. Plants cannot run away when an animal approaches, but they are hard to digest and they frequently contain toxic compounds. Herbivorous vertebrates show an array of specializations to deal with the difficulties of eating plants. These specializations include elaborately sculptured teeth and digestive tracts that provide sites in which symbiotic microorganisms digest compounds that are impervious to the digestive systems of vertebrates.

Reproduction is a critical factor in the evolutionary success of an organism, and vertebrates show an astonishing range of behaviours associated with mating and reproduction. In general, males court females and females care for the young, but these roles are reversed in many species of vertebrates. The forms of reproduction employed by vertebrates range from laying eggs to giving birth to babies that are largely or entirely independent of their parents (precocial young). These variations range across almost all kinds of vertebrates: Many fishes and amphibians produce precocial young and a few mammals lay eggs. At the time of birth or hatching some vertebrates are entirely self-sufficient and never see their parents, whereas other vertebrates (including humans) have extended periods of obligatory parental care. Extensive parental care is found in seemingly unlikely groups of vertebrates—fishes that incubate eggs in their mouths, frogs that incubate eggs in their stomachs, and birds that feed their nestlings a fluid called crop milk that is very similar in composition to mammalian milk.

The diversity of living vertebrates is fascinating, but the species now living are only a small proportion of the species of vertebrates that have existed. For each living species there may be as many as ten extinct species, and some of these have no counterparts among living forms. The dinosaurs, for example, that dominated the Earth for 180 million years are so entirely different from any living animals that it is hard to reconstruct the lives they led. Even mammals were once more diverse than they are now: The Pleistocene saw giants of many kinds—ground sloths as big as modern rhinoceroses, and raccoons and rodents as large as bears. Humans are great pes, close relatives of chimpanzees and gorillas, and much of the biology of humans is best understood in the context of our vertebrate heritage. The number of species of vertebrates probably reached its maximum in the Pliocene and Pleistocene, and has been declining since then. Some parts of that decline can probably be attributed to the effects of humans on species of humans on species they used for food or viewed as competitors, although changes in climate and vegetation have also been instrumental. In the modern world, however, the effects of human activities are paramount, and the fate of other species of vertebrates is very much affected, for good or ill, by human decisions. Our responsibilities to other vertebrates cannot be ignored.

The story of vertebrates is fascinating: Where they originated, how they evolved, what they do, and how they work provide endless intriguing details. In preparing to tell this story we must introduce some basic information: what the different kinds of vertebrates are called and how they are classified, how evolution works, and what the world within which the story of vertebrates unfolded was like.

Different Kinds of Vertebrates

The modern approach to biological classification, which is popularly known as cladistics, recognizes only groups of organisms that are related by common descent, or *phylogeny* (*phyla*=tribe, *genesis*=origin). The application of cladistic methods is making the study of evolution more rigorous than it has been heretofore. The natural groups recognized by cladistics are easier to understand than the artificial groups we are used to, except that we are familiar with the names of the artificial groups and the names of the new groups are sometimes strange. At this point we need to establish a basis for talking about particular animals by naming them, and to relate the old, familiar names to the new, less familiar ones.

Hagfishes and Lampreys (Myxinoidea and Petromyzontoidea)

Lampreys and hagfishes are elongate, scaleless, and slimy and have no internal hard tissues. They are scavengers and parasites and

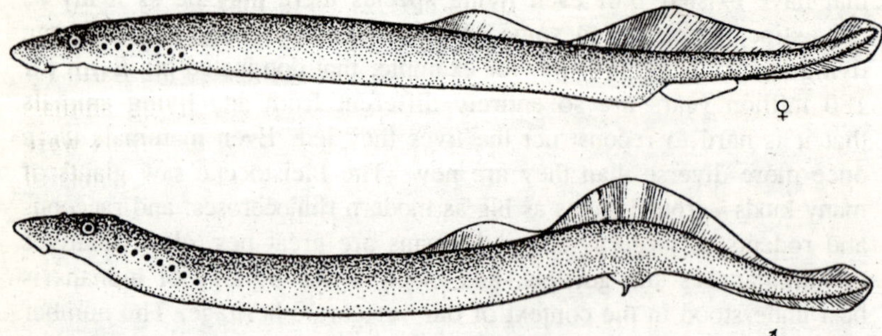

Fig. 1.2. Brook lampreys, Lampetra planeri. Ripe female with anal fin fold, and ripe male.

are specialized for those roles. Hagfishes (about 40 species) are marine and occur on the continental shelf and open ocean at depths around 100 meters, whereas many of the 41 species of lampreys are migratory forms that live in oceans and spawn in rivers.

Hagfishes and lampreys are unique among living vertebrates because they lack jaws, and they occupy an important position in the study of vertebrate evolution. Hagfishes and lampreys have traditionally been grouped as agnathans (*a*=without, *gnath*=jaw) or cyclostomes (*cyto*=round, *stoma*=mouth), but they probably represent two

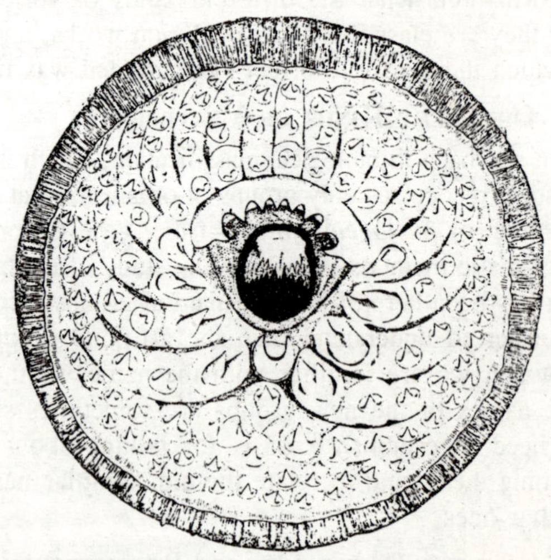

Fig. 1.3. The sucker of Petromyzon marinus showing the dentition in the feeding stage: the outer circular lip, the teeth, and at the centre the tongue, also with teeth.

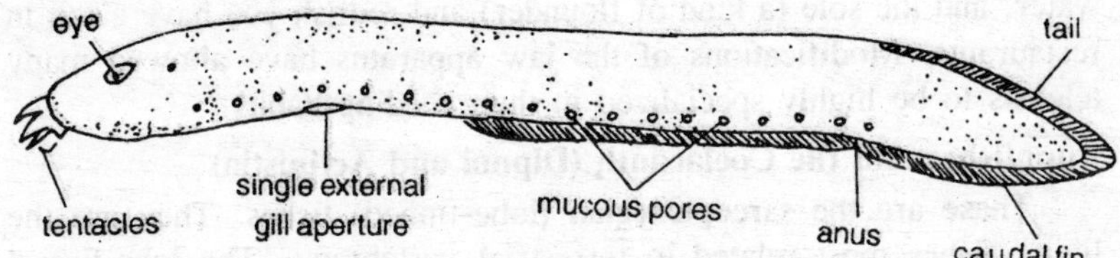

Fig. 1.4. Myxine.

independent evolutionary lineages. Hagfishes lack many of the features that characterize most vertebrates; for example, they have no trace of vertebrae. In contrast, lampreys share many derived characters with jawed vertebrates. The jawless condition of both lampreys and hagfishes, however, is ancestral.

Sharks, Rays, and Ratfishes (Elasmobranchii and Holocephali)

Sharks have a reputation for ferocity that most of the 350 to 400 species would have difficulty living up to. Many sharks are small (15 centimeters or less), and the largest species, the whale sharks, which grows to 10 meters, is a filter feeder that subsists on plankton it strains from the water. The 450 species of rays are dorsoventrally flattened, frequently bottom dwellers that swim with undulations of their extremely broad pectoral fins. Ratfish are bizarre marine fishes with long, slender tails and buck-toothed faces that look rather like rabbits. The name Chondrichthyes (*chondro*=cartilage, *ichthyes*=fish) refers to the cartilaginous skeletons of these fishes.

Bowfin, Gars, and Others (Actinopterygii)

Bony fishes are so diverse that any attempt to characterize them briefly is doomed to failure. Two broad categories can be recognized, the ray-finned fishes (actinopterygians; *actino*=ray, *ptero*=wing or fin) and the lobe-finned or fleshy-finned fishes (sarcopterygians; *sarco*=lobe).

The actinopterygian fishes we have included in this category are called polypterids, holosteans, and neopterygians in various systems of classification. Most have cylindrical bodies, thick scales, and jaws armed with sharp teeth. These fishes seize prey in their mouths with a sudden rush or gulp; they lack the specializations of the jaw apparatus that allow derived bony fishes to use more complex feeding modes.

Teleost Fishes (Actinopterygii)

More than [illegible],000 species of fishes fall into this category and they cover every imaginable range of body sizes, habitats, and habits, from seahorses to the giant ocean sunfish. Most of the familiar fishes are in this category—the bass and panfish you have fished for in fresh water, and the sole (a kind of flounder) and redfish you have eaten in restaurants. Modifications of the jaw apparatus have allowed many teleosts to be highly specialized in their feeding habits.

Lungfishes and the Coelacanth (Dipnoi and Actinistia)

These are the sarcopterygian (lobe-finned) fishes. They are the living fishes most related to terrestrial vertebrates. The lobe-finned fishes are heavy-bodied and slow-moving. The four species of lungfishes live in fresh water, and the coelacanth is marine.

Salamanders, Frogs, and Caecilians (Urodela, Anura, and Gymnophiona)

These three groups of vertebrates are popularly known as amphibians (*amphi*=double, *bios*=life) in recognition of their complex life histories, which often include an aquatic larval form (the larva of a salamander or caecilian and the tadpole of a frog) and a terrestrial adult. All amphibians have bare skins (that is, lacking scales, hair, or feathers) that are important in the exchange of water, ions, and gases with their environment. Salamanders are elongate animals, mostly terrestrial and usually with four legs; anurans (frogs, toads, treefrogs) are short bodied with large heads and large hind legs used for walking, jumping, and climbing; and caecilians are legless aquatic or burrowing animals.

Turtles (Chelonia)

Turtles are probably the most immediately recognizable of all vertebrates. The shell that encloses a turtle has no exact duplicate among other vertebrates, and morphological modifications associated with the shell make turtles extremely peculiar animals. They are, for example, the only vertebrates with the shoulders (pectoral girdle) and hips (pelvic girdle) inside the ribs.

Tuatara, Lizards, and Snakes (Lepidosauria)

These three kinds of vertebrates can be recognized by their scale-covered skin as well as by characteristics of the skull. The two species of tuatara, stocky-bodies animals found only on some islands near New Zealand, and the sole living remnant of a lineage of animals called Sphenodontida that were more diverse in the Mesozoic. In contrast, lizards and especially snakes are now at the peak of their diversity.

Alligators and Crocodiles (Crocodilia)

These impressive animals (the saltwater crocodile has the potential to grow to a length of 7 meters) are in the same lineage (the Archosauria) as dinosaurs and birds. Crocodilians, as they are known collectively, are semiaquatic predators with long snouts armed with numerous teeth. Their skin contains many bones (osteoderms; *osteo*=bone, *derm*=skin) that lie beneath the scales and provide a kind of armor plating. Crocodilians are noted for the parental care they provide for their eggs and young.

Birds (Aves)

The birds are a lineage of dinosaurs that evolved flight in the Mesozoic. Feathers are the distinguishing characteristic of birds. Indeed, some fossils of *Archaeopteryx*, the earliest bird known, were originally classified as dinosaurs because no marks of the feathers were visible. Birds are conspicuous, often vocal, and active during the day (diurnal). As a result they have been studied extensively and much of our information about the behaviour and ecology of terrestrial vertebrates is based on studies of birds.

Mammals (Mammalia)

The living mammals can be traced to an origin in the late Paleozoic from some of the earliest fully terrestrial vertebrates. Extant mammals include about 4500 species, most of which are placental (eutherian) mammals. Their name comes from the placenta, a structure that transfers nutrients from the mother to the embryo and removes the waste products of the embryo's metabolism. Most of the familiar animals of the world are placentals. Marsupials (mammals in which the young is born at a very small size and continues its development in an external pouch on the mother's abdomen) dominate the mammalian fauna only in Australia. Kangaroos, koalas, and wombats are familiar Australian marsupials. The strange monotremes, the duck-billed platypus and the echidnas, are mammals whose young are hatched from eggs.

The Cyclostomes, comprising the lampreys and the hagfishes or myxinoids, are a group of eel-like aquatic vertebrates, not widely known outside zoological circles, for whom they have a special interest as the sole survivors of an extinct, but once flourishing, group of jawless or agnathan Palaeozoic 'fishes' known as the Ostracoderms.

The entirely marine hagfish, once classified as an 'intestinal worm' is a blind scavenger, living on the sea bed in comparatively deep water and attracting the attention of commercial fishermen only because they have a habit of devouring their bait. Their colloquial name 'hag' is an expression of the repulsion that they tend to arouse; an aversion that owes much to their 'soft, naked body, blotched by the sub-cutaneous blood sinuses, the lack of eyes and a characteristic clinging odour'. Since they are said to be rubbery in texture, tasting strongly of fish oils (Jensen, 1966) it is hardly surprising that they have little or no economic value and, except in one locality in Japan, have not been exploited as human food.

Lampreys, although primarily freshwater animals, may spend part of their adult life in the sea, before finally returning to the rivers to spawn and die. Here the egg develops into a blind, larval form – the ammocoete which burrows into the silt banks of the river, where it remains for several years, growing to a maximum length of about eight inches and feeding on microscopic organisms, filtered from the water. During this part of their life cycle, they are seen only rarely, but may be dredged up in large numbers from the mud in which they are hidden, or caught by the use of electrical fishing gear, which causes them to emerge from their burrows. The transformation of the fully grown ammocoete into an adult lamprey occurs during late summer or early autumn and, after metamorphosis, the further development of the young adult or macrophthalmia (so called because of its relatively large eyes) may follow either one of two divergent patterns. In many species – the so called brook lampreys – the adult never feeds and cannot therefore, be any larger than the ammocoete at the end of its larval life. These dwarf lampreys have an adult life span of only 6-9 months, after which they spawn and die. Other lampreys, once their metamorphosis has been completed, become parasitic on fish, either in freshwater or in the sea, attaching themselves to their prey by their toothed oral disc or sucker and feeding on blood or other tissues.

Where river systems remain relatively free from pollution, lampreys and their larvae may occur in very large numbers but, because of their unobtrusive habits, are likely to be detected only by an experienced observer. They are of no interest to the angler (except as bait) and, because of their jawless mouth and parasitic feeding habits, cannot be taken by rod and line. Adult lampreys are usually only seen when they are ascending the rivers on their spawning migration. Even then, they are mainly nocturnal and daytime activity is generally restricted to a very brief period in spring or early summer while spawning is in progress. However, although rarely observed and even then often confused with eels, lampreys have acquired a widespread and traditional reputation as a delicacy and smoked lamprey is particularly valued in the Baltic countries of Eastern Europe, where commercial lamprey fisheries are still of considerable economic importance. This esteem has some scientific backing, since at the start of their migration (when their flesh is most palatable), the high fat content of the muscle tissue rivals that of the eel, while their extraordinarily high levels of vitamin A accounts for their use in Japan as a remedy for night blindness.

In spite of far-reaching differences in their internal organization, lampreys and hagfishes are superficially similar and sinuous creatures, swimming gracefully by eel-like undulatory waves. The largest species of both groups reach lengths approaching one metre and body weights of about one kilogram. In keeping with their more active habits; the median fins are better developed in the lamprey, which usually has two distinct dorsal fins. Moreover, the propulsive tail region behind the cloacal opening is relatively much longer in the lamprey and although both have symmetrical caudal fins, that of the hagfish is more rounded and spatulate in shape. Behind the branchial region the hagfish may have a small ventral fin fold, but this lacks internal cartilaginous supports (fin rays) and rarely projects more than one or two millimetres from the surface. Whereas the hagfish body tends to taper towards the anterior end, the branchial and head region of the lamprey is deeper than the trunk and bears a constant number of seven circular gill ports, the first a short distance behind its prominent paired eyes. In this respect, hagfishes are highly variable. The Pacific hagfish, *Eptatretus stouti* has 12 separate gill openings, placed much further back than in a lamprey, but as many as 15 of these openings may be found in some species. On the other hand in the Atlantic hagfish, *Myxine glutinosa*, there are only five pairs of gills, whose external ducts open at a common pore. Peculiar to the hagfish is a pharyngo-cutaneous duct, usually opening in common with the last gill on the left side and connecting internally with the pharynx. Characteristic of the myxinoids are the large slime glands opening by rows of ducts along the ventro-lateral margins of the body. The specific name *glutinosa* given to the Atlantic hagfish enshrines the ability of these animals to throw off large volumes of slime when they are irritated. For example, when stimulated by handling or by dilute formaldehyde, *Myxine* is capable of producing about half a litre of slime, consisting of a tangled fibrous network, holding imbibed water in its meshes and converting it into a gelatinous mass.

The tapering head of the hagfish carries three pairs of sensory tentacles and between the base of the upper pair is the so-called 'nostril' or nasohypophysial opening, leading internally to a duct, passing backwards below the brain to join the pharynx. Through this system, the respiratory water current passes backwards to the gills, propelled by the movements of a complex scroll-like velum housed in a chamber just in front of the junction with the pharynx. The opening of the mouth is somewhat ventral and overhung by a forwardly projecting rostrum. The eye of *Myxine* is extremely small and degenerate and covered by a layer of muscle so that it cannot be seen from the surface, but in *E. stouti*, where it is rather more developed, its position is marked by a small area of skin, relatively free of pigmentation.

The mouth of the lamprey is surrounded by the expanded sucker-like oral disc, covered on its internal surface by small disc teeth. Above and behind the oral disc is the nasohypophysial opening and behind this is a small unpigmented area of skin marking the site of the pineal organs. Internally, the head region shows marked differences from the hagfish of which the most important is the separation of the food and respiratory passages. The gill pouches open internally into a separate blind-ended respiratory tube or *Wassergang*, whose entrance into the mouth cavity is guarded by a very small valve-like velum. The food passage or oesophagus lies above the water tube and opens separately into the mouth. In the lamprey, the nasohypophysial tract conveys water to the olfactory organ, proceeding backwards to end blindly in a dilated sac above the anterior gill pouches.

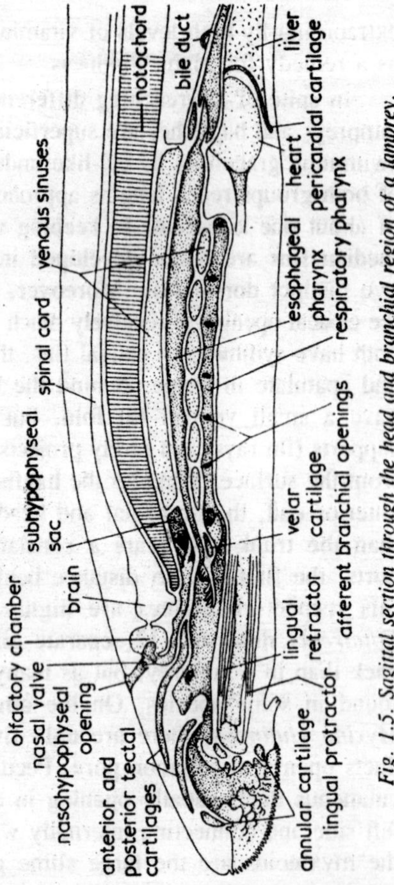

Fig. 1.5. Sagittal section through the head and branchial region of a lamprey.

Although they lack the paired fins of other vertebrates and their skeleton is entirely cartilaginous, the organization of the cyclostome body conforms to the basic vertebrate or, more appropriately, craniate pattern, in which the main axial skeleton consists of a notochord, stretching from below the brain to the tip of the tail, but which in these animals is unconstricted by the development of jointed vertebrae. Immediately below the notochord lies the main distributive artery –

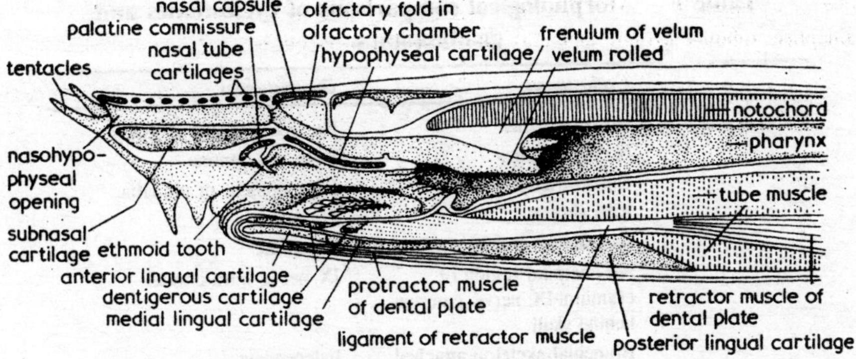

Fig. 1.6. The head of a hagfish.

the dorsal aorta, flanked on either side by the main venous channels – the cardinal veins. Above it is the flattened spinal cord passing forwards to the brain, here housed in an incomplete cranium, with only a membranous roof. Immediately behind the branchial region is the heart. In the lamprey, this is contained within a closed pericardium, whereas in the hagfish, it is open to the general body cavity behind. Through the latter runs a straight and almost uniform intestine, opening together with the kidney ducts at the cloaca.

In its external appearance, internal organization and mode of life, so great is the contrast with an adult lamprey, that for a long time the true larval status of the ammocoete was not appreciated and although Baldner in the 18th century had recognized these 'blind lampreys' as belonging to the same species as the adults, his observations were generally ignored, until their true relationship was finally established by Muller (1856). Lacking the suctorial disc of the adult, the mouth is overhung by the flexible oral hood and on the dorsal surface of the head is a nasohypophysial depression leading, at this stage, to a solid cord of cells and not to a hollow duct. Behind this depression is a translucent spot marking the position of the pineal, but the paired eyes are buried below the skin surface and are not visible from the exterior. The larval pharynx is an undivided tube into which open the seven gill pouches and at whose entrance are the muscular velar paddles, which aid in maintaining the respiratory current On the floor of the pharynx is a ciliated groove and below it lies a complex tubular endostyle opening by a median duct.

Restricted as they were to mainly morphological criteria and impressed by the many apparently primitive characters that were shared by the lampreys and hagfishes, the earlier comparative anatomists had

Table 1 Morphological comparisons of cyclostomes and gnathostomes.

	Cyclostomes	*Gnathostomes*
Skeleton and fins	No paired fins	Pectoral and pelvic fins
	No hard tissues	Bone and dentine
	Unconstricted notochord	Development of arcualia
	Ribs absent	Present
	Incomplete cranial roof	Complete
	No occipital region of cranium–IX nerve emerges behind skull	IX nerve intra-cranial
	Branchial skeleton attached to cranium	Independent
	Visceral arches external to gills	Internal
	Biting or rasping tongue mechanism	Hinged jaws
Muscles	No transverse septum in myotomes	Present
	No internal muscles of accommodation in eye	Present
	4th somite present	Disappears
Gills	Pouch-like gill sacs	
	5-15 gills vagally innervated	5 functional gills
Gut	Velum present	Absent
	Single 'nostril' opening into nasohypophysial tract	Paired nostrils opening into mouth
	No differentiated stomach	Present
	No discrete exocrine pancreas	Present
Vascular system	Extensive sinus system but no true lymphatics	Lymphatic system
	No capillary fenestrations	Present
	No plasma cells	Present
	Single Cuvierian duct	Paired ducts
Nervous system and sense organs	Flattened spinal cord	Cylindrical (except in *Latimeria*)
	Nerve fibres non-medullated	Medullated and non-medullated fibres
	No sympathetic chain	Present
	Post-trematic innervation of gills	Pre- and post-trematic innervation
	No horizontal semicircular canal	Present
Excretory and reproductive systems	Persistence of pronephros (only in larval lamprey)	Embryonic or larval stages only
	No genital ducts	Present

generally favoured a system of classification that grouped them together in a distinct vertebrate Class, the Cyclostomata or Marsipobranchii; the latter referring to their pouch-like gill sacs. Within this class the two groups were given the status of orders or subclasses as the Hyperoartii (lampreys) and Hyperotreti (hagfishes) or as the Petromyzontia and Myxinoidea. As may be seen from the morphological comparisons in Table 1.1, many of the features that distinguish the Cyclostomes from the higher vertebrates (Gnathostomes) relate to the absence in the hagfish or lamprey of structures that have been developed by other vertebrates such as paired fins, hinged jaws, hard tissues, jointed vertebral column, stomach, horizontal semicircular canal, sympathetic nerve chain or genital ducts.

Although in many cases we may be confident that the absence of this or that feature represents a primitive condition that must have existed in the common vertebrate ancestor, in other cases it is possible that structures originally present in the vertebrate line may have been lost secondarily by the cyclostomes. Examples of this are the absence of jointed vertebrae and the failure of the cyclostomes to develop bony structures. Here, although we cannot exclude the possibility that these are primitive conditions (especially in the hagfish), there are some grounds for a belief that the absence of these structures may be the result of secondary reduction from the extensive exoskeleton that existed in the fossil agnathans. Similarly, it has even been argued on the basis of inferences from the fossil cephalaspids that lampreys are descended from ancestors with mobile paired fins. On the positive side are structures peculiar to the cyclostomes, but which in some instances are known to have been present in the related fossil agnathans. These include the relatively large numbers of gill pouches, a nasohypophysial tract with a single external opening and possibly a velum. Above all, the cyclostomes differ radically from the gnathostomes in the arrangement and relationships of their branchial skeleton, which is fused into a continuous structure, placed externally to the gills. Indeed so fundamental is this difference that at one time it was proposed as a basis for dividing the vertebrates into two distinct phyla – the Endobranchiata and the Ectobranchiata – according to whether the gills were endodermal or ectodermal in origin and placed inside or outside the visceral arches.

Yet, in spite of these common 'cyclostome characteristics' and the superficial resemblance between the lampreys and hagfishes, it is increasingly apparent that a considerable gulf separates the two groups;

a separation that becomes more pronounced as we learn more of their physiology and biochemistry. This has led to the general abandonment of the term 'cyclostome' as a taxonomic unit and to the ranking of the hagfishes and lampreys as distinct Classes or subclasses within higher taxa that also include the various groups of fossil agnathans. Some idea of the extent of this divergence may be gauged from the morphological comparisons in Table 1.2, summarizing many of the points that arise in subsequent chapters. To this may be added an almost equally impressive list of biochemical or physiological criteria which are examined in the appropriate sections of this volume. At the same time, in our eagerness to document the phyletic distance that appears to separate these animals, we must not overlook the fact that just as it is possible to discern an 'agnathan grade' of vertebrate organization, so too the cyclostomes undoubtedly share many distinctive biochemical and physiological characteristics marking them off from the higher vertebrates.

Table 1.2 Morphological comparisons of lampreys and hagfishes.

Lampreys	*Hagfishes*
External features	
One, or more usually two dorsal fins with fin rays	At most a small skin fold without fin rays
Oral funnel	Absent
No sensory tentacles	Three pairs of sensory tentacles
Skeleton and muscles	
Neurocranium with incomplete cartilaginous roof	Roof of neurocranium entirely membranous
Rudimentary 'vertebral' elements	Absent
Radial fin muscles	Absent
Parietal muscles meet ventrally	Parietal muscles do not reach ventral surface
Parietal muscle 'boxes' surrounded on all except medial surface by slow fibres	Slow fibres on ventral and lateral surface only
Dual innervation of central fibres	Innervation at one end of fibre only
Nervous system and sense organs	
Dorsal and ventral spinal roots not united	United
Müller and Mauthner neurons present	Absent
Cerebellum present	Absent
Choroid plexuses present	Absent

Well-developed brain ventricles	Brain ventricles reduced
Heart innervated by vagus	Aneural heart
First gill innervated by IX	Vagal innervation of all gills
Posterior branchial nerve forms a hypoglossal	Absent
Functional paired eyes	Eyes degenerate
Extrinsic eye muscles present	Absent
Pineal complex present	Absent
Labyrinth with two semicircular canals	Simple toroidal labyrinth
Ciliated chambers in labyrinth	Absent
Lateral line system present	Absent
Skin photoreceptors innervated by lateralis nerve	Innervation by spinal nerves
Neurulation with formation of solid neural keel	Neurulation involves formation of neural canal
Accessory olfactory organ	Absent
Retinal receptors with synaptic ribbons	Spherical synaptic bodies
Normal ciliary pattern (9+2)	Olfactory and kinocilia of labyrinth lack central filaments (9+0)[2]
Distinct fat column above spinal cord	Absent
Cardiovascular system	
Single dorsal aorta	Paired and median aortae
Single left Cuvierian duct	Single right Cuvierian duct
No accessory hearts	Cardinal, portal and caudal hearts
Afferent and efferent arteries supply hemibranchs of adjacent gill pouches	Each pouch supplied by single afferent
Blood volume not more than 10 %	Blood volume greater than 10%
Gut and respiratory system	
Gill sacs of adult open into separate water tube	Pharynx not divided
Seven gill pouches	Number varies from 5-14
1st gill close to head	Branchial zone posterior
Nasohypophysial opening dorsal	Terminal
Nasohypophysial tract ends blindly	Opens into pharynx
No pharyngo-cutaneous duct	Present
'Protopancreas' in ammocoete	Zymogen cells not concentrated in diverticulum
No bile duct or gall bladder in adult	Present
Ciliated intestine with typhlosole	Not ciliated and without typhlosole

Excretory organs	
Pronephric tubules atrophy at metamorphosis	Persistent pronephros
Differentiated kidney tubule with collecting ducts	Rudimentary tubule only
Reproduction	
Sex dimorphism at maturity	No sex dimorphism
Small isolecithal eggs	Large yolky eggs with horny shell
Holoblastic cleavage	Discoidal cleavage
Prolonged larval stage	No larval stage
Determinate life span–die after single spawning	Eggs produced throughout life

A continually recurring problem is to decide how far the apparently simpler organization of the myxinoid is to be regarded as a primitive condition or alternatively as a direct or indirect result of its habits and mode of life. In the extent of the degeneration in its sensory equipment the hagfish has retreated so far down the road of evolutionary regression that it has been described as little more than a 'vertebrate worm' (Ross, 1963) and there can be little doubt that this degeneration has had wider repercussions on other functional systems. But, although we can have little difficulty in relating the degeneration of the eyes and lateral line system or the reduction in skin pigmentation to their deep water habitat and burrowing habits, there remain many other features, whose apparent simplicity is more difficult to evaluate. Among these are the rudimentary kidney tubule, the simple ring-like labyrinth, the absence of a cerebellum, the undifferentiated state of the adenohypophysis or even the absence of extrinsic eye muscles and their motor nerves. Are we to regard these as primitive features or is their apparent simplicity or total absence a result of evolutionary regression? Where we lack guidance from related fossil forms, the answers to questions such as these can be no more than a balance of probabilities, based on such evidence as we can find from the mode of life of the hagfish and a more detailed knowledge of its physiology and development.

The idea that the agnathans were ancestral to the higher jawed vertebrates is one that is still reflected in the biological literature. This concept of an agnathan-gnathostome evolutionary sequence arose quite naturally from the fact that the jawless vertebrates appear to precede the gnathostomes in the geological record. However, quite apart from the conceptual problems of deriving the hinged jaws and internal visceral skeleton of the gnathostome from a continuous external branchial skeleton of the cyclostome type, many would now regard this as a misleading oversimplification, based on a fundamental misconception of the manner in which new species or higher taxa arise. For adherents of the cladistic view, these are formed by the splitting up of an ancestral species or group resulting in the appearance of two or more sister groups and the disappearance of the ancestral form. Accordingly, throughout this volume, the agnathans and gnathostomes are regarded as sister groups of equal antiquity, derived from an unknown and extinct common ancestor. At the same time this in no way diminishes the evolutionary significance of the cyclostomes. Even though they and their fossil relatives may no longer be thought of as being in the direct line of descent of the higher vertebrates, a close study of their biology may still enable us to make some reasonable inferences on conditions that may have existed in the remote common ancestor of both jawless and jawed vertebrates.

2

Origin of Earth

Changing Surface of the Earth

The fact that the geography and climate of the world has not always been the same began to be known in the nineteenth century, but only in the last 20 years has the full extent of the change become appreciated. Yet in spite of all the new information collected by geophysicists and geologists it is only possible to give a preliminary outline of the facts that the biologist wants to know. To understand why animals and plants have changed, and how they have become spread over the earth, we need knowledge of the distribution and composition of the land, sea and atmosphere at various times in the past, as well as information about the temperature and other physical conditions.

Most changes in climate and geography occur with such long periods that they are without appreciable effect on individual organisms, but may greatly affect the history of the race. The idea of geographical change is made familiar by the fact that coastlines and river courses have altered appreciably in historical times. We often hear stories of destruction of some houses or of a village by the sea, though it may come as a shock to learn that the sea level has changed so much that England and France were connected by land 8000 years ago, and that man-made instruments fished up from the Dogger Bank in the North Sea show that it was an inhabited peat bog 6000 years BC. Such changes in sea level are signs of the changes in extent of ice caps and of 'diastrophic movements', which are major features of long-period geological evolution.

Geologists have long known that there have been great changes in the extent of land and sea. The causes of these are still not fully

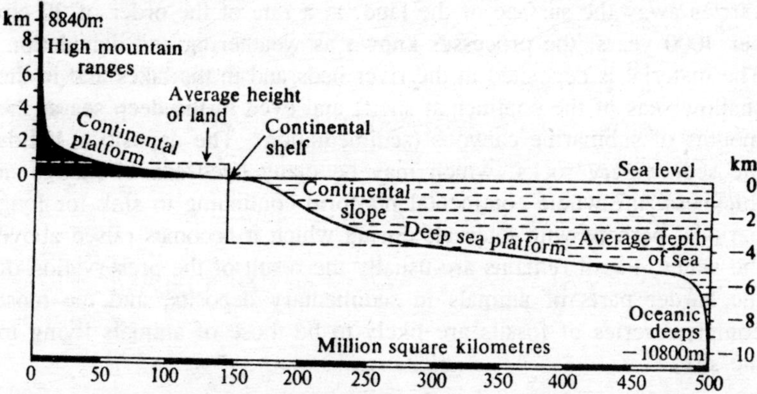

Fig. 2.1. Curve showing the area of the earth's solid surface in relation to the sea level.

known but are probably related to changes in the flow of heat from within the earth, which produces movements of portions of the earth's crust ('plates'), spreading of the sea floor, collisions between continents, and the resulting uplifting of mountains. These periodic changes in the heat flow may be the cause of the repeated elevation and sinking of land masses. Invasion of the edges of the continents by the seas is known as transgression (positive eustasy), the reverse process is regression (negative eustasy). Unfortunately there is no agreement amongst geologists as to whether these processes proceed in any regular cyclical fashion. Changes in sea level may be due to a combination of features. (1) Variations in climate alter the volume of ice that rests upon the land surface. (2) Accumulation of sediment due to weathering alters the balance of land and sea, though this is of lesser importance. (3) Movements of the earth's crust produce variations in the volume of the ocean ridges. (4) Heat flow changes under the ocean ridges may be the most important factor. High heat flow renders basalt more buoyant, hence displacing more sea water onto continents. Following the changes in level the sea leaves more or less of the continental shelf uncovered. Such upward and downward movements have been common and have profoundly influenced the climate. Oceanic climatic influences tend to produce a damp, equable climate, with large areas of marsh and forest. When the land stands higher, or is more laterally extensive, extremes of climate develop, some parts being cold, others forming large, dry interior plains, this is known as greater 'continentality' of climate. Undoubtedly there have been great changes in extent of land and sea and also periods of extensive mountain building (orogenesis). Then the action of frost, wind, and rain breaks up and carries away the surface of the land, at a rate of the order of 30 cm per 4000 years, the processes known as weathering and denudation. The material is deposited in the river beds and in the lakes and in the shallow seas of the continental shelf, and even in the deep sea at the mouths of submarine canyons (sedimentation). The deposition builds the sedimentary rocks, which may be many hundreds of metres in thickness, the whole continental platform continuing to sink for long periods, perhaps with intervals during which it becomes raised above the water. Fossil remains are usually the result of the preservation of the harder parts of animals in sedimentary deposits, and the most complete series of fossils are likely to be those of animals living in the sea.

Components of the Earth

The earth consists of a central dense core, whose outer part is liquid iron, with a radius of about 3470 km, surrounded by a less dense, but more solid mantle of 2870 km thickness. The outermost layer is a thin crust of materials of lesser rigidity and density, which is an average of 5 km thick below the sea floors and 33 km below the continental surfaces. These depths are known from the distribution of the propagation of earthquake shock waves, which changes velocity at the boundary between mantle and crust, known as the Mohorovicic discontinuity (*Moho* for short).

Above the crust are the still lighter water and atmosphere, whose evolution has greatly influenced life and been affected by it. The earliest atmosphere contained reducing gases such as methane, carbon monoxide and ammonia, and the earliest living things, formed more than 3000 Ma ago, were probably anaerobic. The oldest known evidences of life are 'stromatolites' rocks formed from blue-green algae 3.4×10^9 years ago. Such organisms produced free oxygen, which then reacted with methane and ammonia to form carbon dioxide, water, and nitrogen. Water and nitrogen also probably appeared by 'outgassing' from the earth during its chemical evolution. The free oxygen and nitrogen were thus largely the product of organisms, directly or indirectly, during the first 2 billion years of life. By 1.8 billion years ago there was enough oxygen to make red iron oxide deposits and so presumably sufficient for aerobic animal life to begin. However, the earliest animals with skeletons appear only much later, and rather suddenly, in Cambrian time 600 Ma ago.

The composition of the ocean is of special interest to biologists since life first became possible because of conditions in the sea, and then evolved there for perhaps 2000 Ma, that is to say for the greater part of its whole history. Living things still retain in their ionic make-up certain interpreted the blood plasma of vertebrates as a relic of the Palaeozoic sea. The salts in the oceans probably come partly from chemical processes in the mantle, added to by erosion from the land.

Changes of Climate

Evidence of marked changes of climate is the finding in England and other regions now temperate of animal and plant remains appropriate to warmer or colder conditions (corals and woolly rhinoceros, for instance). There is thus every reason to think that

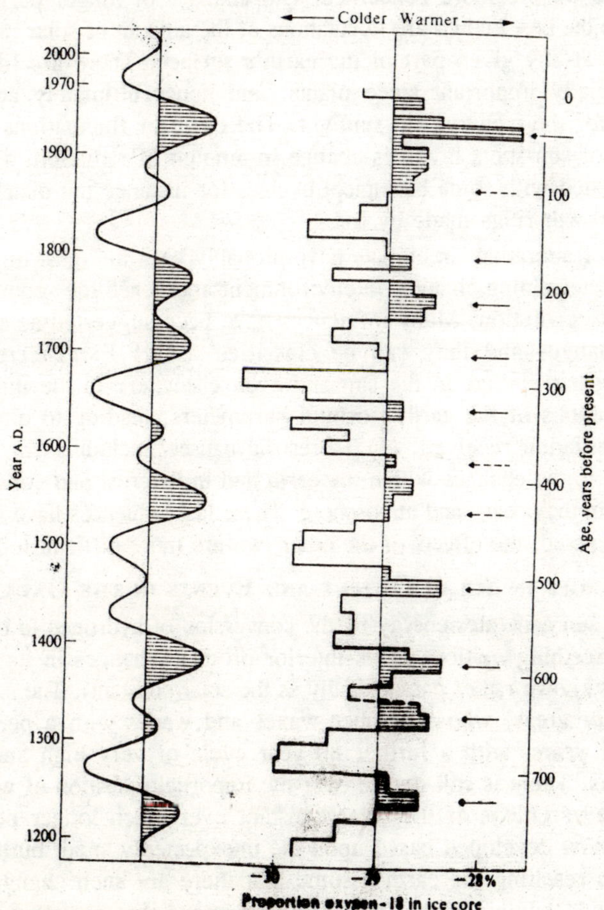

Fig. 2.2. Variations in solar radiation.

great changes from hot to cold and wet to dry conditions have occurred, in conjunction with the changes in latitude and in level of the land.

These fluctuations in geography and climate are obviously of great importance to the biologist. In order to be able to assess the influence of such changes on life we must know more about the rates at which they occur, and careful study shows that some of the climatic changes are rhythmic. Rhythmic changes of climate are, of course, very familiar to us in the cycles of days, months, and years, and the immense importance of these short-period changes for animal and plant life must not be forgotten.

Here we are more concerned with changes of longer periodicity of which the best known are fluctuations of the amount of solar radiation received at any given part of the earth's surface. There are likely to be especially important since plants, and hence ultimately animals, depend for their energy on sunlight. The cycle of fluctuations in the number of sun-spots involves change in amount of radiation, and this is associated with some biological cycles, for instance the distribution of the growth rings made by trees.

Major alterations in climate have probably been the most important factors determining changes in environment and hence the sequence of vertebrate evolution. Many influences have been suggested as causing these changes and they can be classified as (1) Extra-terrestrial, including happenings in the sun and even elsewhere in the universe. (2) Variations of the earth's orbital parameters, leading to alteration of the radiation received. (3) Terrestrial causes, including continental drift due to the changes within the earth and in its crust and consequent changes in the oceans and atmosphere. These last influences have already been discussed, the effects of the other two are more difficult to assess.

Changes in Solar Output and Events in the Galaxy

The sun generates energy by the conversion of hydrogen to helium, and the seething motions in its interior produce changes in its output (which is known rather paradoxically as the solar constant). The sunspots are darker areas, whose number waxes and wanes with a period of about 11 years, with a further 80 year cycle of very high and very low peaks. There is still debate over the important question of whether there are variations of the solar constant over much longer periods. Models now developed based upon the unexpectedly small number of neutrinos reaching the earth assume that there are such changes and these would explain global climatic phenomena that must have been very important for vertebrates, such as the world-wide ice ages.

Further agents of change may be events in more distant parts of the galaxy, such as the explosion of radiate in a few weeks as much energy as ten billion suns. They occur about one every 50 years per galaxy, which means that one near enough to affect the earth takes place every 70 Ma. It is suggested that such events as the extinction of the dinosaurs at the end of the Cretaceous may have been the result of such an explosion. At that time up to 75 per cent of all species of animals became extinct, which is an effect greater than would be expected if half of the world's nuclear stockpile was exploded.

Changes in the Earth's Orbit

The variable orbital parameters of the earth are: (1) The change from circle to ellipse in a cycle of ~93000 years. (2) The axis wobbles, so that the season of closest approach to the sun varies, with a cycle of 'precession' of 21000 years. (3) The tilt of the axis varies in relation to the plane of orbit with a cycle of 4000 years. These three cycles interact to produce variations in the distribution of solar radiation known as the Milankovicic mechanism. These radiation curves agree with the fluctuations of temperature that can be shown to have occurred during the past million years. Presumably there have been similar extensive fluctuations throughout geological history, with marked effects on the flora and fauna.

These various influences, some of them cyclic, interact to produce complicated changes in climate both over the whole globe and

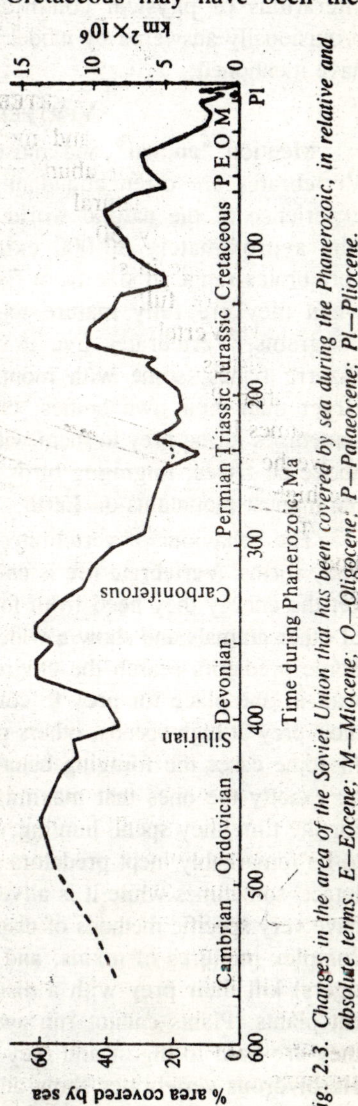

Fig. 2.3. Changes in the area of the Soviet Union that has been covered by sea during the Phanerozoic, in relative and absolute terms. E—Eocene; M—Miocene; O—Oligocene; P—Palaeocene; Pl—Pliocene.

locally. Unfortunately the only record available of the alterations that have taken place throughout the history of the earth is embodied in the material of the crust, which is itself continually changing its position and composition. It is ironic that the appearance of new fossil types often provides the only reliable source of information about the past climatic variation that may have generated them. Questions about global alterations of physical conditions over long periods can only be provisionally answered by model studies such as those of neutrinos we have mentioned.

VERTEBRATE STORY

Mention "animal" and most people will think of a vertebrate. Vertebrates are often abundant and conspicuous parts of people's experience of the natural world. Vertebrates are also very diverse: The approximately 50,000 extant (=currently living) species of vertebrates range in size from fishes that weight as little as 0.1 gram when they are fully mature to whales that weigh nearly 100,000 kilograms. Vertebrates live in virtually all the habitats on Earth: Bizarre fishes, some with mouths so large they can swallow prey larger than their own bodies, cruise through the depths of the sea, sometimes luring prey to them with glowing lights. Some 15 kilometers above he fishes, migrating birds fly over the crest of the Himalayas, the highest mountains on Earth.

The behaviour of vertebrates are as diverse and complex as their body forms. Vertebrate life is energetically expensive, and vertebrates get the energy they need from food they eat. Carnivores eat the flesh of other animals and show a wide range of methods of capturing prey: Some predators search the environment to find prey, whereas others wait in one place for prey to come to them. Some carnivores pursue their prey at high speeds, others pull prey into their mouths by suction. In some cases the foraging behaviours that vertebrates use appear to be exactly the ones that maximize the amount of energy they obtain for the time they spend hunting; in other cases vertebrates can appear to be remarkably inept predators. Many vertebrates swallow their prey intact, sometimes while it is alive and struggling, but other vertebrates have very specific methods of dispatching prey: Venomous snakes inject complex mixtures of toxins, and cats (of all sizes from house cats to tigers) kill their prey with a distinctive bite on the neck. Herbivores eat plants. Plants cannot run away when an animal approaches, but they are hard to digest and they frequently contain toxic compounds. Herbivorous vertebrates show an array of specializations to deal with the difficulties of eating plants. These specializations include elaborately sculptured teeth and digestive tracts that provide sites in which symbiotic microorganisms digest compounds that are impervious to the digestive systems of vertebrates.

Reproduction is a critical factor in the evolutionary success of an organism, and vertebrates show an astonishing range of behaviours associated with mating and reproduction. In general, males court females and females care for the young, but these roles are reversed in many species of vertebrates. The forms of reproduction employed by vertebrates range from laying eggs to giving birth to babies that are largely or entirely independent of their parents (precocial young). These variations range across almost all kinds of vertebrates: Many fishes and amphibians produce precocial young and a few mammals lay eggs. At the time of birth or hatching some vertebrates are entirely self-sufficient and never see their parents, whereas other vertebrates (including humans) have extended periods of obligatory parental care. Extensive parental care is found in seemingly unlikely groups of vertebrates—fishes that incubate eggs in their mouths, frogs that incubate eggs in their stomachs, and birds that feed their nestlings a fluid called crop milk that is very similar in composition to mammalian milk.

The diversity of living vertebrates is fascinating, but the species now living are only a small proportion of the species of vertebrates that have existed. For each living species there may be as many as ten extinct species, and some of these have no counterparts among living forms. The dinosaurs, for example, that dominated the Earth for 180 million years are so entirely different from any living animals that it is hard to reconstruct the lives they led. Even mammals were once more diverse than they are now: The Pleistocene saw giants of many kinds—ground sloths as big as modern rhinoceroses, and raccoons and rodents as large as bears. Humans are great pes, close relatives of chimpanzees and gorillas, and much of the biology of humans is best understood in the context of our vertebrate heritage. The number of species of vertebrates probably reached its maximum in the Pliocene and Pleistocene, and has been declining since then. Some parts of that decline can probably be attributed to the effects of humans on species of humans on species they used for food or viewed as competitors, although changes in climate and vegetation have also been instrumental. In the modern world, however, the effects of human activities are paramount, and the fate of other species of vertebrates is very much affected, for good or ill, by human decisions. Our responsibilities to other vertebrates cannot be ignored.

The story of vertebrates is fascinating: Where they originated, how they evolved, what they do, and how they work provide endless intriguing details. In preparing to tell this story we must introduce some basic information: what the different kinds of vertebrates are called and how they are classified, how evolution works, and what the world within which the story of vertebrates unfolded was like.

Different Kinds of Vertebrates

The modern approach to biological classification, which is popularly known as cladistics, recognizes only groups of organisms that are related by common descent, or *phylogeny* (*phyla*=tribe, *genesis*=origin). The application of cladistic methods is making the study of evolution more rigorous than it has been heretofore. The natural groups recognized by cladistics are easier to understand than the artificial groups we are used to, except that we are familiar with the names of the artificial groups and the names of the new groups are sometimes strange. At this point we need to establish a basis for talking about particular animals by naming them, and to relate the old, familiar names to the new, less familiar ones.

Hagfishes and Lampreys (Myxinoidea and Petromyzontoidea)

Lampreys and hagfishes are elongate, scaleless, and slimy and have no internal hard tissues. They are scavengers and parasites and are specialized for those roles. Hagfishes (about 40 species) are marine and occur on the continental shelf and open ocean at depths around 100 meters, whereas many of the 41 species of lampreys are migratory forms that live in oceans and spawn in rivers.

Hagfishes and lampreys are unique among living vertebrates because they lack jaws, and they occupy an important position in the study of vertebrate evolution. Hagfishes and lampreys have traditionally been grouped as agnathans (*a*=without, *gnath*=jaw) or cyclostomes (*cyto*=round, *stoma*=mouth), but they probably represent two independent evolutionary lineages. Hagfishes lack many of the features that characterize most vertebrates; for example, they have no trace of vertebrae. In contrast, lampreys share many derived characters with jawed vertebrates. The jawless condition of both lampreys and hagfishes, however, is ancestral.

Sharks, Rays, and Ratfishes (Elasmobranchii and Holocephali)

Sharks have a reputation for ferocity that most of the 350 to 400 species would have difficulty living up to. Many sharks are small (15 centimeters or less), and the largest species, the whale sharks, which grows to 10 meters, is a filter feeder that subsists on plankton it strains from the water. The 450 species of rays are dorsoventrally flattened, frequently bottom dwellers that swim with undulations of their extremely broad pectoral fins. Ratfish are bizarre marine fishes with long, slender tails and buck-toothed faces that look rather like rabbits. The name Chondrichthyes (*chondro*=cartilage, *ichthyes*=fish) refers to the cartilaginous skeletons of these fishes.

Bowfin, Gars, and Others (Actinopterygii)

Bony fishes are so diverse that any attempt to characterize them briefly is doomed to failure. Two broad categories can be recognized, the ray-finned fishes (actinopterygians; *actino*=ray, *ptero*=wing or fin) and the lobe-finned or fleshy-finned fishes (sarcopterygians; *sarco*=lobe).

The actinopterygian fishes we have included in this category are called polypterids, holosteans, and neopterygians in various systems of classification. Most have cylindrical bodies, thick scales, and jaws armed with sharp teeth. These fishes seize prey in their mouths with a sudden rush or gulp; they lack the specializations of the jaw apparatus that allow derived bony fishes to use more complex feeding modes.

Teleost Fishes (Actinopterygii)

More than 20,000 species of fishes fall into this category and they cover every imaginable range of body sizes, habitats, and habits, from seahorses to the giant ocean sunfish. Most of the familiar fishes are in this category—the bass and panfish you have fished for in fresh water, and the sole (a kind of flounder) and redfish you have eaten in restaurants. Modifications of the jaw apparatus have allowed many teleosts to be highly specialized in their feeding habits.

Lungfishes and the Coelacanth (Dipnoi and Actinistia)

These are the sarcopterygian (lobe-finned) fishes. They are the living fishes most related to terrestrial vertebrates. The lobe-finned fishes are heavy-bodied and slow-moving. The four species of lungfishes live in fresh water, and the coelacanth is marine.

Salamanders, Frogs, and Caecilians (Urodela, Anura, and Gymnophiona)

These three groups of vertebrates are popularly known as amphibians (*amphi*=double, *bios*=life) in recognition of their complex life histories, which often include an aquatic larval form (the larva of a salamander or caecilian and the tadpole of a frog) and a terrestrial adult. All amphibians have bare skins (that is, lacking scales, hair, or feathers) that are important in the exchange of water, ions, and gases with their environment. Salamanders are elongate animals, mostly terrestrial and usually with four legs; anurans (frogs, toads, treefrogs) are short bodied with large heads and large hind legs used for walking, jumping, and climbing; and caecilians are legless aquatic or burrowing animals.

Turtles (Chelonia)

Turtles are probably the most immediately recognizable of all vertebrates. The shell that encloses a turtle has no exact duplicate among other vertebrates, and morphological modifications associated with the shell make turtles extremely peculiar animals. They are, for example, the only vertebrates with the shoulders (pectoral girdle) and hips (pelvic girdle) inside the ribs.

Tuatara, Lizards, and Snakes (Lepidosauria)

These three kinds of vertebrates can be recognized by their scale-covered skin as well as by characteristics of the skull. The two species of tuatara, stocky-bodies animals found only on some islands near New Zealand, and the sole living remnant of a lineage of animals called Sphenodontida that were more diverse in the Mesozoic. In contrast, lizards and especially snakes are now at the peak of their diversity.

Alligators and Crocodiles (Crocodilia)

These impressive animals (the saltwater crocodile has the potential to grow to a length of 7 meters) are in the same lineage (the Archosauria) as dinosaurs and birds. Crocodilians, as they are known collectively, are semiaquatic predators with long snouts armed with numerous teeth. Their skin contains many bones (osteoderms; *osteo*=bone, *derm*=skin) that lie beneath the scales and provide a kind of armor plating. Crocodilians are noted for the parental care they provide for their eggs and young.

Birds (Aves)

The birds are a lineage of dinosaurs that evolved flight in the Mesozoic. Feathers are the distinguishing characteristic of birds. Indeed, some fossils of *Archaeopteryx*, the earliest bird known, were originally classified as dinosaurs because no marks of the feathers were visible. Birds are conspicuous, often vocal, and active during the day (diurnal). As a result they have been studied extensively and much of our information about the behaviour and ecology of terrestrial vertebrates is based on studies of birds.

Mammals (Mammalia)

The living mammals can be traced to an origin in the late Paleozoic from some of the earliest fully terrestrial vertebrates. Extant mammals include about 4500 species, most of which are placental (eutherian) mammals. Their name comes from the placenta, a structure that transfers nutrients from the mother to the embryo and removes the waste products of the embryo's metabolism. Most of the familiar animals of the world are placentals. Marsupials (mammals in which the young is born at a very small size and continues its development in an external pouch on the mother's abdomen) dominate the mammalian fauna only in Australia. Kangaroos, koalas, and wombats are familiar Australian marsupials. The strange monotremes, the duck-billed platypus and the echidnas, are mammals whose young are hatched from eggs.

EVOLUTION

Evolution is the process that has shaped the vertebrate story, and it is the underlying principle of biology. An understanding of the principles and processes of evolution is essential to appreciating the diversity of vertebrates, because that diversity is the direct result of evolution.

Scientific ideas are shaped by the society and philosophical system in which they form, and in turn they may reshape society and philosophy. That process has been a conspicuous part of the development of evolutionary theory in western societies, most recently in the conflicts over teaching of evolution in schools. Many parts of western though had their origin in Greek philosophy, but evolution was not among them. The Greek concept of nature was a static one, and the idea of evolutionary change in species of plants and animals was foreign to Greek thought. Platonic philosophy was based on the concept of an abstract ideal; earthly manifestations of that archetype (individual animals or plants, for example) were imperfect copies of its form. This attitude was incorporated into Christian theology as the view that an eternal, inviolable essence of every worldly existed in the mind of God. A strict interpretation of the Bible insists that everything on Earth was created by God in its present form, and therefore nothing can change or has changed

This view of life as unchanging was incorporated into biological thought and is still manifested in some biological practices. For example, the Rules of Zoological Nomenclature require that when a new species is named, one specimen of the new species must be designated the type specimen, or holotype of the species. The holotype is deposited in a museum, where it seres as a permanent record of the new species. The original role of the holotype was to show what the species was like. If you had another individual that you thought might be a member of the same species, you could compare it to the holotype and decide. Thus the holotype was a single example that was considered to define the essence of the new species. In effect, it was the designated representative of the Platonic ideal of the species.

Naming a new species still requires a holotype, but the role of the holotype has changed significantly, and in addition to the holotype it is usual to deposit several additional specimens called paratypes or the paratypic series. The paratypes are intended to show the range of variation of the species in such characteristics as body size, colour, and pattern. The holotype now defines the species in those rare cases in which two unnamed species have been confused and both are represented in the paratypic series. The name remains with the species represented by the holotype, and the other species receives a new name.

The recognition that variation within species is biologically important is a major change from the idea that one specimen by itself could define a species, and it reflects the most significant change in biological thought that occurred between the writings of Plato (about 380 B.C.) and the work of Darwin and Wallace in the nineteenth century. In a Platonic view, the variations within a species are the imperfections of different copies of the archetype and they have no significance. To an evolutionary biologist, those variations are the raw material of evolution.

Biological Variation

The idea that no two people (except for identical twins) are exactly the same is a familiar one. Humans use individual variation every day in contexts that range from recognizing friends to finger printing criminal suspects. Individual variation is not confined to morphological variation like facial contours or fingerprints; it extends to any characteristic that can be measured.

These sorts of variation are examples of phenotypic variation. The phenotype of an organism means its form in a very broad sense. For example, phenotype can refer to colour (pale skin versus dark skin), size or shape (tall versus short), or performance capacity (fast versus slow runners). Phenotypes can also be defined on the basis of molecular characteristics: Haemoglobin B is the normal form of the beta chain of the adult human haemoglobin molecule and haemoglobin

S is the form of the beta chain that leads to sickle-cell anemia. The two forms differ in only one amino acid residue: A valine residue appears in haemoglobin S in place of a glutamine residue.

Behavioural phenotypes also can be defined: Guppies are most familiar as aquarium fish, but they occur wild in streams in the New World tropics. In Trinidad guppies occupy stream habitats with relatively few predators and other streams where predators are abundant. Populations in these habitats differ in male colour and in the response of males. The bright colours of male guppies are part of the courtship display, but they also make males more visible to predators. In low-predation areas most male guppies are bright coloured, but in areas of high predation they are plain. Experiments have shown a genetically determined difference in the responses of female guppies from streams with high and low predation to courtship by bright and plain males: Female guppies from low-predation habitats choose to mate with bright-coloured males, whereas females from high-predation areas choose plain males. This response is apparently mediated by the chances that the offspring will survive. Bright-coloured males sire bright-coloured male offspring. In low-predation habitats, bright-coloured male offspring are advantageous because they are attractive to females and run little risk of being eaten by a predator. However, in high-predation habitats bright-coloured male offspring are vulnerable to predation, and females with the genetically determined behavioural trait of mating with plain males leave more surviving offspring than females that mate with bright-coloured males.

These types of phenotypic variation are manifestations of genetic variation. Evolution is defined as change through time in the frequencies of different forms of a gene in the gene pool of a population. Different forms of a gene are called alleles and different alleles produce different phenotypes. The genotype of an individual is its genetic composition. Remember that in sexually reproducing organisms an individual receives one allele from its mother and one from its father. To use the example of sickle-cell anemia, there are three possible genotypes: If an individual receives the normal form of the beta-chain allele from both parents it will be Hb_BHb_B, that is, a homozygous genotype, and it produces the normal beta-chain phenotype. An individual that receives one allele from its mother and a different allele from its father would be Hb_BHb_S, that is, a heterozygote. Heterozygotes for the sickle-cell allele have an advantage in areas where malaria is endemic, apparently because red blood cells that contain haemoglobin S collapse (sickle) when a malaria parasite enters them. This sickling causes a decrease in the concentration of potassium inside the red cell and the eventual death of the parasite. An individual that receives the sickle-cell allele from both parents will have the genotype Hb_sHb_s: These homozygous individuals are at a disadvantage because the haemoglobin molecules sickle whenever the oxygen concentration in the blood drops, as it may, for example, during physical exercise.

Evolution depends on the existence of variation in genotypes among individuals that is (1) manifested in the phenotype, (2) inherited, and (3) associated with differences in reproductive success. Natural selection changes the relative frequency of different alleles in a population, and these changes in alleles are reflected in changes in the frequencies of different phenotypes. It is easier to think of phenotypes than of genotypes in discussions of evolution because phenotypes are visible whereas alleles are not. However, it is alleles that are inherited.

Variation in genotypes results from the combined actions of several genetic processes in sexually reproducing organisms. During meiosis and the formation of gametes (sperm and ova) the pairs of chromosomes that make up the genome of an individual are first duplicated and then separated into gametes that contain only one of each pair of chromosomes. (This is the haploid chromosomal complement.) When two gametes unite to form a zygote (fertilized egg) the diploid chromosome number is regained: Two chromosomes of each pair are present, one from the father and one from the mother. Alleles are not entirely independent entities; their function is affected by other alleles present in the genotype. Consequently, the genetic rearrangement that accompanies sexual reproduction is an important source of phenotypic variation. Mutation is another source of variation, of smaller magnitude than rearrangement but nonetheless important. Mutation can result from errors in copying the genetic code during the initial duplication of chromosomes during meiosis, or it can be caused by events that precede meiosis, such as exposure of the sex cells to ionizing radiation. Other cellular events can cause changes in the linear sequence of genetic loci on chromosomes. Because the action of alleles is influenced by neighbouring alleles, these changes in sequence can affect the phenotype.

Natural Selection

Evolution results from the action of natural selection on different phenotypes, and natural selection works through differential reproductive success, that is, the contribution of different phenotypes to the gene pool of succeeding generations. The term *fitness* is used to describe the relative contribution of different individuals to future generations. If individual A produces 100 offspring that survive to reproduce and individual B produces only 90 offspring, individual B is only 90 percent as successful as individual A. The fitness of the most successful individual (A in this example) is defined as 1.00, and other individuals are defined in relation to A. Thus the fitness of B would be 0.90. That means that the alleles represented in the genotype of individual B would be under-represented in the next generation by 10 percent compared with alleles in the genotype of A. In this example, selection would be said to select against B and the selection coefficient would be 0.10.

Modes of Selection

Natural selection can affect the distribution of phenotypes in a population in three different ways. *Directional selection* discriminates against individuals at one extreme of the variation in a phenotypic character. For example, small individuals might be at a disadvantage compared with individuals of normal or larger-than-normal size. The effect of directional selection is to move the mean value of the character in the direction of the most fit phenotypes from generation to generation.

Stabilizing selection discriminates against individuals that have extreme variation in the phenotypic character in either direction. It favours individuals that have values that lie close to the mean, and it can result in a reduction of the amount of variation within a population, but it does not change the mean value.

Disruptive selection is the reverse of stabilizing selection. That is, individuals at both extremes of the range of variation are favoured

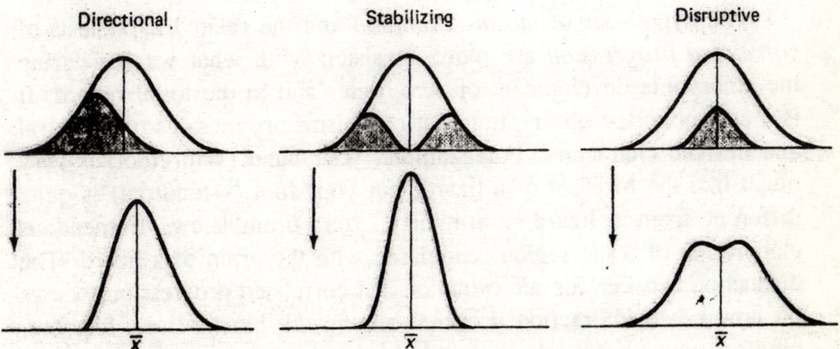

Fig. 2.4. Three kinds of selection: stabilizing, and disruptive.

and individuals near the mean are at a disadvantage. Disruptive selection increases the variation but does not change the mean value.

Note that these descriptions have assumed that only one mode of selection affects the trait being considered. In fact, more than one mode of selection can occur simultaneously, and both the mean value of the trait and the amount of variation can change simultaneously. Natural selection si difficult to demonstrate under natural conditions because many different and sometimes conflicting selective forces operate simultaneously. Nonetheless, some examples are known, and John Endler (1986) has reviewed the evidence for natural selection in the wild.

Evolution of Complex Systems

Organisms are extraordinarily complicated entities, and that complexity is one of the principal reasons they are so fascinating. We try to illustrate that perspective throughout this book by emphasizing the ways in which ecology, behaviour, morphology, and physiology interact in living vertebrates, and by considering how similar interactions in the past may have influenced the evolution of vertebrates.

How do such complex systems evolve? Three general explanations have been proposed: saltation, mosaic evolution, and correlated progression. The *saltation* hypothesis suggested that the mixture of characters that distinguish a mammal from a reptile, for example, could have been acquired all at once as a result of the random effects of genetic mutation and recombination. This hypothesis, which is associated with the early part of the twentieth century, is not supported by either our current understanding of developmental mechanisms or the fossil record.

The hypothesis of *mosaic evolution* and the related hypothesis of *correlated progression* are more consistent with what we see during the embryonic development of vertebrates and in the fossil record. It is a commonplace observation that organisms are mosaics of ancestral and derived characters. For example, your hand (=forefoot) is very much like the forefoot of a lizard, but your foot (=hindfoot) is quite different from a lizard's. Similarly, your brain shows tremendous elaboration of some regions compared with the brain of a lizard. The distinction between mosaic evolution and correlated progression focuses on how much interaction occurred among the blocks (i.e., forefoot, hindfoot, eye, brain) during evolution. Independent evolution of the blocks because of coincidental response to the same new environment is mosaic evolution, whereas an interaction in which changes in one block influence changes in other blocks is correlated progression.

Mosaic evolution and correlated progression are two ends of a continuum rather than being mutually fall at various points along the line. A particularly good example, and one that is treated in more detail later in this book, is the evolution of mammals from early reptilelike animals. That transition involved a host of changes among interdependent systems. Endothermy (i.e., the regulation of body temperature by heat produced by an animal's metabolism), for example, requires a physiological mechanism for heat production (high metabolic rate) and a morphological mechanism for retaining heat (an insulating covering of hair), and neither feature is advantageous unless the other one is already present. This catch-22 situation and the sequence of appearance of the changes in skeletal structure, dentition, maternal care, and brain development that distinguish mammals from reptiles. In the case of mammals, we clearly see a process of correlated progression, in which changes in skeletal structure were associated with changes in locomotion, physiology, reproduction, and parental care.

Inclusive Fitness

We have defined fitness as the relative genetic contribution of different individuals to future generations, and described it in terms of the number of offspring produced by an individual. That is an oversimplification, however, because different individuals of a species have alleles in common, and the more closely related two individuals are the higher will be the proportion of their shared alleles. Siblings, for example, have an average of 50 percent of their alleles. Remember that it is alleles that are transmitted from generation to generation, and you can see that it is possible for an individual to increase its own fitness (that is, to transmit alleles identical to its own) by assisting a related individual to reproduce successfully. For example, when your sibling reproduces, he or she transmits half the alleles you share to that offspring (which is your niece or nephew), and 25 percent of the alleles in the genotype of the offspring are the same as alleles in your genotype (50 percent of alleles shared between siblings times 50 percent of the genotype contributed by each parent = 25 percent of the alleles in the genotype of the offspring). Your own offspring has 50 percent of the alleles in your genotype, twice as many as your niece or nephew. All else being equal, then, two nieces or nephews are the equivalent of one offspring of your own in terms of transmitting your alleles to future generations. The same calculations can be applied to increasingly distant genetic relationships The important point is that reproduction of relatives contributes to the fitness of an individual. In some situations helping relatives to reproduce successfully may be the best way for an individual to increase its own fitness. This principle of inclusive fitness is believed to underlie many of the altruistic behaviours of birds and mammals.

Variation of Evolution

Phenotypic and genotypic variation is the material on which natural selection operates, but not all kinds of variation are subject to selection, and characteristics of the biology of certain species can increase or decrease the importance of selection. In the first place, variation must be heritable if selection is to act on it. That is, variation in phenotypic characters must have an underlying genetic basis, and parents that manifest a particular character must produce offspring that also show that character. Not all variation does have a genetic basis, and many characters that do have a genetic basis are also affected by the environment. The body size of turtles at hatching is an example of this interaction between genetic and environmentally induced variation. Newly hatched turtles receive no parental care: They must dig their way out of nests in the soil, find their way to water, and begin to catch their own food entirely by their own efforts. Probably it is desirable for a female turtle to produce the largest hatchlings she can, because larger hatchlings can dig more strongly, move to water faster, and capture a wider variety of prey than can small individuals. The size of a hatchling turtle is correlated with the size of the egg it came from, and some female turtles lay larger eggs than others. All else being equal, female turtles that lay large eggs should produce larger hatchlings and be more fit than females that lay smaller eggs. However, the availability of water in the nest also affects the body size of hatchlings—nests in moist soil give rise to larger hatchlings than do nests in dry soil. Furthermore, hatchlings from moist nests can also crawl and swim faster than hatchlings from dry nests. This component of variation in body size of hatchlings has no genetic basis and cannot be acted on directly by natural selection. If selection does act on this type of variation (and this sort of selection has not been demonstrated) it would have to be indirect, via a reduction in the fitness of female turtles that chose to nest in dry sites.

Variation that is Subject to Natural Selection

Individual variation is the type of variation one usually thinks of in terms of natural selection—some individuals are larger than others, or faster, or more colourful, or more aggressive than others. The list of characters by which individuals differ can be extended indefinitely, and these differences result from the luck of the draw when gametes combine to form a zygote. The effects of genetic recombination and mutation lead to different genotypes and different phenotypes, and selection can act on those differences.

Individual variation is usually continuous. That is, some animals are small, others are large, and most are somewhere between those extremes. A second type of variation is discontinuous: Instead of having a bell-shaped curve of frequencies of different forms, discontinuous variation can be sorted out into discrete categories. *Polymorphism* (*poly*=many, *morph*=form) is a common type of discontinuous variation. Some individuals are uniformly coloured, some have stripes along the sides of the body, and some have a mottled pattern of dark blotches. These different patterns are genetically determined and offspring inherit the patterns of their parents. One can find individuals with all these patterns in any population of the frogs, but some patterns are more common in one kind of habitat than another, probably because certain patterns are especially cryptic (hard for predators to detect) in particular habitats. In grasslands, for example, the striped patterns occur in high frequency, whereas in forests the unicoloured and mottled patterns are more common. In grassy areas predators see the frogs against a background of grass stems, and the striped pattern may blend well with the straight, light-coloured stems. In forests predators often see the frogs on a background of fallen leaves. There are few sharp, straight lines in a forest and mottled and solid patterns are probably more cryptic than stripes. The differences in frequencies of the patterns in different habitats indicate that natural selection has acted on the gene pool to change the frequencies of alleles producing the patterns.

Sexual dimorphism (*di*=two) is a special case of polymorphism in which males and females of a species differ in secondary sexual characters such as size or colour. Sexual dimorphism results from the different roles that males and females play in reproduction and the differences in the selective pressures that affect the fitness of individuals of the two sexes. The difference in sexual roles begins with the investment of males and females in gametes: Males produce sperm, females produce ova. Sperm are small, energetically cheap, and quick to produce, whereas ova may contain large quantities of yolk and require a substantial investment of time and energy by the female.

As a result of this asymmetry of investment, the reproductive strategies that increase the fitness of males and females can be quite different. Because a male can readily produce more sperm he can usually increase his fitness by mating with as many females as possible—each mating has only a small cost and he can mate repeatedly with little delay between matings. By the time she is ready to mate she has invested time and energy in producing mature ova. Furthermore, in many species the female provides most of the parental care. As a result, it is costly in time and energy for a female to breed repeatedly. Instead, her fitness is increased by mating with the best possible male.

Sexual dimorphism often reflects these differences in the selective forces acting on males and females: In *The Descent of Man and Selection in Relation to Sex*. Darwin pointed out that males are often larger than females and more aggressive. They may engage in ritualized or actual combat with other males, and may form dominance hierarchies in which high-ranking males have opportunities to court females and low-ranking males do not. Darwin noted that females often must be courted repeatedly by a male before they will mate. Many secondary sexual characteristics of males, such as bright colours, the presence of horns or antlers, and vocalization, are examples of sexual dimorphism that are employed in combat or courtship. The role of female choice in determining the reproductive success of individual males has received increased attention recently. The bright-coloured and plain-coloured Trinidad guppies discussed in preceding section provide an example of female choice in determining the reproductive success of males, and additional examples are found in subsequent chapters.

Factors Promoting Speciation

Evolution consists of changes in frequencies of alleles within populations. If two populations of a species accumulate more and more differences in allelic frequencies over time, they become more and more distinct. Some alleles may be lost from one population but retained in the other population, and mutation may introduce new alleles into each population that are not represented in the other. How long can this process continue before the populations are no longer one species but two?

To answer that question we need a definition of a species, and that is a contentious issue among biologists. For sexually reproducing organisms, which include the overwhelming majority of vertebrates, the definition of a *biological species* that was formulated by Ernst Mayr (1942) has been remarkably durable. "Species are groups of interbreeding natural populations that are reproductively isolated from other such groups." Critics of Mayr's biological species definition have pointed out that it does not encompass unisexual organisms, or geographically isolated populations that are prevented from interbreeding by a barrier such as a river or a mountain range. Also, the biological species definition does not include the idea of evolutionary continuity over time.

The *evolutionary species* concept, first defined by George Gaylord Simpson in 1961 and modified by E.O. Wiley in 1978, defines a species as "a single lineage of ancestral-descendant populations which maintains its identity from other such lineages and which has its own evolutionary tendencies and historical fate." All of these definitions of species emphasize that a species is the expression of a genetic system that evolves independently from other species. This view of species stresses the fact that members of a species not only evolve independently of other species but also must be able to survive in the face of competition from members of other species.

The role of isolation in speciation

As long as populations are in contact, alleles can move among them. Even in species with large geographic ranges, gene flow maintains genetic continuity between populations at the extremes of the range of the species via intervening populations. A new allele created by a mutation in one part of the species range can spread through the entire species, and an allele that is lost in one part of the range will eventually be replaced by gene flow from the gene pool of the species. Gene flow is both a constraining and a creative force in evolution. On one hand, gene flow prevents local populations from a accumulating enough genetic differences to evolve into different species, but at the same time gene flow allows superior alleles and combinations of alleles to spread through an entire species.

Allopatric speciation describes a situation in which a physical barrier such as a river or a mountain range prevents gene flow and allows two populations to diverge. Sympatric speciation describes the separation of a single population into two species without such a physical barrier. In this situation, ecological or behavioural differences between subdivisions of the population restrict or eliminate gene flow.

The importance of population size in evolution

How speciation might proceed in large interbreeding populations has perennially puzzled evolutionists. A large population that occupies extensive areas can usually be divided into subpopulations, or demes. Large continental areas are seldom homogeneous, but rather are a series of patchy habitats because of variation in geology and vegetation.

Members of a deme are more likely to breed with members of their own deme than with members of adjacent demes, if for no other reason than proximity. The result is restricted gene flow within the species, and presumably, a reduction in heterozygosity or genetic variation in the deme. If the deme is large the effect of interbreeding will be slight, but if it is small the reduction in genetic variation may be significant. Occasional outbreeding with individuals from other demes will tend to maintain heterozygosity within the deme: An average of one individual exchanged between populations per generation is sufficient to prevent divergence.

Superimposed on this concept of semi-isolated demes with limited outbreeding is the amount or the intensity of natural selection. If, for example, environmental change, predation, or competition for resources is intense within a deme, individuals with the most appropriate phenotypes for the local conditions will be favoured and their genotypes will increase within the population. If gene flow between demes remains low, genetic divergence of demes can occur.

The rate of evolution is most likely to be rapid in small populations, and small breeding populations can result from various circumstances including dispersion of a few adults to a new location, the sudden local isolation of a new deme by some environmental event, or a sudden reduction in the size of a population (a population crash) These periods of diminished population size are called bottlenecks, and they potentially stimulate rapid changes in the frequencies of alleles in the population. Furthermore, the small population resulting from the bottleneck will contain only those alleles present in the individuals that pass through the bottleneck, and these are the only alleles that will be represented in the descendant population. A similar phenomenon, called the founder effect, describes a situation in which a new isolated population is founded by a few individuals. For example, a flock of birds might be blown out to sea by a storm and ultimately land on an island. If conditions on the island are suitable, the waifs could breed and establish a new population that will contain only the alleles represented in the genotypes of the founding individuals. This is the sequence of events that is thought to have produced the bird faunas of many oceanic islands, including Hawaii and the Galapagos Islands.

A classic example of evolution: the Galapagos finches

How geographic isolation promotes speciation is emphasized on oceanic islands, as Darwin himself observed in this examination of the endemic forms of animal life in the Galapagos Islands, which lie

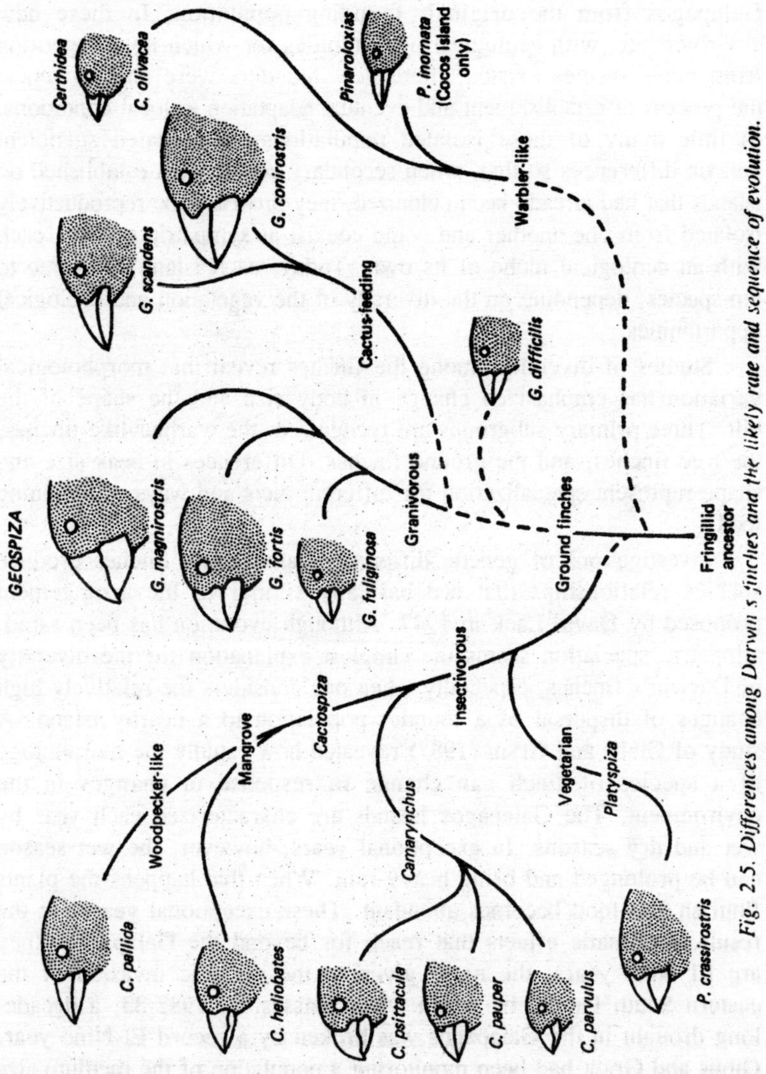

Fig. 2.5. Differences among Darwin's finches and the likely rate and sequence of evolution.

about 1000 kilometers off the coast of Ecuador. Darwin's finches (subfamily Geospizinae) have become the classic example of evolution at the species level. These 14 species have evolved rapidly from a single, ancestral population of seed-eating birds that arrived from South America in relatively recent geological time—perhaps less than 500,000 years ago.

From time to time individuals spread to other islands in the Galapagos from the original, founding population. In these new environments, with ecological opportunities for which no competition from other species existed, these new founders were able to repeat the process of establishment and eventual adaptation to local conditions. In time many of these isolated populations accumulated sufficient genetic differences so that, when secondary contact was established on islands that had already been colonized, they proved to be reproductively isolated from one another and could coexist as sympatric species, each with an ecological niche of its own. Today, each island has three to ten species, depending on the diversity of the vegetation and ecological opportunities.

Studies of diversity among the finches reveal that morphological variation has emphasized change in body size and the shape of the bill. Three primary subgroups are recognized: the warbler-like finches, the tree finches, and the ground finches. Differences in beak size and shape represent specialization for different diets and ways of obtaining food.

Investigations of genetic differences among the finches produce species relationships that are basically similar to the arrangement proposed by David Lack in 1947. Although evolution has been rapid, allopatric speciation seems the simplest explanation for the diversity of Darwin's finches, especially when one considers the relatively high changes of dispersal of a founder population to a nearby island. A study of Gibbs and Grant (1987) revealed how rapidly the morphology of a species of finch can change in response to changes in the environment. The Galapagos Islands are characterized each year by wet and dry seasons. In exceptional years, however, the wet season can be prolonged and bring heavy rain. When that happens the plants flourish and food becomes abundant. These exceptional years are the result of climatic effects that reach for beyond the Galapagos—they are El Nino years, the name given to the periodic invasion of the eastern South Pacific by warm water masses. In 1982-83, a decade-long drought in the Galapagos was broken by a record El Nino year. Gibbs and Grant had been monitoring a population of the medium-size ground finch *Geospiza fortis* on the small island of Daphne Major. The morphology of the birds changed between 1976-77 and 1984-85. Before the El Nino event, the population consisted of large-bodied and heavy-billed birds, but during and following the wet season both body size and bill size diminished. The differences in body and bill size have a genetic basis, and they also relate directly to success in feeding.

In a drought, when food is scarce, the larger birds apparently are favoured because they can crack the large, hard seeds that are the most abundant food for the finches. During wetter times the small birds are apparently successful because small, soft seeds are abundant. Here is a clear example of changing environmental conditions leading to the selection of genetically controlled morphologies that favour survival. The genotype of these finches must carry enough heterozygosity to express either phenotype.

For land-based animals such as finches, oceanic islands are obvious geographic features for the isolation and speciation of populations. The ocean is a barrier to dispersal from a continental source, but it is not an absolute barrier. Habitats on mountain-tops are isolated from each other in much the same way as islands. It is easy to see how a surrounding lowland forest or desert may be an effective barrier to frequent dispersal of animals that are adapted to live only in the highland situation.

Isolation on continents

It has not been to easy to explain how speciation has occurred on continents, particularly in regions where environmental conditions remain uniform over vast areas, such as in the tundra and boreal forests of the Northern Hemisphere or the Amazonian forest of South America. It now appears that a mechanism similar to isolation on islands has been at work repeatedly and that much speciation of land vertebrates on continents can be accounted for by geographic isolation.

The Pleistocene epoch—approximately the last 2 million years of geological history—has been characterized by changes in mean annual temperature and rainfall. Geological and biogeographic evidence supports the conclusion that islandlike refugia of the tundra and taiga ecosystems persisted in northern latitudes even during maximum glaciations, whereas during interglacials these refugia have merged as continuous habitats of great expanse. Thus, species populations that were widely distributed in these uniform habitats during interglacials became fragmented and isolated in refugia during glaciations. This is precisely the sequence of events needed for speciation, and the present distributions of a number of bird and mammal species in boreal North America indicate that they have evolved in this way.

The changes in temperature and precipitation during the Pleistocene have been great enough to have effected changes in most biomes of the world, including those in the tropics. The Amazonian forest has apparently been broken up repeatedly into patches that persisted for a

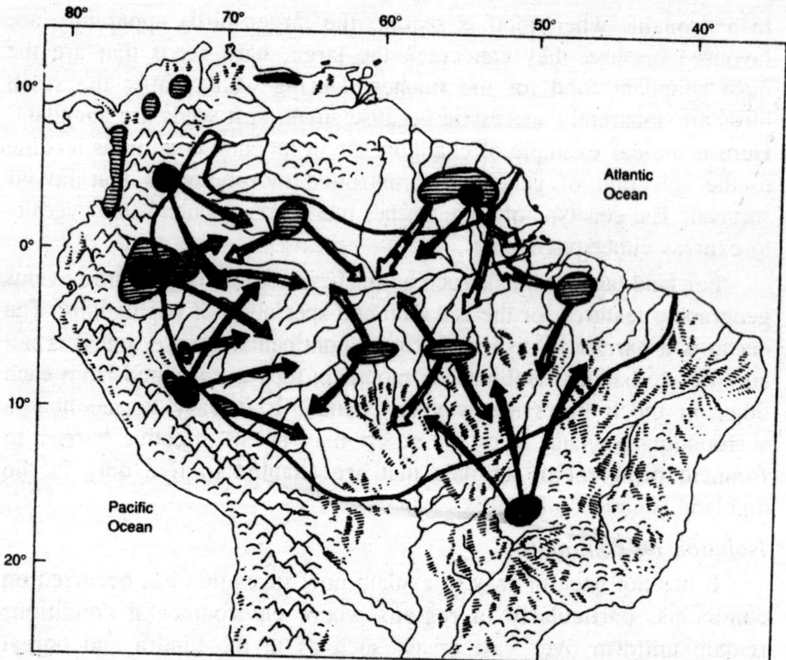

Fig. 2.6. Proposed effects of changes during the Pleistocene on lowland forest vertebrates of the Amazon Basin.

time as isolated forests and then coalesced to form a continuous biome again. Periods of drought have long been considered the cause of this disruption of continuous forest, and a recent interpretation suggests that floods were another factor in the evolution of the Amazonian forest. Similar shifts in the distribution of savanna and forest biomes in Australia, also associated with long-term climatic changes, may account for much recent speciation among birds on that content for much recent speciation among birds on that continent. Most biogeographers and paleoclimatologists now agree that climate and the major biotic associations of plants and animals have fluctuated repeatedly on a worldwide scale during the Pleistocene. The result has been the repeated fragmentation of biomes, alternating with re-formation of more continuous, continental distributions, plus some change in species composition of the biomes. A few years ago ornithologists though that most bird species had evolved in the Pliocene and were several million years old. Now most ornithologists feel that speciation has been very rapid during the Pleistocene, especially among small birds with limited dispersal abilities.

Earth History and Vertebrate Evolution

The phenomena of isolation and secondary contact that are important in the evolution of individual species like Galapagos finches and eastern warblers have also been important in the larger story of vertebrate evolution. The world in which vertebrates have been evolving has changed enormously and repeatedly since the origin of vertebrates in the early Paleozoic, and these changes have affected vertebrate evolution directly and indirectly. An understanding of the sequence of changes in the positions of the continents and the significance of those positions in terms of climates and interchange of faunas is a central part of understanding the vertebrate story. The history of the Earth has occupied four geological *eons*: the Hadean, Archean, Proterozoic, and Phanerozoic. Only the Phanerozoic contains vertebrate life, and it is divided into three geological *eras*: the Paleozoic, Mesozoic, and Cenozoic. These eras are divided into *periods*, which can be further subdivided in a variety of ways. Here we will be considering only the subdivisions called *epochs* within the Cenozoic period.

Movements of land masses have been a feature of Earth's history, at least since the Proterozoic. The direction of vertebrate evolution has been molded in large part by continental drift. By the early Paleozoic a recognizable scene had appeared: The seas had formed, continents floated atop the Earth's mantle, life had become complex, and an atmosphere of oxygen had formed, signifying that the photosynthetic production of food resources had become a central phenomenon of life.

The continents still drift today—North America is moving westward and

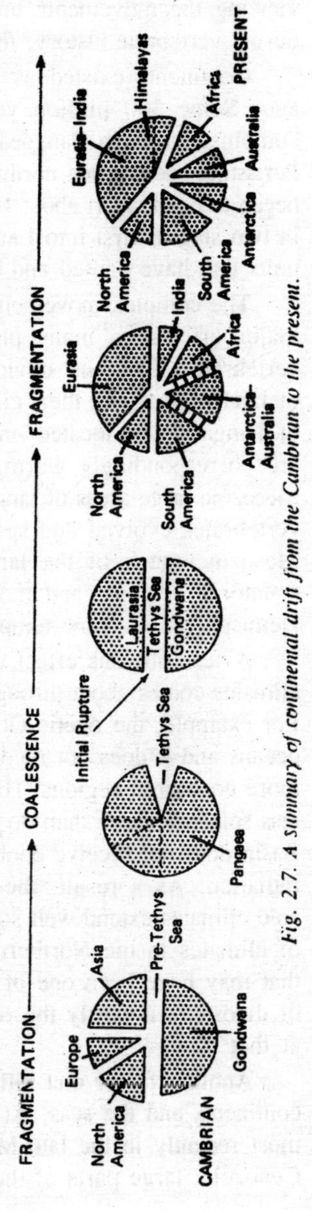

Fig. 2.7. A summary of continental drift from the Cambrian to the present.

Australia northward at approximately 4 centimeters per years. Because the movements are so complex, their sequence, their varied directions, and the precise timing of the changes are difficult to summarize. By viewing the movements broadly, however, a simple pattern unfolds during vertebrate history: *fragmentation—coalescene—fragmentation.*

Continents existed as separate entities over 2000 million years ago. Some 300 million years ago all of these separate continents combined to form Pangaea, birthplace of the terrestrial vertebrates. Persisting and drifting northward as an entity, this single huge continent began to break apart about 150 million years ago. Its separation occurred in two stages: first into Laurasia and Gondwana, then into a series of units that have drifted and become the continents we know today.

The complex movements of the continents through time have had major effects on many phenomena significant to the evolution of vertebrates. The most obvious is the relationship between the location of land masses and their climates. At the start of the Paleozoic much of Pangaea was located on the equator, and climates at the equator are correspondingly warm. During the Paleozoic and much of the Mesozoic large areas of land enjoyed tropical conditions, and terrestrial vertebrates evolved and spread in these tropical regions. By the late Mesozoic much of the land mass of the Earth had moved out of equatorial regions, and most climates in the Northern and Southern Hemispheres are now temperate instead of tropical.

A less obvious effect of the position of continents on terrestrial climates comes about through changes in patterns of ocean circulation. For example, the Arctic Ocean is now largely isolated from the other oceans and it does not receive warm water via currents flowing from more equatorial regions. High latitudes are cold because they receive less solar radiation than do areas closer to the equator, and the Arctic Basin does not receive enough warm water to offset the lack of solar radiation. As a result, the Arctic Ocean is permanently frozen and cold climates extend well southward across the continents. The cooling of climates in the Northern Hemisphere at the end of the Mesozoic that may have been one of the most factors leading to the extinction of dinosaurs is partly the result of the changes in oceanic circulation at that time.

Another factor that influences climates is the relative level of the continents and the seas. At some periods of the history of the Earth, most recently in the late Mesozoic and again in the first part of the Cenozoic, large parts of the continents have been flooded by shallow seas. These *epicontinental seas* extended across the middle of North America and the middle of Eurasia in the Cretaceous and early Cenozoic. Water has a great capacity to absorb heat as environmental temperatures rise and to release that heat as temperatures fall. The heat capacity of water buffers temperature change in areas of land near bodies of water. Areas with maritime climates do not get very hot in summer or very cold in winter, and they are usually moist because water that evaporates from the sea falls as rain on the land. *Continental climates*, which characterize areas of land far from the sea, are usually dry with cold winters and hot summers. The draining of the epicontinental seas at the end of the Cretaceous was a second factor that probably contributed to the demise of the dinosaurs by making climates in the Northern Hemisphere more continental.

In addition to changing climates, continental drift has formed and broken land connections between the continents. Isolation of different lineages of vertebrates on different land masses has produced dramatic examples of the independent evolution of similar types of organisms. These are well shown by mammals in the Cenozoic, a period when the Earth's continents were more widely separated than they ever have been before or since. Saber-tooth carnivores evolved among placental mammals in the Northern Hemisphere and among marsupials in South America, and grazing animals were represented by entirely different lineages of placental mammals in North America and South America and by kangaroos and wombats (both marsupials) in Australia.

Much of evolutionary history appears to depend on whether or not a particular lineage was in the right place at the right time. This stochastic (random, chance) element of evolution is assuming increasing prominence as more detailed information about the time of extinction of old groups and radiation of new groups is suggesting that competitive replacement of one group by another is not the usual mechanism of large-scale evolutionary change. The movements of continents and their effects on climates and the isolation or dispersal of animals are taking an increasingly central role in our understanding of vertebrate evolution.

3

Origin of Life

About a century ago the question, How did life begin? Which has interested men throughout their history, reached an impasse. Up to that time two answers had been offered: one that life had been created supernaturally, the other that it arises continually from the nonliving. The first explanation lay outside science; the second was now shown to be untenable. For a time scientists felt some discomfort in having no answer at all. Then they stopped asking the question.

Recently ways have been found again to consider the origin of life as a scientific problem – as an event within the order of nature. In part this is the result of new information. But a theory never rises of itself, however rich and secure the facts. It is an act of creation. Our present ideas in this realm were first brought together in a clear and defensible argument by the Russian biochemist A.I. Oparin in a book called *The Origin of Life*, published in 1936. Much can be added now to Oparin's discussion, yet it provides the foundation upon which all of us who are interested in this subject have built.

The attempt to understand how life originated raises a wide variety of scientific questions, which lead in many and diverse directions and should end by casting light into many obscure corners. At the center of the enterprise lies the hope not only of explaining a great past event – important as that should be – but of showing that the explanation is workable. If we can indeed come to understand how a living organism arises from the nonliving, we should be able to construct one – only of the simplest description, to be sure, but still recognizably alive. This is so remote a possibility now that one scarcely dares to acknowledge it; but it is there nevertheless.

One answer to the problem of how life originated is that it was created. This is an understandable confusion of nature with technology. Men are used to making things; it is a ready thought that those things not made by men were made by a superhuman being. Most of the cultures we know contain mythical accounts of a supernatural creation of life. Our own tradition provides such an account in the opening chapters of *Genesis*. There we are told that beginning on the third day of the Creation, God brought forth living creatures—first plants, then fishes and birds, then land animals and finally man.

Theories of Origin

Four theories have been advanced to account for the existence of the varied kinds of animals and plants on earth to-day—theories in some respects diametrically opposed to one another, in other respects somewhat in accord. They are:

Theory of Eternity of Present Conditions

The first theory argues for the unchangeableness of the universe, holding not only that organisms have been unalterable throughout their existence, but that they have always existed and will continue to exist in the same unchanging state throughout eternity. This was apparently the belief of very few authorities, for one finds almost no allusion to it in the literature of science, although Hutton wrote: "The result of this physical enquiry is that we find no vestige of a beginning - no prospect of an end." Whether this should be interpreted as a statement that the world has neither beginning nor end is, however, open to question.

Theory of Special Creation

The second theory, that of Special Creation, or Creationism, is the literal interpretation of the Mosaic account of creation set forth in the first chapter of Genesis - a simple story, beautifully told, derived from the Hebrew tradition and well suited to the state of knowledge of the times and of the people for whom it was written. This account, strictly interpreted, has been the teaching, not alone of the Hebrew, but of the Christian church authorities for many centuries, although the increase of zoological knowledge made it harder and harder to reconcile with observed facts, until it was replaced by the doctrine of Evolution.

Suarez

One of the greatest advocates of the Special Creation doctrine during Christian times was Father Suarez (1548-1617), a Spanish Jesuit priest, who taught emphatically that "the world was made in six natural days. On the first of these days the *materia prima* was made out of nothing, to receive afterwards those 'substantial forms' which moulded it into the universe of things; on the third day, the ancestors of all living plants suddenly came into being, full-grown, perfect, and possessed of all the properties which now distinguish them; while, on the fifth and sixth days, the ancestors of all existing animals were similarly caused to exist in their complete and perfect state, by the infusion of their appropriate material substantial forms into the matter which had already been created. Finally, on the sixth day, the *anima rationalis* – that rational and immortal substantial form which is peculiar to man – was created out of nothing, and 'breathed into' a mass of matter which, till then, was mere dust of the earth, and so man arose. But the species man was presented by a solitary made individual, until the Creator took out one of his ribs and fashioned it into a female" (Huxley).

So profound was Suarez' influence upon European Catholic thought that his teaching continued to be the only orthodox belief in Europe until the middle of the nineteenth century. In a similar manner John Milton (1608-1674) influenced Protestant thought in England by the wondrously written story of the creation in *Paradise Lost*.

Some advocates of the theory claimed that none of the forms had changed in the several thousand years which had elapsed since the beginning; but that the latter-day descendants were in every way precisely similar to the original pair when they issued from the hands of their Creator. Other keen observers, like Linnaeus, thought that all the species of one genus constituted at the creation but one form, *ab initio unam constituerint speciem;* their number being subsequently increased through intercrossing with other species, and the hybrids thus produced forming additional species to those originally created. Linnaeus also held that certain forms had lost their pristine character through degeneracy – the result of climate and environment.

Theory of Spontaneous Generation or Abiogenesis

According to this theory, life has originated from non-living organic matter *abiogenetically* (*Gr. A, not; bios, life; genesis,* origin) from time to time. Greek philosophers of prechristian era, like *Thales*, *Anaximander*, *Anaximenes*, *Xenophanes*, *Empedocles*, *Plato*, *Aristotle* etc. are the followers of this theory.

According to *Epicurus* (342-271 B.C.) worms and numerous other animals were generated from the soil or manure by the action of moist warmth of sun and air. *Anaxagoras* (510-428 B.C.) thought that life come in tiny seeds (spermeia) with the rain water to fructifly the earth. According to *Aristotle* (384-322 B.C.) living creatures are born from like species no doubt, but they also arise spontaneously.

Thus, it was regarded a normal thing for worms, larvae of bees, wasps, mites, glow-worms and other insects to be formed from due, rotting dung and mud, from dry wood, from sweat from meat Flies, moths, butterflies, midges, dungbeetles, fleas, bugs and lice arise from field humus, from moulds and dung, from decaying wood and fruits, (impurities) of vineger and from old wool. *Aristotle* was of the opinion that not only insects and worms but also other highly organized creatures could be spontaneously formed, for example, crabs and various molluscs

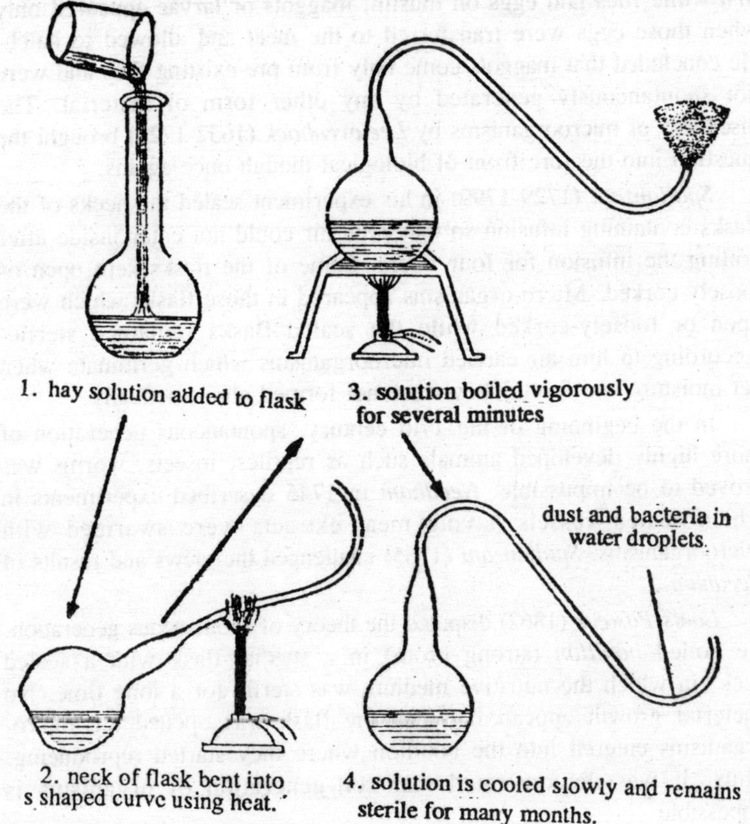

Fig. 3.1. Pasteur's experiment with bend-necked flask to disproove Theory of Sponganeous Generation.

from foul earth and mud; eels and many other fish from the mud of lakes, from sand and from decaying water, plants; even frogs and salamanders could arise from moist earth.

Belief in abiogenesis was so strong that *Van Helmont* (1652) claimed the formation of mice in 21 days from dirty, sweat-soaked shirt put in wheat barn in the dark.

Till the middle of the 17th century, the theory of spontaneous generation was widely accepted. But later on several scholars viz., *Francesco Redi* in 17th century and *Abbe Spallanzani* in 18th century disproved the idea of spontaneous generation.

Redi (1621-1691) described a series of experiments in which he placed meat or fish (eel) under clean muslin coverings and demonstrated that while flies laid eggs on muslin, maggots or larvae appeared only when those eggs were transferred to the meat and allowed to hatch. He concluded that maggots come only from pre-existing flies and were not spontaneously generated by any other form of material. The discovery of microorganisms by *Leeuwenhock* (1632-1723) brought the question into the fore-front of biological though once agains.

Spallanzani (1729-1799) in his experiment sealed the necks of the flasks containing infusion so that even air could not enter inside after boiling the infusion for four hours. Some of the flasks kept open or loosely-corked. Micro-organisms appeared in those flasks which were open or loosely-corked while the sealed flasks remained sterile. According to him air carried microorganisms which germinate when get moisture and food. They were not formed abiogenetically.

In the beginning of the 17th century, spontaneous generation of more highly developed animals such as reptiles, insects, worms was proved to be impossible. *Needham* in 1745 described experiments in which sealed vessels having meat extracts were swarmed with microorganisms. *Spallanzani* (1765) challenged the views and results of *Needham*.

Louis Pasteur (1862) disputed the theory of spontaneous generation. He boiled *bouillon* (strong broth) in a special flask with a sealed neck, in which the nutritive medium was sterile for a long time. No bacterial growth appeared. When the flask was opened the micro-organisms entered into the bouillon where they started reproducing. Thus, it was demonstrated that self-generation of organisms is impossible.

Experiments performed by *Spallanzani* and *Pasteur* proved beyond doubt the fallacy of spontaneous generations.

Theory of Catastrophism

Cuvier

A new complication arose through the discovery of older faunas, the remains of which were preserved in the form of fossils and which seemed to represent creatures whose existence antedated that of the living types. Cuvier (1769-1832), one of the founders of the science of Paleontology, became interested in the bones which lay buried in the gypsum quarries in the hill of Montmartre within the present limits of the city of Paris. His studies of these forms, and especially his reconstructions of their skeletons, showed the great anatomist that he was dealing with extinct animals which had no existing representatives. Cuvier also had, because of his official position in the Jardin des Plantes, the opportunity to study hosts of specimens from all parts of the earth, and as a result of his research, gave to the world a new theory, that of Catastrophism or Catalclysm, to account for the extinction of these forms. He is generally accredited with the belief that the cataclysms were world-wide and that the slaughter of the older fauna necessitated the creation of a new one to take its place. That belief, however, was held by later scholars of the same school, but apparently not by Cuvier.

What Cuvier believed was that the catastrophes were local, "sudden revolutions, such as subsidences of the earth's crust, followed by invasions by the sea of continents once dry;" while "other revolutions resulting in the upheaval of mountain chains have again cast back the waters and allowed, on the foundation of the dried bottom of the sea, the constitution of continental soils favourable to the expansion of new terrestrial faunas; these new faunas are not created on the spot, but come from distant regions, their migration from which has become possible owing to temporary bridges between continents" (Depéret). Cuvier's belief has a great deal of truth in it, except that the "revolutions," with resulting climatic change and consequent extinctions and immigrations, have been rapid only in proportion to the length of geologic time, but very, very slow as morals note the flight of years.

D'Orbigny

Further knowledge of historical geology led to an expansion of the catastrophic belief far beyond the teaching of Cuvier, and postulated are-creation following each cataclysm and corresponding to the principal geologic periods. Alcide d'Orbigny (1802-1857), writing in the year 1848, expounded this theory as follows:

"The first creation shows itself in the Silurian stage. After its annihilation through some geological cause or other, a second creation took place a considerable time after in the Devonian stage, and, twenty-seven times in succession, *distinct creations* have come to re-people the whole earth with its plants and animals after each of the geological disturbances which destroyed everything in living nature. Such is the fact, certain but incomprehensible, which we confine ourselves to stating, without endeavoring to solve the superhuman mystery which envelops it" (Depéret).

Cosmozoic Theory or Theory of Panspermia

According to *Richter*, the protoplasm reached the earth in the form of sports or germ or other simple particle with cosmic duct from other planet.

Arrehenius proposed *cosmic panspermia theory*, according to him the life existed throughout the Universe and their spores etc. could freely travel through the space from one planet to other.

According to *Helmholtz*, the life was brough on the earth with falling meteorites.

However, this theory has been rejected due to two reasons. First, due to intense cold, extreme dryness and intense radiation of interstellar space. Even the most resistant living spores cannot withstand exposure to interstellarspace. Moreover, this theory does not answer the basic question of 'origin of life' upon earth.

Modern Concepts on the Origin of Life

Scientists are now of the opinion that spontaneous generation is not possible at present on Earth and that specific conditions are necessary for the appearance of life.

T.H. Huxley (1869) and *John Tyndall* (1874) asserted that life could be generated from inorganic chemicals. But their ideas were vague as the knowledge of biochemistry was not available at that time.

In 1920s, a new interest in the origin of life arose from independent speculation of two biologists, *A.I. Oparin* and *J.B.S. Haldane*. Both these scientists got the ideas from the new biochemistry that was founded by Sir *F.G. Hopkins*. They were of the opinion that the evolution is purely chemical—the gradual transmutation of inorganic compounds into organic ones.

After 1920s, the subject of the origin of life attracted ever increasing interest. *Pirie* coined the term *Biopoiesis* for the study of the origin of life.

Oparin first of all gave the scientific account of the origin of life in his book entitled. "*The Origin of Life*", which was published in 1936. His main postulates are as follows:

Origin of earth

The earth is presumed to have originated about 5-6 billion years ago. There are two hypothesis for its origin (i) *Planetesimal hypothesis* believes that earth is originated as a part broken off from the modern mass of sub (ii) *Nebular hypothesis*. According to this hypothesis the earth is originated by gradual condensation of interstellar dust from which entire solar system is supposed to be formed. At that time temperature was extremely high which served as basis of primary origin of organic compounds on the earth.

Presence of elements on earth

In the begining, the earth was fiery spinning ball of hot gases and vapours of elements. Gradually, through hundreds of millions of years, the gases condensed into a molten core and different element got stratified according to their density. The original temperature of the earth was very high, about 5000-6000°C. Due to this high temperature carbon elements existed in three major forms, dicarbon (C_2), cynogen (CN), methane (CH_4), metal carbides, carbon dioxide etc. Oxygen was not existing in free state but oxides of àluminium, boron, hydrogen etc. were present. Nitrogen existed in combination with metals to form nitrides.

The earth cooled gradually and some of the gases liquefied and some solidified. Steam condensed into water and resulted in rain. The water again returned to atmosphere because of superheated earth. This cycle continued for millions of years and resulted in the cooling of earth. Due to cooling down of earth first ocean came in existence. The oceanic water contained methane, ammonia of atmosphere in dissolved condition.

Theory of Organic Evolution

Evolution is the gradual development from the simple unorganized condition of primal matter to the complex structure of the physical universe; and in like manner, from the beginning of organic life on the habitable planet, a gradual unfolding and branching out into all the varied forms of beings which constitute the animal and plant kingdoms. The first is called Inorganic, the last Organic Evolution, or descent with modification.

Early Greek Theories

Organic Evolution is often imagined to be a nineteenth century contribution to biologic science, whereas the idea is itself the product of an evolution of thought and is the fruition of no fewer than twenty-four centuries of speculation and research. The germ of the evolutionary idea had its inception with the Greeks, whose wonderful fertility of mind has so enriched the world, the first writer to deal with the problem, Anaximander, living five and a half centuries before the Christian era. Empedocles (495-435 B.C.) may be called the father of Evolution, though the Evolution that he taught implied no succession of related animals, gradually improving in successive generations, but a series of attempts on the part of nature to produce more perfect forms, the unfit being eliminated. He is the first to show the possibility of the origin in the fittest forms through chance rather than through design.

Another Greek, Democritus (460-?357 B.C.), went further than Empedocles in that he taught the adaptations of single structures and organs, whereas the latter applied the idea to entire organisms. But by far the most notable figure in Greek philosophy was Aristotle (384-322 B.C.), whose versatility as a writer upon all aspects of human knowledge was remarkable. In view of the limited opportunities for observation possible in those days when the teeming host of microscopic forms as well as the extinct creatures were utterly unknown, the deductions of Aristotle, even where he appears to retrogress from the truth, are highly logical. He did not believe in Special Creation, nevertheless he postulates an intelligent design as the primary cause of the changes which has been wrought in nature, and the central thought in his evolutionary theory, if such it was, is an *internal perfecting tendency* impelling organisms to greater and greater perfection. As a result of this, he saw a complete gradation in nature from the mineral to the plant, the plant-like animal, the animal with senses and hence locomotor powers, and finally man.

Aristotle considered life a function of the organism, not a separate principle, and had an understanding of adaptations and of heredity, even of the atavistic heredity wherein an ancestral trait reappears in a later descendant after having lain dormant for several generations. Osborn says of him:

Aristotle's argument for "operation of natural law, rather than of chance, in the lifeless and in the living world, is a perfectly logical one, and his consequent rejection of the hypothesis of the Survival of the Fittest, a sound induction from his own limited knowledge of Nature... If he had accepted Empedocles' hypothesis [of the origin of the fittest through chance rather than through design] he would have been the literal prophet of Darwinism."

To summarize, then, the Greeks offered as causes of evolutionary change three explanations:

1. Intelligent design,
2. The operation of natural laws implanted by intelligent design.
3. The operation of natural causes due to laws of change—no evidence of design, even in origin.

Middle Ages

And now, for hundreds of years, owing largely to the repressive measures of the church authorities, though some, like Saint Augustine, would have taught otherwise, the progress of the evolutionary idea virtually ceased until the coming of the philosophers Bacon, Descartes, Leibnitz, and Kant, and the naturalists Linnaeus, Buffon, Erasmus and Charles Darwin, E. Geofisms which lead, during their larval state, a planktonic existence. These forms are as a rule extremely small and have feeble powers of locomotion, generally by means of cilia. Nevertheless they are so numerous that the upper strata of the oceans are literally crowded with them and they form a great source of food supply for the more aggressive forms of life. They pass their short existence floating about in the sea, in swarms. Sooner or later, however, they sink to the bottom, and if they fall upon the proper sort of substratum, they develop into the benthonic adult; but if they fall upon an unfavourable bottom, or if food supply is scarce, they perish.

The necessity of some means of dispersal or for the repopulation of an area wherein accident has destroyed the original inhabitants is imperative, and in sedentary adults can only be attained by this means. Mero-planktonic larvae are found in every group of aquatic sedentary benthonic animals. The young of the sedentary scale insects, on the other hand, which are active for a brief time, are vagrant benthos. Germs, spores, and many seeds like those of the maple and dandelion constitute about all that can possibly be included in aerial mero-plankton.

Pseudo-plankton is a term proposed for organisms such as the sargassum or gulf sea-weed which is normally or in early life an attached benthonic oganism but which becomes planktonic. The meaning of the term has been extended to include plants or animals living as sedentary or vagrant benthos upon floating objects, such as the algae, hydroids, or bryozoans, which may be attached to the floating sargassum, and the crustaceans, molluscs, or other animals which dwell among them. In many instances the pseudo-planktonic existence of these forms is due to accident, but on the other hand it seems to be habitual with certain forms, which, like the goose-barnacle, *Lepas*, rarely occur except attached to floating objects such as timber or the bottoms of ships, especially when the latter are derelict. Many of the animals found on floating sargassum seem to be characteristic of it in this condition, as they do not occur when it is attached.

The Task

To make an organism demands the right substances in the right proportions and in the right arrangement. We do not think that anything more is needed – but that is problem enough.

The substances are water, certain salts – as it happens, those found in the ocean – and carbon compounds. The latter are called *organic* compounds because they scarcely occur except as products of living organisms.

Organic compounds consist for the most part of four types of atoms: carbon, oxygen, nitrogen and hydrogen. These four atoms together constitute about 99 per cent of living material, for hydrogen and oxygen also form water. The organic compounds found in organisms fall mainly into four great classes: carbohydrates, fats, proteins and nucleic acids. The illustrations on this and the next three pages give some notion of their composition and degrees of complexity. The fats are simplest, each consisting of three fatty acids joined to glycerol. The starches and glycogens are made of sugar units strung together to form long straight and branched chains. In general only one type of sugar appears in a single starch or glycogen; these molecules are large, but still relatively simple. The principal function of carbohydrates and fats in the organism is to serve as fuel – as a source of energy.

The *nucleic acids* introduce a further level of complexity. They are very large structures, composed of aggregates of at least four types of unit – the nucleotides – brought together in a great variety of proportions and sequences. An almost endless variety of different nucleic acids is possible, and specific differences among them are believed to be of the highest importance. Indeed, these structures are thought by many to be the main constituents of the genes, the bearers of hereditary constitution.

Variety and specificity, however, are most characteristic of the proteins, which include the largest and most complex molecules known. The units of which their structure is built are about 25 different amino acids. These are strung together in chains hundreds to thousands of units long, in different proportions, in all types of sequence, and with the greatest variety of branching and folding. A virtually infinite number of different proteins is possible. Organisms seem to exploit this potentiality, for no two species of living organism, animal or plant, possess the same proteins.

Organic molecules therefore form a large and formidable array, endless in variety and of the most bewildering complexity. One cannot think of having organisms without them. This is precisely the trouble, for to understand how organisms originated we must first of all explain how such complicated molecules could come into being. And that is only the beginning. To make an organism requires not only a tremendous variety of these substances, in adequate amounts and proper proportions, but also just the right arrangement of them. Structure here is as important as composition - and what a complication of structure! The most complex machine man has devised - say an electronic brain-is child's play compared with the simplest of living organisms. The especially trying thing is that complexity here involves such small dimensions. It is on the molecular level; it consists of a detailed fitting of molecule to molecule such as no chemist can attempt.

The Possible and Impossible

One has only to contemplate the magnitude of this task to concede that the spontaneous generation of a living organism is impossible. Yet here we are - as a result of spontaneous generation. It will help to digress for a moment to ask what one means by "impossible."

With every event one can associate a probability - the chance that it will occur. This is always a fraction, the proportion of times the event occurs in a large number of trials. Sometimes the probability is apparent even without trial. A coin has two faces; the probability of tossing a head is therefore 1/2. A die has six faces; the probability of throwing a deuce is 1/6. When one has no means of estimating the probability beforehand, it must be determined by counting the fraction of successes in a large number of trials.

Our everyday concept of what is impossible, possible or certain derives from our experience: the number of trials that may be encompassed within the space of a human lifetime, or at most within recorded human history.

But even within the bounds of our own time there is a serious flaw in our judgement of what is possible. It sounds impressive to say that an event has never been observed in the whole of human history.

We should tend to regard such an event as at least "practically" impossible, whatever probability is assigned to it on abstract grounds. When we look a little further into such a statement, however, it proves to be almost meaningless. For men are apt to reject reports over very improbable occurrences. Persons of good judgement think it safer to destrust the alleged observer of such an event than to believe him. The result is that events which are merely very extraordinary acquire the reputation of never having occurred at all. Thus the highly improbable is made to appear impossible.

To give an example: Every physicist knows that there is a very small probability, which is easily computed. The event requires no more than that the molecules of which the table is composed, ordinarily in random motion in all directions, should happen by chance to move in the same direction. Every physicist concedes this possibility; but try telling one that you have seen it happen.

A final aspect of the problem is very important. When we consider the spontaneous origin of a living organism, this is not an event that need happen again and again. It is perhaps enough for it to happen once. The probability with which we are concerned is of a special kind; it is the probability that an event occur at least once. To this type of probability a fundamentally important thing happens as one increases the number of trials. However improbable the event in a single trial, it becomes increasingly probable as the trials are multiplied. Eventually the event becomes virtually inevitable. For instance, the chance that a coin will not fall head up in a single toss is ½. The chance that no head will appear in a series of tosses is ½ × ½ × ½ . . . . as many times over as the number of tosses. In 10 tosses the chance that no head will appear is therefore ½ multiplied by itself 10 times, or 1/1,000. Consequently the chance that a head will appear at least once in 10 tosses is 999/1,000. Ten trials have converted what started as a modest probability to a near certainty.

The same effect can be achieved with any probability, however small, by multiplying sufficiently the number of trials. Consider a reasonably improbable event, the chance of which is 1/1,000. The chance that this will not occur in one trial is 999/1,000. The chance that it won't occur in 1,000 trials is 999/1,000 multiplied together 1,000 times. This fraction comes out to be 37/100. The chance that it will happen at least once in 1,000 trials is therefore one minus this number - 63/100 - a little better than three chances out of five. One thousand trials have transformed this from a highly improbable to a highly probable event. In 10,000 trials the chance that this event will occur at least once comes out to be 19,999/20,000. It is now almost inevitable.

It makes no important change in the argument if we assess the probability that an event occur at least two, three, four or some other small number of times rather than at least once. It simply means that more trials are needed to achieve any degree of certainty. Otherwise everything is the same.

In such a problem as the spontaneous origin of life we have no way of assessing probabilities beforehand, or even of deciding what we mean by a trial. The origin of a living organism is undoubtedly a stepwise phenomenon, each step with its own probability and its own conditions of trial. Of one thing we can be sure, however: whatever constitutes a trial, more such trials occur the longer the interval of time.

The important point is that since the origin of life belongs in the category of at least once phenomena, time is on is side. However improbable we regard this event, or any of the steps which it involves, given enough time it will almost certainly happen at least once. And for life as we know it, with its capacity for growth and reproduction, once may be enough.

Time is in fact the hero of the plot. The time with which we have to deal is of the order of two billion years. What we regard as impossible on the basis of human experience is meaningless here. Given so much time, the "impossible" becomes possible, the possible probable, and the probable virtually certain. One has only to wait: time itself performs the miracles.

Organic Molecules

This brings the argument back to its first stage: the origin of organic compounds. Until a century and a quarter ago the only known source of these substances was the stuff of living organisms. Students of chemistry are usually told that when, in 1828, Friedrich Wöhler synthesized the first organic compound, urea, he proved that organic compounds do not require living organisms to make them. Of course it showed nothing of the kind. Organic chemists are alive; Wöhler merely showed that they can make organic compounds externally as well as internally. It is still true that with almost negligible exceptions all the organic matter we know is the product of living organisms.

The almost negligible exceptions, however, are very important for our argument. It is now recognized that a constant, slow production of organic molecules occurs without the agency of living things. Certain geological phenomena yield simple organic compounds. So, for example, volcanic eruptions bring metal carbides to the surface of the earth, where they react with water vapour to yield simple compounds of carbon and hydrogen. The familiar type of such a reaction is the process used in old-style bicycle lamps in which acetylene is made by mixing iron carbide with water.

Recently harold Urey, Nobel laureate in chemistry, has become interested in the degree to which electrical discharges in the upper atmosphere may promote the formation of organic compounds. One of his students, S.L. Miller, performed the simple experiment of circulating a mixture of water vapour, methane (CH_4), ammonia (NH_3) and hydrogen - all gases believed to have been present in the early atmosphere of the earth - continuously for a week over an electric spark. The circulation was maintained by boiling the water in one limb of the apparatus and condensing it in the other. At the end of the week the water was analyzed by the delicate method of paper chromatography. It was found to have acquired a mixture of amino acids! Glycine and alanine, the simplest amino acids and the most prevalent in proteins, were definitely identified in the solution, and there were indications it contained aspartic acid and two others. The yield was surprisingly high. This amazing result changes at a stroke our ideas of the probability of the spontaneous formation of amino acids.

A final consideration, however, seems to me more important than all the special processes to which one might appeal for organic synthesis in inanimate nature.

It has already been said that to have organic molecules one ordinarily needs organisms. The synthesis of organic substances, like almost everything else that happens in organisms, is governed by the special class of proteins called enzymes - the organic catalysts which greatly accelerate chemical reactions in the body. Since an enzyme is not used up but is returned at the end of the process, a small amount of enzyme can promote an enormous transformation of material.

Enzymes play such a dominant role in the chemistry of life that it is exceedingly difficult to imagine the synthesis of living material without their help. This poses a dilemma, for enzymes themselves are proteins, and hence among the most complex organic components of the cell. One is asking, in effect, for an apparatus which is the unique property of cells in order to form the first cell.

This is not, however, an insuperable difficulty. An enzyme, after all, is only a catalyst; it can do not more than change the *rate* of a chemical reaction. It cannot make anything happen that would not have happened, though more slowly, in its absence. Every process that is catalyzed by an enzyme, and every product of such a process, would occur without the enzyme. The only difference is one of rate.

Once again the essence of the argument is time. What takes only a few moments in the presence of an enzyme or other catalyst may take days, months or years in its absence; but given time, the end result is the same.

Indeed, this great difficulty in conceiving of the spontaneous generation of organic compounds has its positive side. In a sense, organisms demonstrate to us what organic reactions and products are *possible*. We can be certain that, given time, all these things must occur. Every substance that has ever been found in an organism displays thereby the finite probability of its occurrences. Hence, given time, it should arise spontaneously. One has only to wait.

It will be objected at once that this is just what one cannot do. Everyone knows that these substances are highly perishable. Granted that, within long spaces of time, now a sugar molecule, now a fat, now even a protein might form spontaneously, each of these molecules should have only a transitory existence. How are they ever to accumulate; and, unless they do so, how form an organism?

We must turn the question around. What, in our experience, is known to destroy organic compounds? Primarily two agencies: decay and the attack of oxygen. But decay is the work of living organisms, and we are talking of a time before life existed. As for oxygen, this introduces a further and fundamental section of our argument.

It is generally conceded at present that the early atmosphere of our planet contained virtually no free oxygen. Almost all the earth's oxygen was bound in the form of water and metal oxides. If this were not so, it would be very difficult to imagine how organic matter could accumulate over the long stretches of time that alone might make possible the spontaneous origin of life. This is a crucial point, therefore, and the statement that the early atmosphere of the planet was virtually oxygen-free comes forward so opportunely as to raise a suspicion of special pleading.

Apparently something similar was true also for another common component of our atmosphere - carbon dioxide. It is believed that most of the carbon on the earth during its early geological history existed as the element or in metal carbides and hydrocarbons; very little was combined with oxygen.

This situation is not without it irony. We tend usually to think that the environment plays the tune to which the organisms must dance. The environment is given; the organism's problem is to adapt to it or die. It has become apparent lately, however, that some of the most important features of the physical environment are themselves the work of living organisms. Two such features have just been named. The atmosphere of our planet seems to have contained no oxygen until organisms placed it thereby the process of plant photosynthesis. It is estimated that at present all the oxygen of our atmosphere is renewed by photosynthesis once in every 2,000 years, and that all the carbon dioxide passes through the process of photosynthesis once in every 300 years. In the scale of geological time, these intervals are very small indeed. We are left with the realization that all the oxygen and carbon dioxide of our planet are the products of living organisms, and have passed through living organisms over and over again.

Forces of Dissolution

In the early history of our planet, when there were no organisms or any free oxygen, organic compounds should have been stable over very long periods. This is the crucial difference between the period before life existed and our own. If one were to specify a single reason why the spontaneous generation of living organisms was possible once and is so no longer, this is the reason.

We must still reckon, however, with another destructive force which is disposed of less easily. This can be called spontaneous dissolution – the counter-part of spontaneous generation. We have noted that any process catalyzed by an enzyme can occur in time without the enzyme. The trouble is that the processes which synthesize an organic substance are reversible: any chemical reaction which an enzyme may catalyze will go backward as well as forward. We have spoken as though one has only to wait to achieve synthesis of all kinds; it is true to say that what one achieves by waiting is *equilibria* of all kinds – equilibria in which the synthesis and dissolution of substances come into balance.

In the vast majority of the processes in which we are interested the point of equilibrium lies far over toward the side of dissolution. That is to say, spontaneous dissolution is much more probable, and hence proceeds much more rapidly, than spontaneous synthesis. For example, the spontaneous union, step by step, of amino acid units to form a protein has a certain small probability, and hence might occur over a long stretch of time. But the dissolution of the protein or of an intermediate product into its component amino acids is much more probable, and hence will go ever so much more rapidly. The situation we must face is that of patient Penelope waiting for Odysseus yet much worse: each night she undid the weaving of the preceding day, but here a night could readily undo the work of a year or a century.

How do present-day organisms manage to synthesize organic compounds against the forces of dissolution? They do so by a continuous expenditure of energy. Indeed, living organisms commonly do better than oppose the forces of dissolution; they grow in spite of them. They do so, however, only at enormous expense to their surroundings. They need a constant supply of material and energy merely to maintain themselves, and much more of both to grow and reproduce. A living organism is an intricate machine for performing exactly this function. When, for want of fuel or through some internal failure in its mechanism, an organism stops actively synthesizing itself in opposition to the processes which continuously decompose it, it dies and rapidly disintegrates.

Forces of Integration

At present we can make only a beginning with this problem. We know that it is possible on occasion to protect molecules from dissolution by precipitation or by attachment to other molecules. A wide variety of such precipitation and "trapping" reactions is used in modern chemistry and biochemistry to promote synthesis. Some molecules appear to acquire a degree of resistance to disintegration simply through their size. So, for example, the larger molecules composed of amino acids – polypeptides and proteins – seem to display much less tendency to disintegrate into their units than do smaller compounds of two or three amino acids.

Again, many organic molecules display still another type of integrating force – a spontaneous impulse toward structure formation. Certain types of fatty molecules – lecithins and cephalins – spin themselves out in water to form highly oriented and well-shaped structures – the so-called myelin figures. Proteins sometimes orient even in solution, and also may aggregate in the solid state in highly organized formations. Such spontaneous architectonic tendencies are still largely unexplored, particularly and they may occur in complex mixtures of substances, and they involve forces the strength of which has not yet been estimated.

What we are saying is that possibilities exist for opposing *intramolecular* dissolution by *intermolecular* aggregations of various kinds. The equilibrium between union and disunion of the amino acids that make up a protein is all to the advantage of disunion, but the aggregation of the protein with itself or other molecules might swing the equilibrium in the opposite direction: perhaps by removing the protein from access to the water which would be requried to disintegrate it or by providing some particularly stable type of molecular association.

In such a scheme the protein appears only as a transient intermediate, an unstable way-station, which can either fall back to a mixture of its constituent amino acids or enter into the formation of a complex structural aggregate: amino acids $\leftrightarrows$ protein $\rightarrow$ aggregate.

Such molecular aggregates, of various degrees of material and architectural complexity, are indispensable intermediates between molecules and organisms. We have no need to try to imagine the spontaneous formation of an organism by one grand collision of its component molecules. The whole process must be gradual. The molecules form aggregates, small and large. The aggregates add further molecules, thus growing in size and complexity. Aggregates of various kinds interact with one another to form still larger and more complex structures. In this way we imagine the ascent, not by jumps or master strokes, but gradually, piecemeal, to the first living organisms.

First Organisms

Where may this have happened? It is easiest to suppose that life first arose in the sea. Here were the necessary salts and the water. The latter is not only the principal component of organisms, but prior to their formation provided a medium which could dissolve molecules of the widest variety and ceaselessly mix and circulate them. It is this constant mixture and collision of organic molecules of every sort that constituted in large part the "trials" of our earlier discussion of probabilities.

The sea in fact gradually turned into a dilute broth, sterile and oxygen-free. In this broth molecules came together in increasing number and variety, sometimes merely to collide and separate, sometimes to react with one another to produce new combinations, sometimes to aggregate into multimolecular formations of increasing size and complexity.

What brought order into such complexes? For order is as essential here as composition. To form an organism, molecules must enter into intricate designs and connections; they must eventually form a self-repairing, self-constructing dynamic machine. For a time this problem of molecular arrangement seemed to present an almost insuperable obstacle in the way of imagining a spontaneous origin of life, or indeed the laboratory synthesis of a living organism. It is still a large and mysterious problem, but it no longer seems insuperable. The change in view has come about because we now realize that it is not altogether necessary to *bring* order into this situation; a great deal of order is implicit in the molecules themselves.

The epitome of molecular order is a crystal. In a perfect crystal the molecules display complete regularity of position and orientation in all planes of space. At the other extreme are fluids - liquids or gases - in which the molecules are in ceaseless motion and in wholly random orientations and positions.

Lately it has become clear that very little of a living cell is truly fluid. Most of it consists of molecules which have taken up various degrees of orientation with regard to one another. That is, most of the cell represents various degrees of approach to crystallinity - often, however, with very important differences from the crystals most familiar to us. Much of the cell's crystallinity involves molecules which are still in solution - so-called liquid crystals - and much of the dynamic, plastic quality of cellular structure, the capacity for constant change of shape and interchange of material, derives from this condition. Our familiar crystals, furthermore, involves only one or a very few types of molecule, while in the cell a great variety of different molecules come together in some degree of regular spacing and orientation - *i.e.*, some degree of crystallinity. We are dealing in the cell with highly mixed crystals and near-crystals, solid and liquid. The laboratory study of this type of formation has scarcely begun. Its further exploration is of the highest importance for our problem.

In a fluid such as water the molecules are in very rapid motion. Any molecules dissolved in such a medium are under a constant barrage of collisions with water molecules. This keeps small and moderately sized molecules in a constant turmoil; they are knocked about at random, colliding again and again, never holding any position or orientation for more than an instant. The larger a molecule is relative to water, the less it is disturbed by such collisions. Many protein and nucleic acid molecules are so large that even in solution their motions are very sluggish, and since they carry large numbers of electric charges distributed about their surfaces, they tend even in solution to align with respect to one another. It is so that they tend to form liquid crystals.

We have spoken above of architectonic tendencies even among some of the relatively small molecules; the lecithins and cephalins. Such molecules are insoluble in water yet possess special groups which have a high affinity for water. As a result they tend to form surface layers, in which their water-seeking groups project into the water phase, while their water-repelling portions project into the air, or into an oil phase, or unite to form an oil phase. The result is that quite spontaneously such molecules, when exposed to water, take up highly oriented positions to form surface membranes, myelin figures and other quasicrystalline structures.

Recently several particularly striking examples have been reported of the spontaneous production of familiar types of biological structure by protein molecules. Cartilage and muscle offer some of the most intricate and regular patterns of structure to be found in organisms. A fibre from either type of tissue presents under the electron microscope a beautiful pattern of cross striations of various widths and densities, very regularly spaced. The proteins that form these structures can be coaxed into free solution and stirred into completely random orientation. Yet on precipitating, under proper conditions, the molecules realign with regard to one another to regenerate with extraordinary fidelity the original patterns of the tissues.

We have therefore a genuine basis for the view that the molecules of our oceanic broth will not only come together spontaneously to form aggregates but in doing so will spontaneously achieve various types and degrees of order. This greatly simplifies our problem. What it means is that, given the right molecules, one does not have to do everything for them; they do a great deal for themselves.

Oparin has made the ingenious suggestion that natural selection, which *Darwin* proposed to be the driving force of organic evolution, begins to operate at this level. He suggests that as the molecules come together to form colloidal aggregates, the latter begin to compete with one another for material. Some aggregates, by virtue of especially favourable composition or internal arrangement, acquire new molecules more rapidly than others. They eventually emerge as the dominant types. Oparin suggests further than considerations of optimal size enter at this level. A growing colloidal particle may reach a point at which it becomes unstable and breaks down into smaller particles, each of which grows and redivides. All these phenomena lie within the bonds of known processes in nonliving systems.

The Sources of Energy

We suppose that all these forces and factors, and others perhaps yet to be revealed, together give us eventually the first living organism. That achieved, how does the organism continue to live?

It has already noted that a living organism is a dynamic structure. It is the site of a continuous influx and outflow of matter and energy. This is the very sign of life, its cessation the best evidence of death. What is the primal organism to use as food, and how derive the energy it needs to maintain itself and grow?

For the primal organism, generated under the conditions we have described, only one answer is possible. Having arisen in an oceanic broth of organic molecules, its only recourse is to live upon them. There is only one way of doing that in the absence of oxygen. It is called fermentation: the process by which organisms derive energy by breaking organic molecules and rearranging their parts. The most familiar example of such a process is the fermen-tation of sugar by yeast, which yields alcohol as one of the products. Animal cells also ferment sugar, not to alcohol but to lactic acid. These are two examples from a host of known fermentations.

The yeast fermentation has the following over-all equation: $C_6H_{12}O_6 \rightarrow 2CO_2 + 2C_2H_5OH + \text{energy}$. The result of fragmenting 180 grams of sugar into 88 grams of carbon dioxide and 92 grams of alcohol is to make available about 20,000 calories of energy for the use of the cell. The energy is all that the cell derives by this transaction; the carbon dioxide and alcohol are waste products which must be got rid of somehow if the cell is to survive.

The cell, having arisen in a broth of organic compounds accumualted over the ages, must consume these molecules by fermentation in order to acquire the energy it needs to live, grow and reproduce. In doing so, it and its descendants are living on borrowed time. They are consuming their heritage, just as we in our time have nearly consumed our heritage of coal and oil. Eventually such a process must come to an end, and with that life also should have ended. It would have been necessary to start the entire development again.

Fortunately, however, the waste product carbon dioxide saved this situation. This gas entered the ocean and the atmosphere in ever-increasing quantity. Some time before the cell exhausted the supply of organic molecules, it succeeded in interventing the process of photosynthesis. This enabled it, with the energy of sunlight, to make its own organic molecules; first sugar from carbon dioxide and water, then, with ammonia and nitrates as sources of nitrogen, the entire array of organic compounds which it requires. The sugar synthesis equation is: $6\ CO_2 + 6\ H_2O + \text{sunlight} \rightarrow C_6H_{12}O_6 + 6\ O_2$. Hence 264 grams of carbon dioxide *plus* 108 grams of water *plus* about 700,000 calories of sunlight yield 180 grams of sugar and 192 grams of oxygen.

This is an enormous step forward. Living organisms no longer needed to depend upon the accumulation of organic matter from past ages; they could make their own. With the energy of sunlight they could accomplish the fundamental organic synthesis that provide their substance, and by fermentation they could produce what energy they needed.

Fermentation, however, is an extraordinarily inefficient source of energy. It leaves most of the energy potential of organic compounds unexploited; consequently huge amounts of organic material must be fermented to provide a modicum of energy. It produces also various poisonous waste products - alcohol, lactic acid, acetic acid, formic acid and so on. In the sea such products are readily washed away, but if organisms were ever to penetrate to the air and land, these products must prove a serious embarrassment.

One of the by-products of photosynthesis, however, is oxygen. Once this was available, organisms could invent a new way to acquire energy, many times as efficient as fermentation. This is the process of cold combustion called respiration: $C_6H_{12}O + 6O_2 \rightarrow 6CO_2 + 6H_2O + energy$. The burning of 180 grams of sugar in cellular respiration yields about 700,000 calories, as compared with the approximately 20,000 calories produced by fermentation of the same quantity of sugar. This process of combustion extracts all the energy that can possibly be derived from the molecules which it consumes. With this process at its disposal, the cell can meet its energy requirements with a minimum expenditure of substance. It is a further advantage that the products of respiration - water and carbon dioxide - are innocuous and easily disposed of in any environment.

Life's Capital

It is difficult to overestimate the degree to which the invention of cellular respiration released the forces of living organisms. No organism that relies wholly upon fermentation has ever amounted to much. Even after the advent of photosynthesis, organisms could have led only a marginal existence. They could indeed produce their own organic materials, but only in quantities sufficient to survive. Fermentation is so profligate a way of life that photo-synthesis could do little more than keep up with it. Respiration used the material of organisms with such enormously greater efficiency as for the first time to leave something over. Coupled with fermentation, photosynthesis made organisms self-sustaining; coupled with respiration, it provided a surplus. To use an economic analogy, photosynthesis brought organisms to the subsistence level; respiration provided them with capital. It is mainly the capital that they invested in the great enterprise of organic evolution.

The entry of oxygen into the atmosphere also liberated organisms in another sense. The sun's radiation contains ultraviolet components which no living cell can tolerate. We are sometimes told that if this radiation were to reach the earth's surface, life must cease. That is not quite true. Water absorbs ultraviolet radiation very effectively, and one must conclude that as long as these rays penetrated in quantity to the surface of the earth, life had to remain under water. With the appearance of oxygen, however, a layer of ozone formed high in the atmosphere and absorbed this radiation. Now organisms could for the first time emerge from the water and begin to populate the earth and air. Oxygen provided not only the means of obtaining adequate energy for evolution but the protective blanket of ozone which alone made possible terrestrial life.

This is really the end of our story. Yet not quite the end. Our entire concern in this argument has been to bring the origin of life

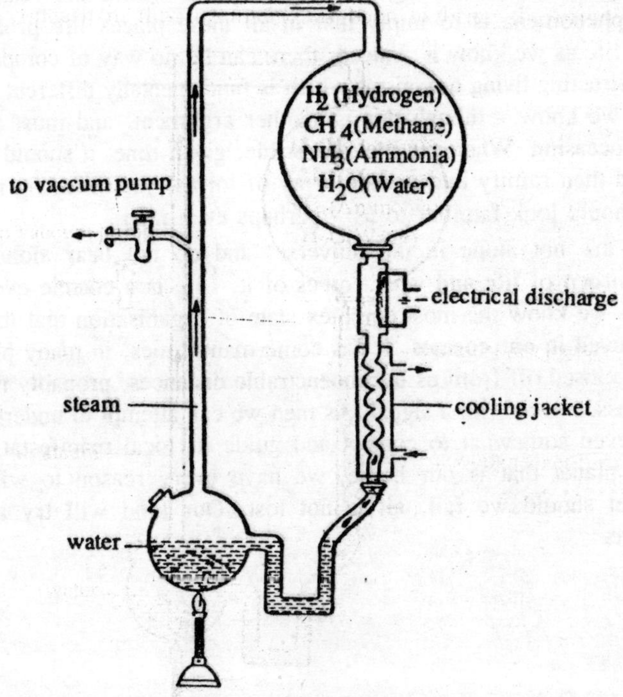

Fig. 3.2. Spark discharge apparatus.

within the compass of natural phenomena. It is of the essence of such phenomena to be repetitive, and hence, given time, to be inevitable.

This is by far our most significant conclusion – that life, as an orderly natural event on such a planet as ours, was inevitable. The same can be said of the whole of organic evolution. All of it lies within the order of nature, and apart from details all of it was inevitable.

Astronomers have reason to believe that a planet such as ours – of about the earth's size and temperature, and about as well-lighted – is a rare event in the universe. Indeed, filled as our story is with improbable phenomena, one of the least probable is to have had such a body as the earth to begin with. Yet though this probability is small, the universe is so large that it is conservatively estimated at least 100,000 planets like the earth exist in our galaxy alone. Some 100 million galaxies lie within the range of our most powerful telescopes, so that throughout observable space we can count apparently on the existence of at least 10 million planets like our own.

What it means to bring the origin of life within the realm of natural phenomena is to imply that in all these places life probably exists – life as we know it. Indeed, there can be no way of composing and constructing living organisms which is fundamentally different from the one we know – though this is another argument, and must await another occasion. Wherever life is possible, given time, it should rise. It should then ramify into a wide array of forms, yet including many which should look familiar to us – perhaps even men.

We are not alone in the universe, and do not bear alone the whole burden of life and what comes of it. Life is a cosmic event – so far as we know the most complex state of organisation that matter has achieved in our cosmos. It has come many times, in many places – places closed off from us by impenetrable distances, probably never to be crossed even with a signal. As men we can attempt to understand it, and even somewhat to control and guide its local manifestations. On this planet that is our home, we have every reason to wish it well. Yet should we fail, all is not lost. Our kind will try again elsewhere.

4

CHORDATE ORGANIZATION

The chordata occupy a greater variety of habitats and show more complicated mechanisms of self-maintenance than any other group in the whole animal kingdom. They and the arthropods and the pulmonate molluscs have fully solved the problém of life on the land—which they now dominate. This domination is achieved by most delicate mechanisms for resisting desiccation, for providing support, and for conducting many operations that are harder in the air than in water. By even more wonderful devices the body temperature is raised and kept uniform and thus all reactions accelerated. Finally, use is made of this high metabolism for the development of the nervous system into a most delicate instrument, allowing the animal not only to change its response to a given stimulus from moment to moment, but also to store up and act upon the fruits of past experience.

Besides these more developed types of chordate that dominate the land and air there are also great numbers of extremely successful aquatic and amphibious types. The frog is often referred to as a somewhat lowly and unsuccessful animal, but frogs and toads are found all over the world. The sharks and bony fishes share with the squids and whales the culminating ecological position in the food chains of the sea, while the bony fishes are the only animals that have achieved considerable size and variety in fresh water. Among the still more lowly chordates the sea-squirts take a very important, though not dominant, position among the animal and plant communities that occupy the sea bottom, but they have never entered fresh water.

One could continue indefinitely with particulars of the amazing types produced by this most adaptable phylum. Yet through all their variety of structure the chordates show a considerable uniformity of general plan, and there can be no doubt that they have all evolved from a common ancestor of what might be called a 'fish-like' habit. In the very earlier stages only the larva was fish-like, and the life-history probably also included a sessile adult stage, such as the tunicates still show today. This bottom-living phase was then eliminated by paedomorphosis, the larvae becoming the adults. Therefore the essential organization of a chordate is that of a long-bodied, free-swimming creature. All the other types can be derived from such an ancestor, though in some cases only by what is often called 'degeneration'.

Classification of Chordates

We may conveniently divide the Phylum Chordata into four Subphyla:

Subphylum 1: Hemichordata
Balanoglossus; *Cephalodiscus*; *Rhabdopleura*
Subphylum 2: Cephalochordata (=Acrania)
Branchiostoma (Amphioxus)
Subphylum 3: Tunicata (=Urochordata)
Ciona, sea-squirts
Subphylum 4: Vertebrata (=Craniata)
Superclass 1: Agnatha
Class 1: Cyclostomata. Lampreys and hag fishes
Class 2: Cephalaspidomorphi. *Cephalaspis*
Class 3: Pteraspidomorphi. *Pteraspis*
Class 4: Anaspida. *Birkenia*, *Jaymoytius*
Superclass 2: Gnathostomata
Class 1: Placodermi. *Acanthodes*
Class 2: Elasmobranchii. Dogfishes, skates, and rays
Class 3: Actinopterygii. Bony fishes
Class 4: Crossopterygii. Lung fishes
Class 5: Amphibia
Class 6: Reptilia
Class 7: Aves
Class 8: Mammalia

Amphioxus, a Generalized Chordate

It has long been realized that despite their great variety all these types show certain common features, often referred to as the *typical chordate characters*. It is better to regard these not as a list of isolated 'characters' but as the signs of a certain pattern of organization that is characteristic of the group. There is much reason to suppose that this basic chordate organization was that of a free-swimming marine animal, probably feeding by the collection of minute particles. We are fortunate in having still alive a little animal, amphioxus, the lancelet, which possesses nearly all of these features. Study of amphioxus will go a long way to show the basic plan on which all later chordates are built, and indeed, gives us a strong indication of what the early chordates may have been like.

Though it can swim freely through the water, amphioxus is essentially a burrowing animal, and many of its special features are connected with this habitat. It lives in the sand, at small depths, and has been found round the edges of all the oceans of the world. Evidently, in spite of its simplicity, it is a successful type. It is found on British coasts and, indeed, the first individual described was sent (preserved) from Cornwall to the German zoologist P.S. Pallas, who supposed it to be a slug and in 1774 called it *Limax lanceolatus*. It was first figured and given the name *Amphioxus* (two-ended) *lanceolatus* by William Yarrell in 1836. However, the name *Branchiostoma* (=gill mouth) had been given in 1834 by O.G. Costa and by the rules of priority this is the official name of the genus. We may keep amphioxus as a common name. Some twenty or more species of *Branchiostoma* are recognized, and in addition there is a group of about six species referred to the genus *Asymmetron*. These resemble *Branchiostoma* in general organization, but they have gonads only on the right side.

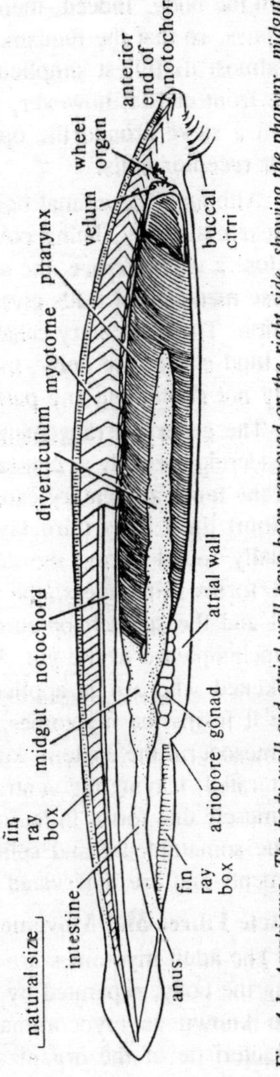

Fig. 4.1. Amphioxus. The body wall and atrial wall have been removed on the right side, showing the pharynx, midgut with its diverticulum and intestine.

The adult *Branchiostoma lanceolatum* is usually rather less than 5 cm long and has the typical fish-like organization, whose main external features are related to the methods of locomotion and feeding. The body is elongated, and flattened from side to side. The skin has no pigment, and the muscles can be easily seen as a series of blocks, the myotomes, serving to bend the body into folds. As the name implies, the body is pointed at both ends; there is no recognizable head separated from the body. Indeed, there are no separate eyes, nose, or ears, and no jaws, so that the fundamental plan of chordate organization appears in almost its fullest simplicity from one end of the body to the other. The front end is, however, marked by a series of buccal cirri, which form a sieve around the opening of the oral hood and are provided with receptor cells.

Although the animal has a large number of gill slits these do not appear externally, being covered by lateral folds of the body, which enclose a ventral space, the atrium, opening posteriorly by an atriopore. These metapleural folds give the body a triangular shape in transverse section. The alimentary canal opens posteriorly by an anus, in front of the hind end of the body, thus leaving a definite tail—a region of the body not containing any part of the alimentary canal.

The general arrangement of the organs can best be understood by considering the body as consisting of two tubes, the outer skin (ectoderm) and the inner alimentary canal (endoderm), with a space between (the coelom) lined by a third layer (the mesoderm). This arrangement is actually found during the course of development. The mesoderm at first forms thin layers, the somatopleure applied to the outer body wall and the splanchnopleure applied to the outer body wall and the splanchnopleure to the gut. Very soon the inner layer becomes much thickened where it is applied to the nerve cord and notochord, and here it forms the myotomes, or muscle blocks. In this dorsal part of the mesoderm the coelom, known here as the myocoele, soon becomes obliterated, leaving the ventral splanchnocoele around the gut. Besides the muscle that forms in the myotomes, non-myotomal muscles develop in the somatopleure and splanchnopleure. These are not divided into segments and are innervated by the dorsal nerve roots.

Muscle Fibres and Movement

The adult myotomes are blocks of striated muscle fibres, running along the body, separated by sheets of connective tissue, the myosepta (also known as myocommas). This repetition or segmentation is characteristic of the organization of all chordates. The myosepta do not run straight down the body from dorsal to ventral side but are >>-shaped. However, each muscle fibre runs straight from before backwards, and the contraction of the whole myotome therefore bends the body.

The muscle fibres are in the form of flat plates or lamellae. Each plate stretches between the myosepta and contains longitudinally orientated cross-striated myofibrils. The muscles are connected with the nerve cord in an unusual way, not by motor nerve fibres but by thin processes of the muscle cells themselves, which were thought by earlier workers to be ventral roots. Each passes medially to join the spinal cord, and there makes a cholinergic synapse with axons of central neurons. There are two types of muscle lamella. Those near the lateral surface are narrow, contain irregular myofibrils, may mitochondria, and glycogen. Their 'ventral root' fibres are thin. The other lamellae are wide, composed mainly of myofilaments and have thick ventral root fibres. The two kinds probably correspond to the slow (sarcoplasm-rich) and fast muscle fibres of higher chordates.

Amphioxus can swim equally well forwards or backwards. Movement is produced by serial contraction of the myotomes resulting in transverse motion of the body inclined at varying angles in such a way as to result in forward propagation. In forward swimming each myotome contracts after that in front of it—the effect being to produce an ~-bend that moves backwards through the water as the animal moves forward. In backward swimming the wave begins at the tail and moves forward.

For our present purpose the point is that the contraction is serial, that is to say, it depends on the breaking up of the longitudinal muscle into blocks. It was probably the need for division of the musculature that led to the development of the segmentation, and this, affecting primarily the muscles, has come to influence a great part of chordate organization.

Contraction of the longitudinally arranged muscle fibres will only produce a sharp bending of the body if there is no possibility of shortening of the whole. To prevent telescoping, an incompressible and elastic rod, the notochord, runs down the centre of the body. It is often stated that this is a 'supporting structure', but, of course, an animal such as a fish in water needs no 'support'. Nor is the notochord a lever to which muscles are attached, as they are to the bones of many higher forms. No muscles pull on it directly, though the myoseptas are attached to its sheath. Its function is to prevent the shortening of the body that would otherwise be the result of contraction of longitudinal muscles. For this purpose it has developed into a unique 'hydroskeleton' whose stiffness is variable under the control of the nervous system.

The notochord is composed mainly of a series of flattened plates surrounded by a fibrous sheath of Muller's cells. The plates are arranged in a regular manner with their flat surfaces in the transverse plane of the body. Each plate develops as a highly vacuolated cell, the nuclei being later pushed aside alternately to the dorsal or ventral edge. This structure is well suited by the turgidity of its cells, enclosed in the sheath, to resist forces tending to shorten the body. The plates are in fact muscle cells, each crossed by striations showing isotropic and anisotropic bands in polarized light. Electron micrographs show thick and thin filaments, the former being striped with a period of 14.5 nm, similar to that of the paramyosin of the catch muscle of some molluscs such as clams (*Pecten*). The plates contract under electrical stimulation, increasing the internal pressure and stiffness of the notochord. Tails from these notochordal muscles pass to the nerve cord and make cholinergic synapses with the axons of notochordal motoneurons. Probably adjustments of stiffness are made to allow swimming in either direction, the trailing end always shows the greater oscillations. The capacity to move in either direction may be important for burrowing in sand as is the fact that the cord extends from the very tip of the head to the end of the tail, projecting, that is to say, beyond the level of the myotomes.

Amphioxus lives and feeds in the sand and probably does not often swim free in the water since the body is not adapted for fast directional movements though it can maintain 60 cm/s for 50 s but then stops suddenly. This is 13 body lengths (L) per second compared with 10 L/s in the trout. There is also a slow mode of swimming (1-3 L/s), probably using the slow muscle. Lancelets may move for distances of 1 km or more, following changes in the sandy bottom, but movements are not well orientated. There are no elaborate fins such as those of later fishes, which ensure static stability like the feathers on an arrow, or are movable, to allow active control of the direction of swimming. Indeed fins would be an obstacle to burrowing. There is a low dorsal ridge, which continues behind as a small caudal fin. There are no definite paired fins, but the metapleural folds might perhaps be considered comparable to the lateral fin folds from which all vertebrate limbs are probably derived. They are distended with coelomic fluid, and, together with the dorsal ridge, probably serve to protect the body during the rapid dives by means of which the creature enters the sand. The larvae of lampreys swim in a similar way.

Skeletal Structures

Around the notochordal sheath is a further layer of gelatinous material containing fibres. There are no cells within this material but it is secreted by cells around the outside, which retain the epithelial arrangement of the mesoderm from which they were derived. This connective tissue continues as a sheath around the nerve cord and above this into a series of structures known as fin-ray boxes, which support the dorsal ridge. These boxes are more numerous than the segments and each contains some 'cartilage'. The relationship of these structures to the fin supports of vertebrates is obscure. Other skeletal rods occur in the cirri around the mouth and in the gill bars.

Skin

The epidermis differs from that of vertebrates in being very thin, composed of a single layer of cells, ciliated in the very young larva and with the outer border highly cuticularized in the adult. It is not known whether this cuticle contains a substance similar to the keratin produced by the many-layered skin of later forms. There are receptor cells but no glands or chromatophores in the skin.

Fig. 4.2. Amphioxus. V.S. of body-wall.

Below the epidermis is a fibrous cutis, and below this again a gelatinous material containing fibres, the subcutis. Both these layers are secreted by scattered cells having some similarity to the fibroblasts of higher forms. They contain a system of cutaneous canals, with endothelial lining.

Mouth and Pharynx and the Control of Feeding

Amphioxus obtains its food by extracting small particles from a stream of water, which it draws in by means of cilia. In all animals that use cilia for this purpose a very large surface is provided (e.g. lamellibranchs, ascidians), and the pharynx and gill bars of amphioxus occupy more than one-half of the whole surface area of the body. Special arrangements are made for the support and protection of this ciliated surface, the wall of the pharynx being so greatly subdivided that it needs the protection of an outer layer, the atrium.

The mouth lies covered by an oral hood whose edges are drawn out into buccal cirri, provided with sense cells some of which are mechanoreceptors. When stimulated by touching they elicit a velar reflex by which the tentacles come together and close the inhalent aperture and the velar ring sphincter closes. When feeding the cirri are curved to form a funnel-like sieve preventing the entry of large particles. Around the mouth itself there is a further ring of sensory tentacles, the velum. The oral hood contains a complex set of ciliated tracts, the 'wheel organ' of Muller, and this plays a part in sweeping the food particles into the mouth. Near its centre is a groove, Hatschek's pit, formed as an opening of the left first coelomic sac to the exterior.

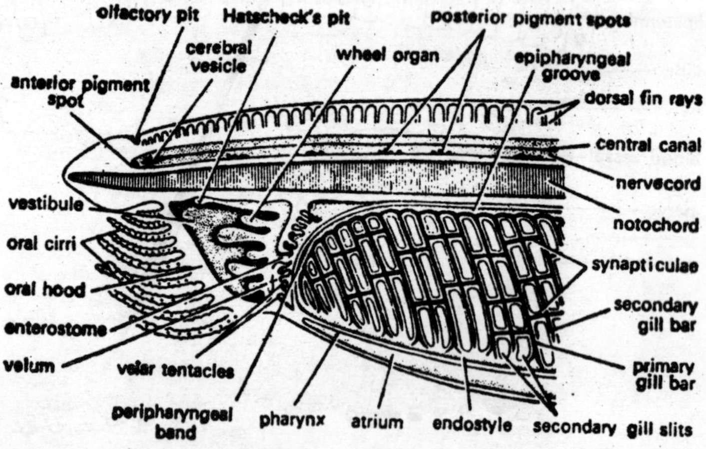

Fig. 4.3. Amphioxus. Lateral view of anterior end.

The main operation of food collection is performed by the pharynx, a large tube, flattened from side to side, whose walls are perforated by nearly 200 oblique vertical slits, the number increasing as the animal gets older. The slits are separated by bars containing skeletal

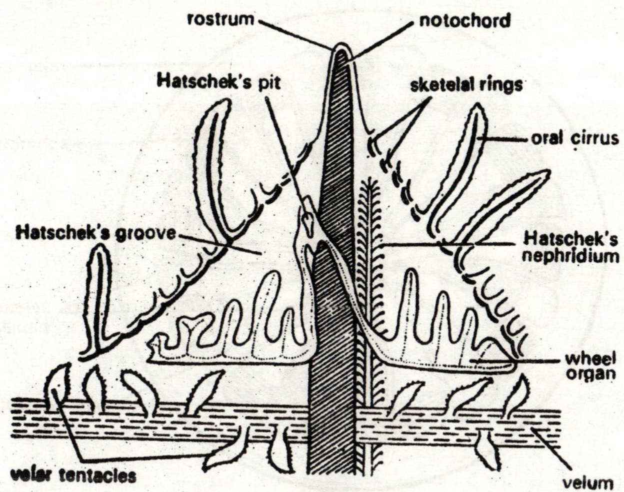

Fig. 4.4. Amphioxus. Ventral view of anterior end showing Hatscheck's pit and groove.

rods and further subdivision is provided by cross-bars (synapticulae). Since the bars slope diagonally many of them are cut in a single transverse section, but it must be remembered that they are essentially the vertical portions of the main walls of body and pharynx, where these have not been perforated by a gill slit. Such a portion of the body wall must contain a coelomic space and this can in fact be seen in the original or primary gill bars. However, an increase of the ciliary surface is produced by downgrowth of secondary or tongue bars from the upper margin, dividing each primary slit; these secondary bars contain no coelom. The coelomic spaces in the primary bars communicate above and below with continuous longitudinal coelomic cavities.

There are cilia on the sides and inner surfaces of the gill bars, the lateral ones being mainly responsible for driving the water outwards through the atrium and thereby drawing the feeding current of water in at the mouth. There is a complicated plexus of nerve cells and fibres in the walls of the pharynx and changes in the rhythm of the cilia occur, but details of the nervous connections are not known. In the floor of the pharynx lies the endostyle, containing columns of ciliated cells, alternating with mucus-secreting cells, which produce sticky threads in which food particles become entangled. Various currents then draw the sticky material along until it reaches the midgut. The frontal cilia of the gill bars produce an upward current, driving the mucus from the endostyle into a median dorsal epipharyngeal groove,

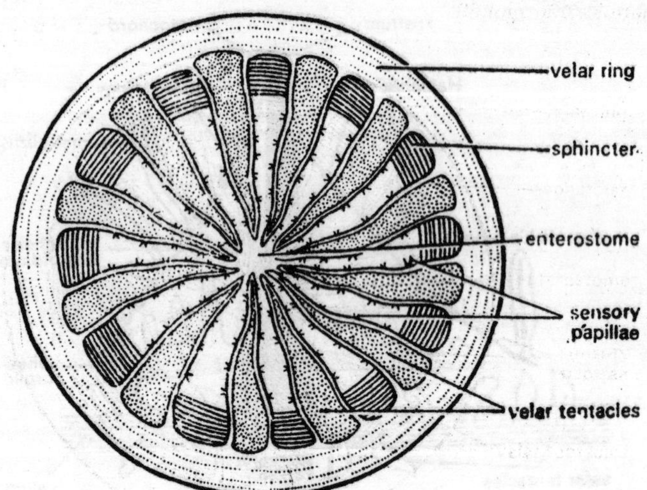

Fig. 4.5. Amphioxus. Velum.

in which it is conducted backwards. The cilia of the endostyle also move mucus along the peripharyngeal ciliated tracts, behind the velum, to join the epipharyngeal groove. Radioactive iodine is concentrated by the cells of one of the columns of the endostyle and secreted with the mucus. There may be regarded as the precursors of thyroid cells. They serve to produce iodinated mucoproteins, which are then absorbed farther down the gut. Homogenates of amphioxus show that mono- and di-iodotyrosine are present, as well as tri-iodothyronine (T_3) and thyroxine (T_4). Unlike higher vertebrates T_3 is more abundant T_4. Metamorphosis of axolotls was produced by implants of the pharynx of about 50 amphioxus, but not by the tails. There is no evidence that these iodine compounds have an endocrine activity in the animal itself.

The pharynx narrows at its hind end to open dorsally into a region best known as the midgut, the name stomach being inappropriate. A large midgut diverticulum reaches forward from this region on the right-hand side of the pharynx. From its position this organ is often called the liver, but it is the seat of the production of digestive enzymes. Zymogen cells, similar to those of the midgut, are found in its walls. Some of these appear to be protein secretors, with rough endoplasmic reticulum and secretory granules. Others contain much glycogen and lipid and may be compared to liver cells. Concentration of mitochondria at both apical and basal locations suggest transport through the cells of the caecum. Its strong dorsal and ventral ciliation maintains in it a circulation of food materials and secretion, and its cells are capable of phagocytosis as well as secretory activity. Amphioxus thus combines intracellular with extracellular digestion, doubtless in connection with its microphagous habit. Particles placed in the diverticulum are swept backwards and join the main food cord that passes through the midgut.

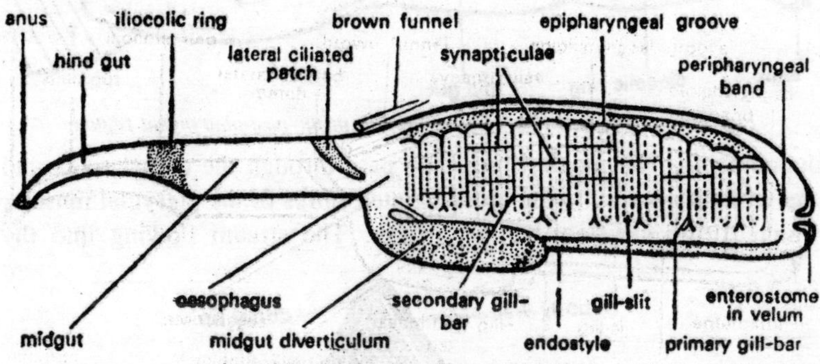

Fig. 4.6. Amphioxus. Alimentary canal.

The hind end of the midgut is marked by a specially ciliated region, the ileo-colon ring, whose cilia rotate the cord of mucus and food. The movement is transmitted to the portion of the food cord in the midgut and presumably assists in the taking up of the enzymes that emerge from the diverticulum. Extracellular digestion takes place in the midgut and the enzymes responsible have been studied by Barrington (1938). The pH of the contents varies from 6.7 to 7.1. An amylase is present in extracts of the diverticulum, midgut, and hindgut, but not in those of the pharynx. Lipase and protease are present in the same regions, the latter having an optimum action at about pH 8.0, being, that is to say, a tryptic type of enzyme. There is no sign of any protease with an acid optimum, similar to the pepsin of higher forms.

Behind the ileo-colon ring the intestine runs as a straight hindgut to the anus. Absorption of food takes place here, and perhaps also in the midgut, apparently partly by intracellular digestion, since ingested carmine particles are taken into the cells.

The feeding current is regulated by the rate of beat of the cilia and the degree of contraction of the inhalent and exhalent apertures. The walls of the atrium contain an elaborate system of afferent and efferent nerve fibres. The receptor cells are ciliated and have no axon but are in contact with processes of large nerve cell bodies, lying beneath the atrial epithelium and sending axons in by way of the

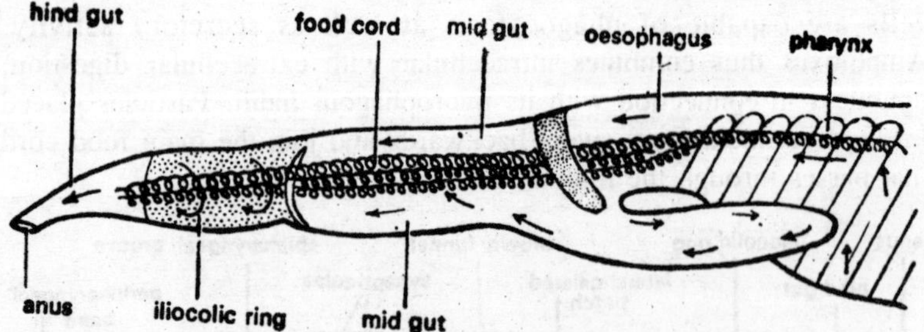

Fig. 4.7. Amphioxus. Path of food current in the post-pharyngeal region.

dorsal roots. The motor fibres also pass through the dorsal roots and run without synapse to the cross-striated fibres of the pterygial muscle, which forms the floor of the atrium. The stream flowing into the

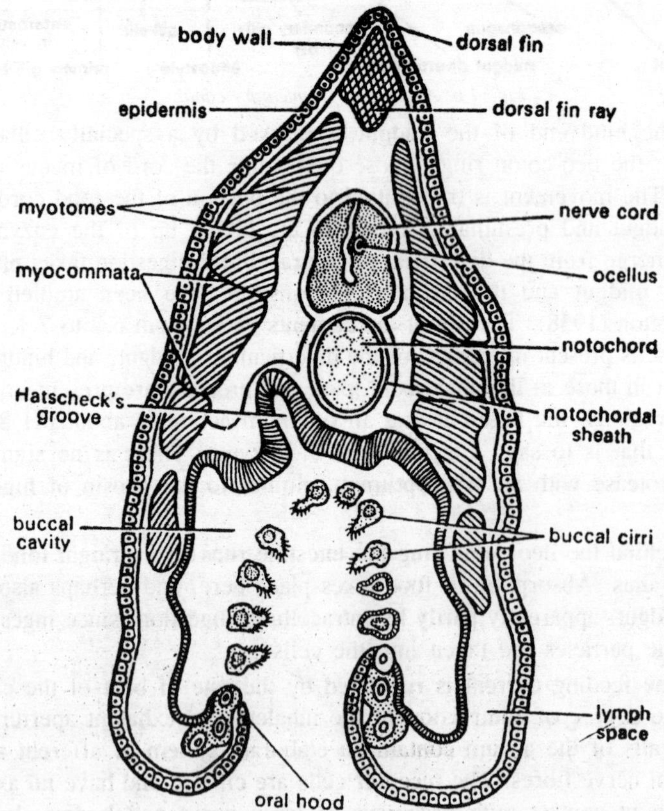

Fig. 4.8. T.S. of Amphioxus through oral hood.

pharynx is tested by the receptors of the velum and atrium, and if noxious material is present, the water is expelled by closing the atriopore and contracting the pterygial muscle, producing a 'cough'. The system can distinguish between suspensions of food material and inorganic particles. When sufficient food has been taken, collection is suspended until it has been digested.

The atrial nervous system probably regulates spawning as well as feeding. It has been compared with the close similarities. The nerve cells in it are receptors and there is no sign of the peripheral synapse on the efferent pathway that is so characteristic of the true autonomic system. The plexus in the wall of the gut also consists of sensory cells. The atrial system is developed in relation to filter feeding and has perhaps been completely lost in higher forms that propel the food along the gut by the action of muscles and have developed the autonomic nervous system to control them.

Circulation

The blood vessels of amphioxus show the fundamental plan on which the circulation of all chordates is based. Slow waves of contraction occur in various separate parts in such a way as to drive the blood forwards in the ventral vessels, backwards in the dorsal ones. Below the hind end of the pharynx there is a large sac, the sinus venosus, into which blood from all parts of the body is collected. From this there proceeds forwards a large endostylar artery (truncus arteriosus or ventral aorta) from which spring vessels carrying blood up the branchial arches. At the base of each primary bar there is a little bulb, functioning as a branchial heart. From the gill bars blood is collected into paired dorsal aortae, which join behind the pharynx. From the paired and median aortae blood is carried to the system of lacunae that supplies the tissues. There are no true capillaries. From the lacunae blood is collected into veins, the most important of which are the caudals, cardinals, and a plexus on the gut. The cardinals are a pair of vessels in the dorsal wall of the coelom, and they collect blood from the muscles and body wall. They lead to the sinus venosus by a pair of vessels, ductus Cuvier, which pass ventrally and across the coelom to join the sinus venosus on the floor of the gut. The caudal veins join the plexus on the gut, from which blood si collected by a large subintestinal vein running on to the liver; from here another plexus leads to the sinus venosus.

Contractions arise independently in the sinus venosus, branchial bulbs, subintestinal vein, and elsewhere. The rhythms are very slow

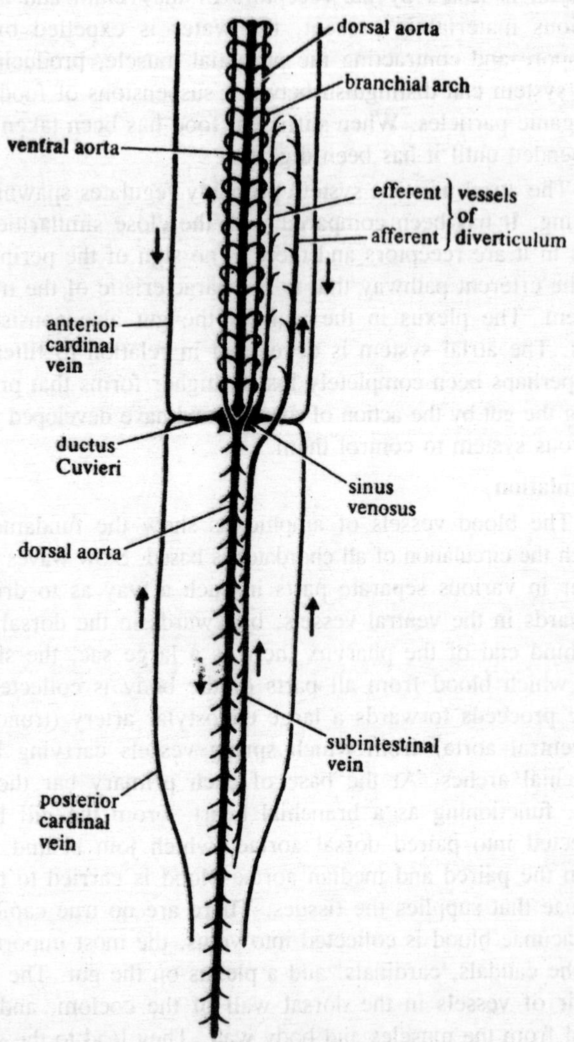

Fig. 4.9. Diagram showing the circulation of amphioxus.

(once in two minutes), irregular, and apparently not co-ordinated by any control system.

The blood is colourless and is not known to contain any respiratory pigment. It contains no cells. Presumably the tension of dissolved oxygen acquired by simple solution is sufficient for the small energy needs of the animal, which spends most of its life at rest. It is by no means certain that any oxygenation of the blood takes place in the gills.

Orton (1913) has suggested that since these, through their cilia, do much of the work of the body, the blood actually leaves the gills less rich in oxygen that when it enters them. Oxygenation probably takes place chiefly in the lacunae close to the skin perhaps especially those of the metapleural folds.

Excretory System

It used to be thought that one of the most mysterious features about the organization of amphioxus was the presence of flame cells, comparable with those found in platyhelminthes, molluscs, and annelids. These so-called nephridia lie above the pharynx. To each primary gill bar there corresponds a sac, opening by a pore to the atrium and studded with numerous elongated flame cells, which were formerly supposed to be solenocytes. These 'flame cells' do not open internally, but are in close contact with special blood vessels (glomeruli) whose walls separate the 'flame cells' from the coelomic epithelium. Assuming that there are 200 of these nephridia, each with 500 'solenocytes' 50 μm long Goodrich (1902), who provided the most accurate classical information about these organs, showed that the total length available for excretion is no less than 5m. It is assumed that excretion takes

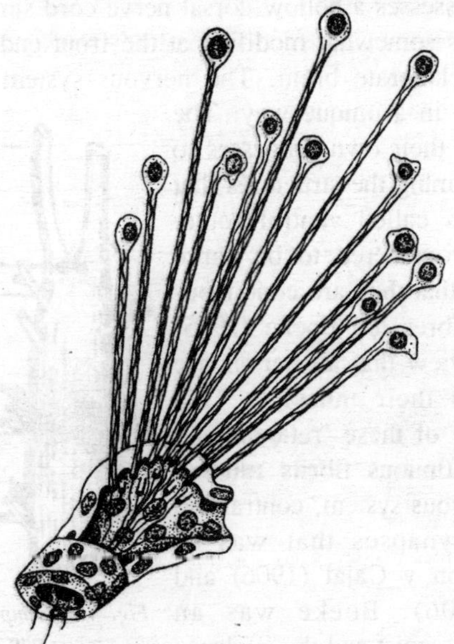

Fig. 4.10. Solenocytes of amphioxus showing the nuclei, long flagella, and the openings into the main excretory canal which leads to the atrium.

place by diffusion through the flame cell wall, the liquid being driven down the tube by cilia. Coloured particles injected into the bloodstream are not excreted by the nephridia.

Electron microscopy has shown that these 'flame cells' are in fact modified coelomic epithelial cells. Indeed they are quite similar to the podocytes that line the renal capsule of vertebrates. They have therefore been given the cumbrous name of cyrtopodocytes. The feet of these cells covering the glomerular blood vessels are joined by a slit-membrane. There is no basement membrane between blood and coelomic spaces as there is in the soloenocytes of polychaetes. So electron microscopy has solved one of the major problems that worried such excellent comparative anatomists as Goodrich.

The brown funnels are blind sacs at the front of the atrium, invaginating into the epibranchial coelom. They are probably receptor organs. Some parts of the atrial wall may perform excretory functions. Masses of cells in the atrial floor, the atrial glands, contain granules that may be excretory but perhaps have been taken up from the food current. In the gonads, especially the testes, there are large yellow masses, containing uric acid, which are extruded with the gametes.

Nervous System

Amphioxus possesses a hollow dorsal nerve cord similar to that of vertebrates. This is somewhat modified at the front end, but it is not enlarged into an elaborate brain. The nervous system is connected with the periphery in a unique way. The muscle fibres send their own processes to the nerve cord, forming the structures that all earlier workers called ventral roots. The fibres were considered to be nerve fibres and the fact that they are continuous with the muscle fibres led Boeke (1935) and others to the view that all nerves are 'continuous' with their muscles. This supported the view of these 'reticularists' that there are continuous fibrils running throughout the nervous system, contrary to the concept of synapses that was so essential for Ramon y Cajal (1906) and Sherrington (1906). Boeke was an accomplished microscopist and the evidence of continuity that he observed in various

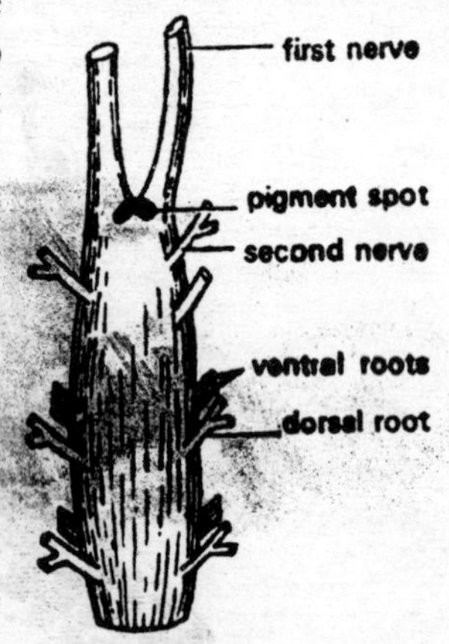

Fig. 4.11. Amphioxus. Anterior part of central nervous system.

tissues was often well founded, as in this case. Only with the coming of the electron microscope was it finally proved that most nerve fibres are separated by a membrane from the sarcoplasm. The dorsal roots run out between the myotomes and carry all the afferent fibres of the segment and motor fibres for the non-myotomal muscles of the ventral part of the body including the 'trapezius' muscle of the floor of the atrium, also the muscles around the anus and the genital sacs.

The fibres of the peripheral nerves differ from those of vertebrates in that they have no myelin sheath. The nerve trunks are surrounded by an epineurium with connective tissue cells but there seem to be no Schwann cells accompanying the nerve fibres.

The afferent fibres of the dorsal roots are unique among chordates in that the cell bodies are not collected into spinal ganglia but mostly lie within the central nervous system. They form two continuous columns along the cord. Their peripheral processes make the subepithelial plexuses of the skin and centrally they end in connection with the motoneurons or with various interneurons. In addition, on the head and tail there are peripheral receptor cells, sending fibres centrally, also complicated encapsulated organs in the metapleural folds. There are some large multipolar sensory nerve cells just beneath the atrial epithelium and very many sensory cells in the walls of the diverticulum and midgut. These cells have many branched dendrites and an axon that runs through a dorsal root to the spinal cord.

The spinal cord has only a narrow lumen and its elements are arranged as in vertebrates, namely, ependyma close to the canal, cell layer ('grey matter'). The glia is mostly in the form of ependyma, with cell bodies near the central canal and end-feet on the outer membrane. There are no blood vessels in the cord. The neurons are not arranged in horns as they are in vertebrates. The somatic motor neurons have a curious form, with a broad process attached to the central canal and a terminal bush in the region of ending of the muscle tails. The 'dendrites' are apparently collaterals of the main trunk. The visceral motor neurons are of more conventional form with dendrites and an axon in the dorsal root.

The most conspicuous cells are the giant Rohde cells, which lie dorsally in the anterior and posterior parts but are absent from about the thirteenth to thirty-ninth segments. Each of these cells has many dendrites, branching in the region of entry of the dorsal root fibres, and a single axon, which runs backwards in the front part of the body, forwards in the hind, passing in each case for the whole length of the cord. These axons probably make connection with the somatic motor cells. The most anterior Rohde cell is the largest and sends a median giant fibre ventrally for the length of the cord close to the visceromotor cells, which probably produce the 'coughing' movements of the atrium. This axon is the fastest in the animal, conducting at 5m/s.

Amphioxus responds to all stimuli by movements of 'flight'. There are no isolated or local movements; the effect of any stimulus such as touch on the side of the body is to produce waves of myotomal contraction. These may, however, vary from strong waves going the whole length of the body to single rapid twitches. The Rohde cells participate in the spread of these waves. It seems likely that the arrangement ensures that touch on the anterior part of the body, normally exposed when feeding, produces backward movement (i.e. withdrawal into the sand) but touch on the hind part the reverse movement of emergence and escape.

At the front end the central canal is enlarged to form a cerebral vesicle. The whole neural tube is hardly wider here than in the region of the spinal cord and there is no thickening of the walls, which are indeed mostly formed of a single layer of ciliated epithelial cells. This is a striking indication of the lack of cephalization of the animal. The brain may be divided into four regions. The most anterior receives the first dorsal roots, carrying fibres from the receptors of the oral hood and tentacles, from a depression known as Kolliker's pit and from a pigmented eye spot ('macula'). The second region, between nerves 2 and 4, contains large dorsal cells and many small ones. The third region, between roots 4 and 6, also contains many large dorsal cells with descending axons. The fourth part of the brain, between nerves 6 and 7, is mainly composed of longitudinal tracts. In the ventral part of the third region is the infundibular composed of tall cells with long cilia, which beat in the opposite direction to those of the rest of the vesicle. From them fibres run backwards down the cord. The organ is also the site of origin of Reissner's fibre. This is a thread of non-cellular material, present in all vertebrates at the centre of the neural canal. It is secreted at the front end and then passed backwards and is often collected and absorbed in a sac at the hind end of the spinal cord. In vertebrates it arises from secretory ependymal cells of the subcommissural organ, lying dorsally in the diencephalon. The infundibular organ of amphioxus is clearly not exactly similar, yet the Reissner's fibres are clearly comparable; an interesting problem in homology.

A further complication is that the cells of the infundibular organ contain material that stains with the Gomori method, and is similar to the neurosecretory material found in the fibres of the hypophysial tract. The organ thus seems to occupy a central position in the control system as a receptor, originator of nerve fibres, and of two sorts of secretion. There may be much to be learned from this about the origin and significance of the control systems of the diencephalon.

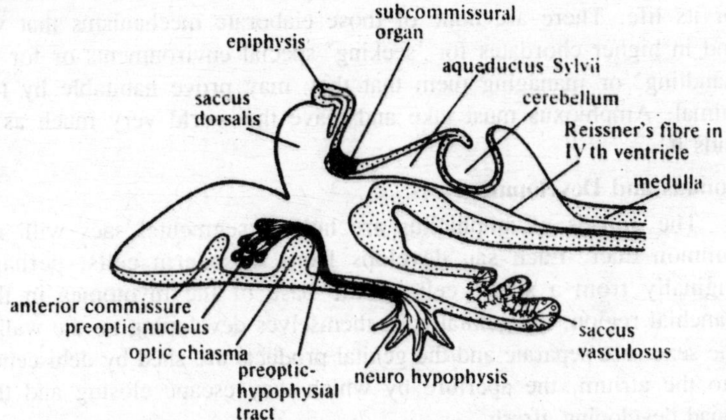

Fig. 4.12. Diagram of the anterior end of the nervous system of a fish, Polypterus.

The brain has largely inhibitory influences on the cord. A suspended lancelet shows periods of spontaneous swimming and these last longer after removal of the brain. A decerebrate animal responds more readily than a normal one to mechanical stimulation or to light, and remains active for longer. Stimulation of the brain can suppress the activation produced in Rohde cells by ascending volleys. On the other hand at the beginning of swimming, electrical activity starts in the brain 200 ms before the middle region of the cord.

Adjacent to the central canal there are rows of photoreceptor cells. Each has a partial covering of pigment and an irregular array of photoreceptive membranes. Their connections and functioning are not known in detail. The 'macula' in the brain consists of a collection of these cells but the lowest threshold for movement responses is to illumination of the cord, not the head.

Amphioxus is therefore provided with receptor and motor systems that serve to keep it in its sedentary position, able to collect food from the current that it makes by the cilia. There are mechanisms that help it to make appropriate movements of escape when it is touched or when the body (but not head) is illuminated. The touch receptors of the buccal cirri produce rejection of large particles and those of the velum are chemoreceptors. The infundibular organ may be some forms of gravity or pressure receptor. By means of these receptor organs and its simple movements of swimming, burrowing, and closing the oral hood, the animal is maintained, probably chiefly by trial and error (phobotactic) behaviour, in an environment suitable for its life. There are none of those elaborate mechanisms that we find in higher chordates for 'seeking' special environments or for so 'handling' or managing them that they may prove habitable by the animal. Amphioxus must take and leave the world very much as it finds it.

Gonads and Development

The gonads of amphioxus are hollow segmental sacs with no common duct. Each sac develops from mesoderm cells, perhaps originally from a single cell, at the base of the myotomes in the branchial region, the genital cells themselves developing on the walls. The sexes are separate and the genital products are shed by dehiscence into the atrium, the aperture by which they escape closing and the gonad developing afresh.

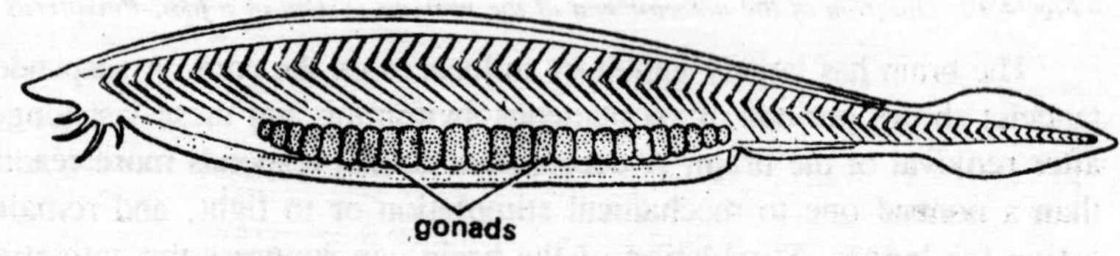

Fig. 4.13. Amphioxus. Diagram showing the location of gonads.

Extrusion of the gametes in amphioxus occurs at Naples in spring, on warm evenings following stormy weather. Fertilization is external and development then occurs free in the water. Numerous eggs are produced and they are small but yolky. Complex flowing movements take place in them after fertilization, and cleavage is then rapid and complete, producing a blastula composed of a dome of somewhat smaller and a floor of rather larger cells. These latter then invaginate to make the archenteron, opening by a wide blastopore, which later becomes the anus. At about this stage the gastrula becomes covered with flagella, by which it rotates within the egg case.

The creature now elongates and its dorsal side flattens and eventually sinks in to form the neural tube. At about this time the dorsal side of the inner layer begins to fold near to the front end, in

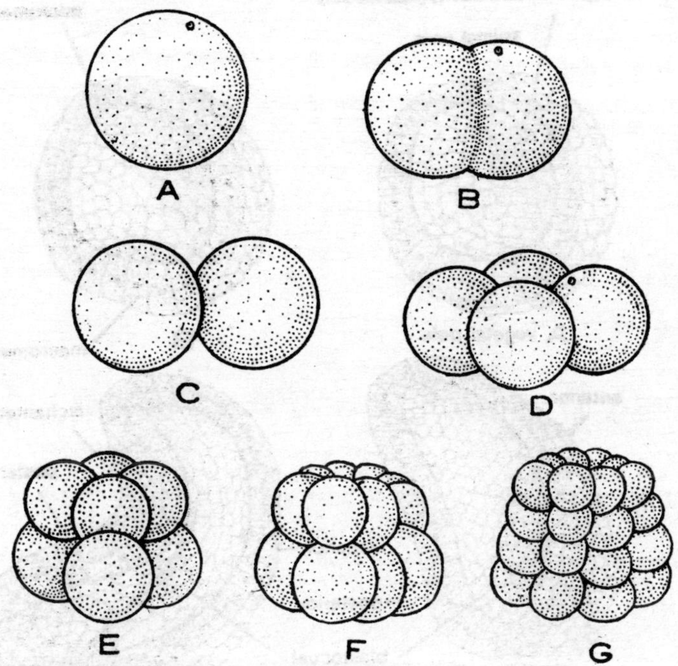

Fig. 4.14. Amphioxus. Holoblastic cleavage up to 32-cell stage.

such a way as to make a pair of lateral pouches. The walls of these pouches are the further mesoderm and the cavity is the coelom. As in other early chordates, therefore, the coelom is continuous at first with the archenteron. The roof of the archenteron also arches up dorsally and forms the notochord, the gut wall being completed by the approximation of the edges of the remaining portion of the inner layer, which is now the definitive gut wall or endoderm.

The analysis of the processes of development now enables us to say something of the forces by which these formative foldings and cell movements are produced. The formation of the neural tube, mesoderm and notochord and the completion of the gut roof all involve an upward movement of cells towards the mid-dorsal line. This process of 'convergence' is a very marked feature of the development of all chordates.

As the animal elongates, further mesodermal pouches are produced, each separating completely from the endoderm and from its neighbours. The cells of each pouch push down ventrally on either side of the gut, the outer ones applying themselves to the body wall to form the somatopleure, the inner to the gut wall as splanchnopleure. The inner

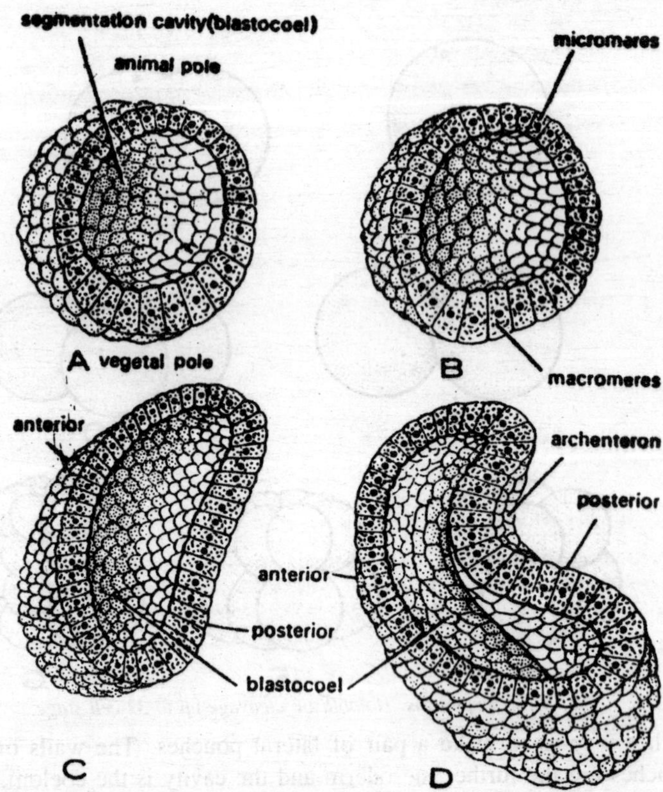

Fig. 4.15. Amphioxus. Blastula and its change in polarity during gastrulation.

to the mesoderm on either side of the nerve cord thickens to form the myotome, and a tongue of cells growing up between this and the nerve cord forms the sheaths of the latter and probably also the fin-ray boxes and other 'mesenchymal' tissues. The upper part of the coelomic cavity, the myocoele, becomes separated from the ventral splanchnocoele. Whereas the former becomes almost completely obliterated, the latter expands to form the adult coelom, the cavities between the adjacent sacs breaking down.

While this differentiation of the mesoderm has been proceeding the animal has elongated into a definitely fish-like form. The neural tube is a small dorsal canal, opening by an anterior neuropore and continuous behind through a neurenteric canal with the gut. The larva hatches when only two gill slits have been formed and swims at the sea surface by means of its ciliated epidermis, turning on its axis from right to left as it proceeds with the front end forwards.

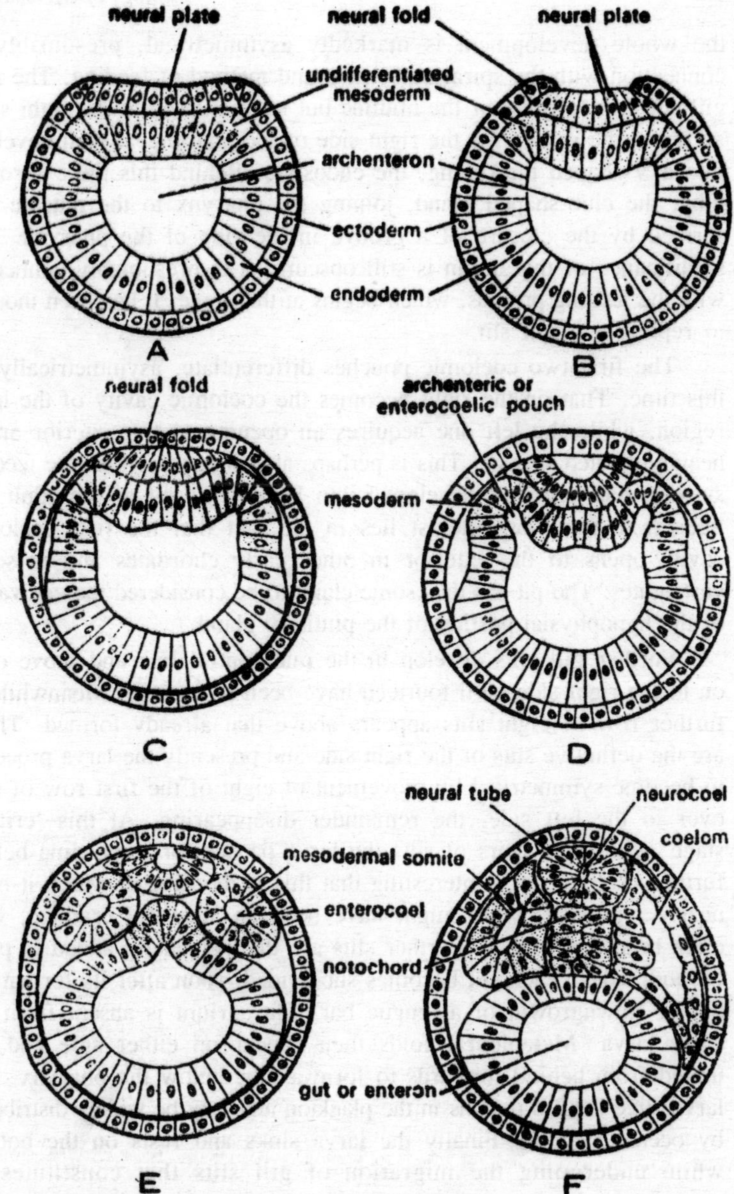

Fig. 4.16. Transverse sections through the Amphioxus embryo. A, B—Gastrula. C, F—Post-gastrula stage.

The mouth now appears as a circular opening and then moves over to the left side and becomes very large. From this time onward the whole development is markedly asymmetrical, presumably in connection with the spiral movement and method of feeding. The first gill slit also forms near the midline but moves up on to the right side. At about the same time the right side of the pharyngeal wall develops into a V-shaped thickening, the endostyle. Behind this there forms a tube, the club-shaped gland, joining the pharynx to the outside and formed by the closure of a groove in the side of the pharynx. The significance of this organ is still obscure; it is presumably connected with the feeding process, which begins at this stage. It has been thought to represent a gill slit.

The first two coelomic pouches differentiate, asymmetrically, at this time. That on the right becomes the coelomic cavity of the head region, while the left one acquires an opening to the exterior and a heavily ciliated surface. This is perhaps also connected with the feeding systems and becomes developed into Hatschek's pit of the adult. Its interest to the morphologist lies in the fact that the first coelomic cavity opens to the exterior in other early chordates and in some vertebrates. The pit has thus some claim to be considered the equivalent of the hypophysial portion of the pituitary gland.

Further gill slits develop in the mid-ventral line and move over on to the right side until fourteen have been so formed. Meanwhile, a further row of eight slits appears above that already formed. These are the definitive slits of the right side and presently the larva proceeds to become symmetrical by movement of eight of the first row of slits over to the left side, the remainder disappearing. At this 'critical stage' with eight pairs of slits the larva pauses for some time before further changes. It is interesting that this is the time at which it most nearly represents what might have been an ancestral craniate, with eight branchial arches. Further slits are then gradually added in pairs on both sides. Each slit becomes subdivided, soon after its formation, by the downgrowth of a tongue bar. The atrium is absent from the early larva. Metapleural folds then appear on either side and are united from behind forwards to form a tube below the pharynx. The larvae live for 2-6 months in the plankton and may be widely distributed by ocean currents. Finally the larva sinks and rests on the bottom while undergoing the migration of gill slits that constitutes its metamorphosis.

The development of amphioxus, like its adult organization, shows us many features of the plan that is typical of all chordates and was presumably present in the earliest of them. Thus the cleavage,

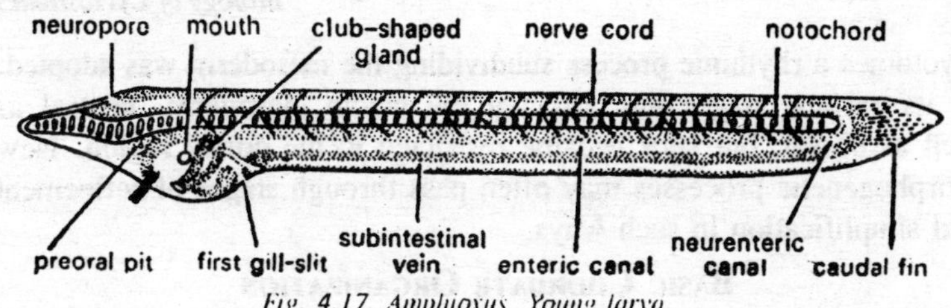

Fig. 4.17. Amphioxus. Young larva.

invagination, and mesoderm formation recall those of echinoderms and other forms similar to the ancestors of the chordates, and also show a pattern from which all later chordate development can be derived. Unfortunately we cannot pursue this study as far as we should like because of the difficulty of investigating the development of amphioxus. Modern embryologists aim at tracing the morphogenetic movements by which the organism is built, and ultimately at discovering the forces responsible for these processes. We still remain ignorant of the details of these movements, and can only guess that the system of cell activities by which an amphioxus is built represents quite closely the original set of morphogenetic processes of vertebrates.

There are, of course, some special features connected with the method of life of the larva, and especially with its asymmetry. The strange sequence of gill formation, the immense left-sided larval mouth, perhaps the club-shaped gland, and Muller's organ, may show considerable modifications of relatively recent date. However, the earliest chordates probably fed by means of cilia and were planktonic, so we must not too hastily assume that even these asymmetrical features are novelties.

The division of the mesoderm of amphioxus into a series of sacs presents an interesting problem. The segmentation of the mesoderm of vertebrates is restricted to the dorsal region. In the lowest chordates, as in their pre-chordate ancestors, there are three coelomic cavities, but it is probable that the many segments of vertebrates arose in order to provide a set of muscles able to contract in a serial manner for the purpose of swimming. Their segmentation would thus be a relatively late development, not related to the segmentation of annelids, which divides the whole body into rings. Accordingly the ventral part of the vertebrate coelom usually remains unsegmented. But in amphioxus (and in the lamprey) it is subdivided from its first appearances and only becomes continuous later. The best interpretation of this condition is to suppose that in order to provide a series of myotomes a rhythmic process subdividing the mesoderm was adopted. In its earliest stages this affected the whole mesoderm, ventral as well as dorsal, but later became restricted to the dorsal region. New morphogenetic processes may often pass through stages of refinement and simplification in such ways.

Basic Chordate Organization

Amphioxus provides us, then, with a valuable example of a chordate that retains the habit of ciliary feeding, which was probably that of the earliest ancestors of our phylum. No doubt in connection with this, and the bottom-living habit, there are many specializations; the enormously developed pharynx with its atrium, the asymmetry, and so on; but the general arrangement of the body is almost diagrammatically simple, and it may well be that amphioxus shows us a stage very like that through which the ancestors of the craniates evolved. Perhaps next the larva remained longer in the plankton and became mature there. The larvae of some acraniates shown signs of such a change.

This might give rise to a suspicion that amphioxus is not an ancestral type but a simplified derivative of the vertebrates, perhaps a paedomorphic form. It possesses, however, sufficient peculiar features to make this view unlikely. Neoteny might explain the regular segmentation, separate dorsal and ventral roots, and other features, but can hardly account for the method of obtaining food, or for the condition of the skin. It may be, therefore, that amphioxus shows us approximately the condition of the early fish-like chordates, living in the lower Ordovician some 500 million years ago, and that it has undergone relatively little change in all the time since.

5

VERTEBRATE ANCESTRY

The history of life, like human history, has been marked by certain great development rising above the general level of events that collectively make up the record. These outstanding evolutionary developments are in the nature of revolutions, affecting profoundly the phylogenetic trends that follow them, just as great historical revolutions, like the American Revolt against Britain or the French revolutions, have affected the subsequent histories of the peoples concerned with them, A better comparison might be with the peaceful revolution in human arts and industries, such as those brought about by the development of the printing press or the application of steam power to machinery.

Similarly animal life has also a significant history. The first vertebrates known in the fossil record are indicated, by scales that have been found in Ordovician freshwater sediments in Colorado. These remains are fragmentary and give no clue as to the nature of the animals represented by them; yet they are sufficiently well preserved to be studied. When placed under a microscope, they show a bony structure. Therefore, it is obvious from these scales, unsatisfactory and tantalizing though they may be as evidence of the first vertebrates in the geological record, that in Ordovician times there was armored vertebrates living in shallow rivers and lakes. The next evidence of vertebrate life comes from England in rocks of Middle Silurian ago. Two genera, *Jamoytius* and *Thelodus*, are represented by completed skeletons which, because of the manner of their preservation, are not easy to interpret. *Jamoytius* is of special interest because it appears to be a very primitive jawless vertebrate, perhaps, occupying a position close to the ancestry of the lamprey and its relatives.

The few, rather enigmatic fossil remains of *Jamoytius* show that this animal was small, elongated and tubular shaped. It had a terminal, sectorial mouth, and on each side of the head region, there was a row of circular gill openings, behind the eye. There was a propulsive tail fin with a long lower lobe and a shorter, deep upper lobe, and possibly there were lateral fin folds and a long dorsal fin, for maintaining balance. *Jamoytius* and *Thelodus* are found together in marine sediments. One of the great events or revolutions in the history of the vertebrates was the appearance of the jaws. The importance of this evolutionary development can hardly be overestimated, for it opened to the vertebrates new lines of adaptation and new possibilities of evolutionary advancement that expanded immeasurably the potentialities of these animals.

The jawless vertebrates were definitely restricted as to their adaptations for different modes of life, and it is possible that the ostracoderms of late Silurian and Devonian times had explored and virtually exhausted the evolutionary possibilities and the ecological niches open to animals of this type. Animal without jaws can evolve as bottom feeders, as did some of the ostracoderms, or they can develop movable plates around the mouth opening that serve after a fashion as weak "jaws," as did some other ostracoderms. Or they can become parasites, like the modern lampreys and hagfishes. Yet even under the most favourable conditions they are denied the possibilities of development that are open to animals with jaws; they simply do not have the structural mechanisms to exploit the opportunities that are available to jawed animals. So it is that the appearance of jaws marked

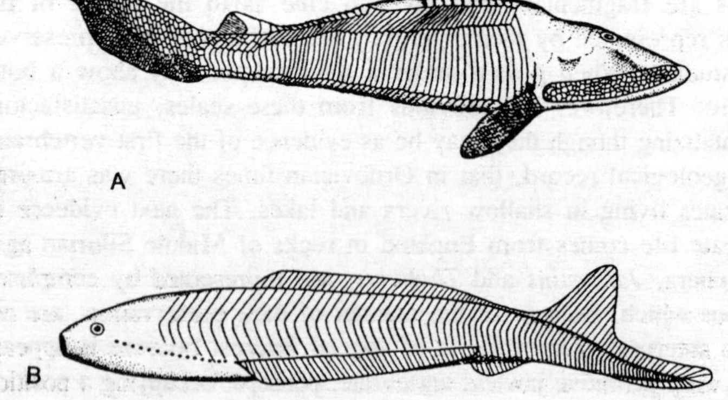

Fig. 5.1. Fossil ostracoderms. A—Hemicyclaspis, a member of the Osteostraci; B—Jamoytius, two members of the Anaspida.

a major turning point in vertebrate evolution. The appearance of the jaws in vertebrates was brought about by a transformation of anatomical elements that originally had performed a function quite different from the function of food gathering. Here we see the working of a process that has taken place innumerable times, in the course of animal evolution; indeed, much of the progress from earlier to later forms of animals has been brought about through the transformation of structures from one function to another.

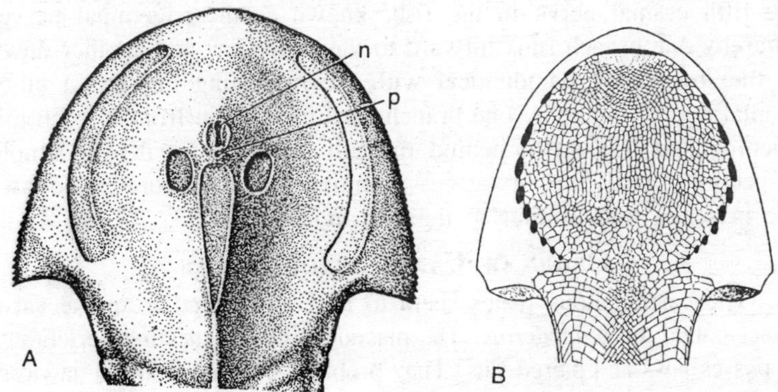

Fig. 5.2. A—Dorsal and B—ventral views of the head region of a fossil ostracoderm of the Cephalaspis type. Dorsally are seen openings for the paired eyes, median eye (p), and a median slit (n) for nostril and hypophyseal sac.

The origin and evolution of the jaws are an excellent example of this evolutionary principle. The jaws were derived originally from gill arched. It will be remembered that the ostracoderms had a large number of gills, as many as ten in *Cephalaspis*, and it seem probable that the presence of comparatively numerous gills with cartilaginous or bony supports was typical of the primitive vertebrates. At an early stage in the history of the vertebrates at least one and probably two of the original anterior gill 'arches" were eliminated, while another arch, probably the third one in the series, was changed from a gill support into a pair of jaws. This transformation actually was not as radical a shift as on first sight it may appear to have been. Each gill support or arch in the primitive vertebrate was formed by a series of several bones, arranged somewhat in the fashion of a V turned on its side, with the point directed posteriorly. Imagine several such lazy V's in series, thus : > > >. Imagine also the first of the these (morphologically the third of the original gill arches) supplied with teeth, and hinged at the point of the V, and you have the primitive vertebrate jaws, in alignment with the gill arches behind them.

It is immediately apparent that the transformation of a gill arch into jaws was a natural evolutionary development, perhaps the simplest possible to the basic problem of developing a pair of vertebrate jaws. There are various facts that support such an origin for the jaws in the vertebrates. For instance, a study of the embryological development of certain modern fishes indicates this origin very strongly. Moreover, the arrangement of the nerves in the head region of sharks shows that the jaws are in series with the gill apparatus. Thus the branching of the fifth cranial nerve in the fish, known as the trigeminal nerve, whereby one branch runs forward to the upper jaw and another down to the lower jaw, is identical with the branching of certain other cranial nerves, whereby one branch runs forward in front of each gill opening and one down behind the gill opening. Finally, a simple inspection reveals that in various primitive jawed vertebrates the jaws are in series with similar to the gill arches.

Origin of Cartilaginous Fishes

The cartilaginous fishes seem to have originated from the early fishes called the *placoderms*. The placoderms were the first vertebrates to posses jaws and paired fins. They probably evolved from the jawless, bottom-dwelling, filter feeding, ostracoderms in early Devonian period i.e., about 350 million yeas ago. However, no connections link between the jawless and the jawed fishes is available in the fossil record. The placoderms were small, mainly freshwater fishes. They retained the body armour of the ancestral ostracoderms. The armour consisted of large plates over the head and anterior part of the trunk and scales elsewhere. The endoskeleton had considerable ossification. The visceral arches were, however, cartilaginous. The notochord persisted throughout life. The first pair of gill-slits were not reduced. The second pair of gill arches were unaffected by jaws.

Paired fins were present and unpaired fins included one or two dorsal fins, a caudal fin and an anal fin. The tail was heterocercal. The jaws developed from the first pair of gill, arches in front of the first gill-slits. Their formation affected neither the first pair of gill-slits nor the second pair of gill-arches. In later fishes, both these structures changed due to posterior expansion of the jaws. Appearance of jaws caused a revolution in the vertebrate feeding as now the fishes could leave filter feeding and resort to other modes of feeding and spread into a variety of habitats. Many of them become predaceous and this led to the development of a more efficient muscular system for quick locomotion. Efficient locomotion in turn required and resulted

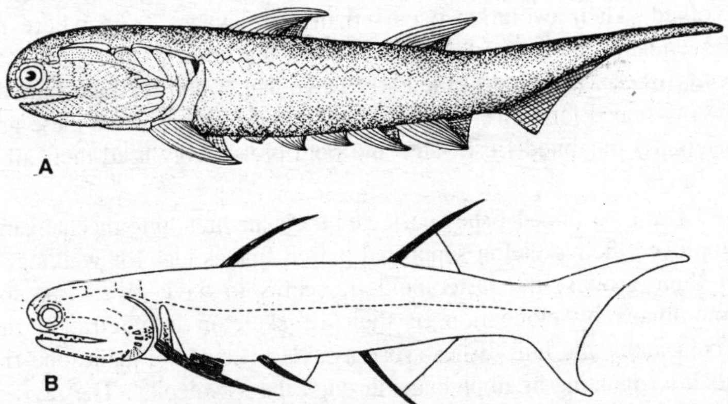

Fig. 5.3. Acanthodians, primitive jawed fishes probably related to the modern Osteichthyes. A—Climatius, a form with heavy scales and very large fin spines; B—Ischnacanthus, one with greatly reduced armor and thin spines.

in an improvement in the respiratory, circulatory, sensory and nervous system. The paired fins may have first appeared in the form of lateral folds, such as those found in the primitive ostracoderm *Jamoytius*, These folds may have broken up into a series of small paired fins such as are found in certain primitive placoderms e.g., *Climatius*.

The anterior and posterior pairs were larger than other, and ultimately developed into the pectoral and pelvic fins respectively. The paired fins between these appeared, some paleontology do not believe in the above "fin fold" theory of the origin of paired fins, and hold that these numerous paired fins may have originated as separate structures. Specific placoderm ancestors of higher fishes are not known. It is, however, possible that the cartilaginous fishes may have evolved from the stegoselachian stock and the bony fishes from the acanthodian stock. Evolution of both the groups occurred in freshwater, but many from later took to marine life. The cartilaginous fishes are known in the fossil records from the mid-Devonian period and the bony fishes from the early Devonian period. This indicates that the bony fishes are oeder than the cartilaginous fishes. This inference is further supported by the fact that the possession of bone is a primitive character and its absence in the cartilaginous fishes is a case degeneration

Origin of Bony Fishes

Bony fishes, as mentioned earlier, arose from the early jawed fishes, the placoderms, which themselves originated from the jawless ostracoderms, specific placoderm ancestor of bony fishes is not known, but it seem possible that they may have evolved from the acanthodian stock. They probably appeared before the cartilaginous fishes had evolved. Their evolution occurred in fresh water, from where many descendants later invaded the sea. The earliest fossil record of the bony fishes was found in the early Devonian, when the lobe-finned and the ray-finned forms were already well-defined groups. Both these groups originally inhabited freshwater and both presumably used their air-sacs as Jungs.

The lobe-finned fishes perfected their air-breathing mechanism and employed their skeleton-supported paired fins as legs for walking. One of their groups, the Osteolepidoti, seems to have given rise to the amphibians by elongation of their fin-skeleton into tetrapod limbs. Coelacanths and lung-fishes arose as side lines of the main lobe-finned stock terminating in amphibians through the osteolepids. The ray-finned fishes started using air-sacs as hydrostatic organs and evolved through palaeoniscoid forms into Chondrostei, Holostei and Peleostei.

Evolution of Fishes

In a general way it is evident from the preceding summary of structure and classification that the Shark-like fishes or Elasmobranchs are the most primitive of existing fishes, and that the bony fishes can be derived from these by modifications of the various organs; also that the Ganoids are the more primitive of the bony fishes, and the Teleostei the most modified. Among the Ganoids, the "fringe-finned" forms or Crossopterygii are the most primitive, being nearest to the Elasmobranchs in the structure of the paired fins. The Dipnoi, or lung-fishes, on the other hand, might be supposed to be the latest stage in the evolution of fishes, especially on account of the advance stage of evolution shown by the lungs in adaptation to atmospheric respiration; but on the other hand in structure the Dipnoi show unmistakable affinities to the Crossopterygii.

In order, however, to discuss the evolution of fishes we must consider the evidence afforded by the chronological succession of fossil forms as well as the evidence of the structure of existing forms; we must study palaeontology as well as comparative anatomy, and include under the latter term the study of development or embryology. We will proceed therefore to give a brief outline of the most important facts concerning the palaeontology of fishes. There is no evidence of the derivation of fishes with jaws from those with sectorial mouths (Cyclostomes) resembling the lampreys and hags. Minute fossils known as Conodonts occur in the most ancient stratified rocks called by geologists palaeozoic, from the Lower Silurian to the Carboniferous

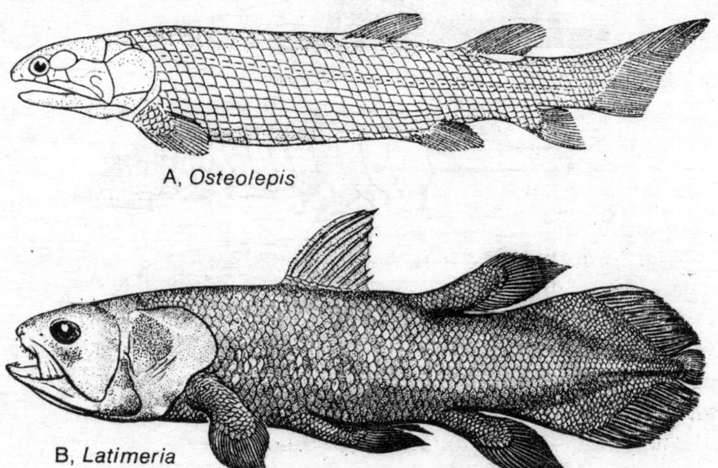

Fig. 5.4. Crossopterygians. A—Typical Devonian form; B—the only living coelacanth.

Limestone, and have been compared to the teeth of the Cyclostomes; but these fossils in microscopic structure do not resemble such teeth, and nothing can be said about them.

Small fossils about two inches long found by *Dr. Traquair* in the Caithness flag-stones near Thurso appear to have the structure of Cyclostomes, but they have a calcified internal skeleton, while the skeleton of existing Cyclostomes is soft and cartilaginous. Indeed, in deposits belonging to the Lower Silurian and Devonian, in Russia, England, and North America, minute, slender, pointed horny bodies, bent like a hook, with sharp opposite margins, have been found and described under the name of *Conodonts*. More frequently they possess an elongated basal portion, in which there is generally a larger tooth with rows of similar but smaller denticles on one or both sides of the larger tooth, according as this is central or at one end of the base. In other examples, there is no prominent central tooth, but a series or more or less similar teeth is implanted on a straight or curved base.

Modifications of these arrangements are very numerous, and many Palaeontologists entertain still doubts whether the origin of these remains is not rather from Annelids and Mollusks than from Fishes. The first undeniable evidence of a fish, or, indeed, of a vertebrate animal, occurs in the *Upper Silurian* Rocks, in a bone-bed of the Downton sandstone, near Ludlow. It consists of compressed, slightly curved, ribbed spines, of less than two inches in length (*Onchus*); of small shagreen-scales (*Thelodus*); the fragment of a jaw-like bar with

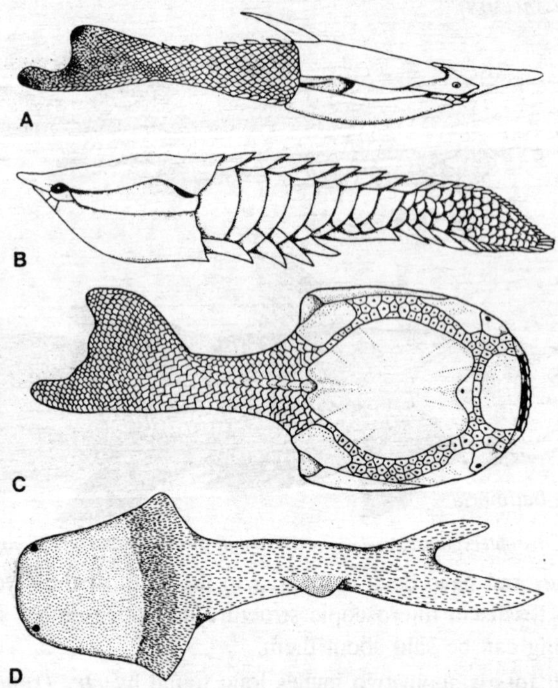

Fig. 5.5. Fossil ostracoderms. A—Pteraspis; B—Anglaspis; C—Drepanaspis; D—Logania.

pluricuspid teeth (*Plectrodus*); the cephalic bucklers of what seems to be a species of (*Pteraspis*); and, finally, the coprolitic bodies of phosphate and carbonate of lime, including recognisable remains of the Mollusks and Crinoids inhabiting the same waters. But no vertebra or other part of the skeleton has been found. The spines and scales seem to have belonged to the same kind of fish, which probably was a Plagiostome.

It is quite uncertain whether or not the jaw (if it be the jaw of a fish) belonged to the buckler-bearing I *teraspis*, the position of which among Ganoids, with which it is generally associate, is open to doubt. No detached undoubted tooth of a Plagiostome or Ganoid scale has been discovered in the Ludlow deposits; but so much is certain that those earliest remains in Palaeozoic rocks belonged to fishes closely allied to forms occurring in greater abundance in the succeeding formation, the Devonian, where they are associated with undoubted Palaeichtyes, Plagiostomes as well as Ganoids. These fish-remains of the *Devonian* or *Old Red Sandstone*, can be determined with greater certainty. They consist of spines or the so called *Ichthyodorulites*, which show sufficiently distinctive characters to be referred to several genera, one of them, *Onchus*, still surviving from the Silurian epoch.

All these spines are believed to be those of Chondropterygians, to which order some pluricuspid teeth (*Cladodus*) from the Old Sandstone in the vicinity of St. Petersburg have been referred likewise. The remains of the Ganoid fishes are in a much more perfect state of preservation, so that it is even possible to obtain a tolerably certain idea of the general appearance and habits of some of them, especially of such as were provided with hard carapaces, solid scales, and ordinary or bony fin-rays. A certain proportion of them, as might have been expected, remind us, with regard to external form, of Teleosteous fishes rather than of any of the few still existing Ganoid types; but it is contrary to all analogy and to all palaeontological evidence to suppose that those fishes were, with regard to their internal structure, more nearly allied to Teleosteans than to Ganoids. If they were not true Ganoids, they may be justly supposed to have the essential characters of Palaeichtyes.

Other forms exhibit even that remote geological epoch so unmistakably the characteristics of existing Ganoids, that no one can entertain any doubt with regard to their place in the system. In none of these fishes is there any trace of vertebral segmentation. The Palaeichtyes of the Old Red Sandstone, the systematic position of which is still obscure, are the *Cephalaspidae* from the Lower Old Red Sandstone of Great Britain and Eastern Canada; *Pterichthys*, *Coccosteus*, and *Dinichthyas*: genera which have been combined in one group—*Placodermi*; and *Acanthodes* and allied genera, which combined numerous branchiostegals with chondropterygian spines and a shagreen-like dermal covering. Among the other Devonian fishes (and they formed the majority) two types may be recognised, both of which are unmistakably Ganoids.

The first approaches the still living *Polypterus*, with which some of the genera like *Diplopterus* singularly agree in the form and armature of the head, the lepidosis of the body, the lobate pectoral fins, and the termination of the vertebral column. Other genera, as *Holoptychius*, have cycloid scales; many have two dorsal fins (*Holoptychius*), and, instead of branchiostegals, jugular scutes; other one long dorsal confluent with the caudal (*Phaneropleuron*). In the second type the principal characters of the *Dipnoi* are manifest, and some of them, for example *Dipterus*, *Paloedaphus*, *Holodus*, approach so closely the Dipnoi which still survive, that the differences existing between them warrant a separation into families only.

Devonian fishes are frequently found under peculiar circumstances, enclosed in the so-called *nodules*. These bodies are elliptical fattened pebbles, which have resisted the action of water in consequence of their greater hardness, whilst the surrounding rock has been reduced to detritus by that agency. Their greater density is due to the dispersion in their substance of the fat of the animal which decomposed in them. Frequently, on cleaving one of these nodules with the stroke of the hammer, a fish is found embedded in the centre. At certain localities of the Devonian, fossil fishes are so abundant that the whole of the stratum is affected by the decomposing remains emitting a peculiar small when newly opened, and acquiring a density and durability not possessed by surata without fishes. The flagstones of Caithness are a remarkable instance of this.

The fish-remains of the *Carboniferous* formation show a great similarity to those of the preceding. They occur throughout the series, but are very irregularly distributed, being extremely scarce in some countries, whilst in others entire beds (the so-called bone-beds) are composed of ichthyolites. In the ironstones they frequently form the nuclei of nodules, as in the Devonian. Of Chondropterygians the spines of *Onchus* and others still occur, with the additions of teeth indicative of the existence of fishes allied to the Cestracion-type (*Cochliodus*, *Psammodus*): a type which henceforth plays an important part in the composition of the extinct marine fish faunae. Another extinct Selachina family, that of Hobodontes, make its appearance, but is known from the teeth only. Of the Ganoid fishes, the family *Palaeoniscidae* (Traquair) is numerously represented; others are Coelacanths (*Coelacanthus*, *Rhizodus*), and *Saurodipteridue* (*Megalichthys*). None of these fishes have an ossified vertebral column, but in some (*Megalichthys*) the outer surface of the vertebrae is ossified into a ring; the termination of their tail is heterocercal.

The carboniferous *Uronemus* and the Devonian *Phaneropleurone* are probably generically the same; and the Devonian *Dipnoi* are continued as, and well represented by, *Ctenodus*. In the Carboniferous and Permian strata some remarkable forms of the shark type have been obtained in recent year in a very well-preserved condition. The most primitive and earliest of these is *Cladoselache*, found in Lower Carboniferous strata in Ohio. Its mouth is at the end of the snout. It is from two to six feet long. The paired fins are attached horizontally

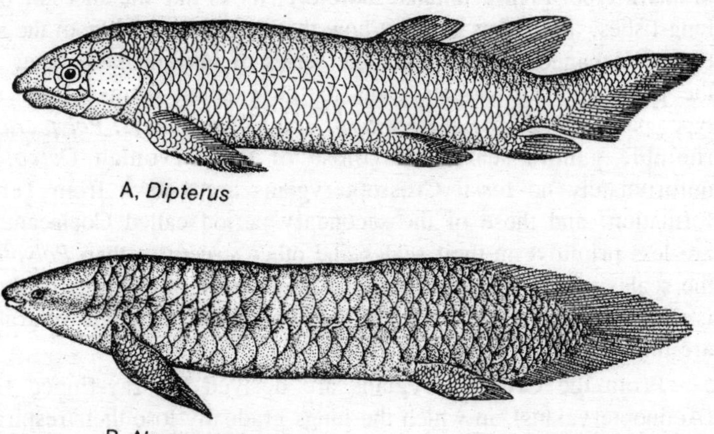

Fig. 5.6. Lungfishes. A—An ancient Devonian fossil type; B—Neoceratodus of Australia.

by a wide base, and have the appearance of enlarged portions of an originally continuous fold. The tail is short and high vertically, with an extremely heterocercal structure; that is to say, the termination of the trunk is bent upwards almost at right angles and the ventral fin-rays are very long.

The Acanthodians already mentioned extend from the Silurian to the end of the Permian and therefore go back to an earlier period than *Cladoselache*; but *Professor Bashford Dean* considers nevertheless that the latter represents the ancestral form of the Acanthodians. *Pleuracanthus*, which occurs in the Carboniferous and Permian both in Europe and North America, on the other hand, differs in many respects from sharks, and in these same features approximates to the Dipnoi. It was a shark in the following features: the gillslits opened separately on the surface; the jaws consisted of the same primitive cartilages as in sharks, claspers occurred in the male. On the other hand, the paired fins show a transition to the featherlike or pinnate structure which is in reality not primitive but secondary, the basal cartilages being separated from the body at the posterior end, and the radial cartilages extending round to the posterior, originally the internal, side, so that a structure is produced which has a central axis and rays along each side.

But the feature most similar to the lung-fish was that the dermal denticles had disappeared from the skin of the trunk, and on the head had developed into dermal bones arranged like those of the lung-fish. The tail retained the primitive symmetry and was not of the heterocercal of shark type. *Pleuracanthus*, therefore, looks like the ancestor of the lung-fishes, and seems to show how these were derived from the shark type. We cannot trace a complete series of intermediate stages from the palaeozoic Crossopterygians to the existing forms of the same type, namely the African *Polypterus* and *Calamochthys*. *Polypterus* has rhombic ganoid scales like those of the Devonian *Osteolepis*; unfortunately no fossil Crossopterygians are known from Tertiary formation, and those of the secondary period called Coelacanthidae are less primitive in their scales and other structures than *Polypterus*; the scales of Coelacanthidae are cycloid, the terminal part of the tail is disappearing and the basals of the two dorsal and single ventral fins are united into a single, forked bone.

From the Crossopterygians are derived the ray-finned fishes (Actinopterygians), in which the lungs gradually lose their respiratory function and paired character, and become more and more completely adapted to serve as an air-bladder which diminishes the specific gravity of the fish. In the evolution of these forms we can trace a gradual return to the sea, for among existing fishes the more primitive forms live more or less completely in fresh water, and the most recent and most modified are typically marine. The fins of Actinopterygians are characterized by the reduction of the basal lobe and its internal skeleton, the dermal rays arising from the surface of the body.

The oldest form is *Chirolepis* which occurs in the Lower Old Red Sandstone together with the earliest Crossopterygians. This genus belongs

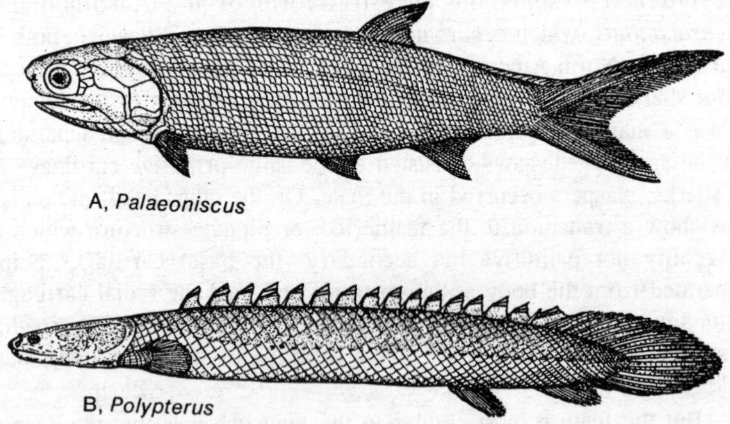

Fig. 5.7. Primitive ray-finned fishes. A—An early palaeoniscoid; B—a living representative of the ancient palaeoniscoids, with modified fin structure.

to the group or branch of the Chondrostei and is the first of the great family Palaeoniscidae which developed an abundance of species and individuals in the Corboniferous and Permian formations and gradually diminished in the Secondary period till the end of the Jurassic. These fishes had a complete armour of ganoid scales, a large heterocercal tail, and a single dorsal fin in the middle of the back. The dermal bones of the head were external and were simply the ganoid plates of that region. As the Palaeoniscidae are disappearing the family Chondrosteidae comes into existence, distinguished by the degeneration of the scales on the body; specimens occur in the Lias or Lower Jurassic, and these forms must be regarded as directly descended from the Palaeoniscidae.

The Chondrosteidae, on the other hand, are the ancestors of the existing Sturgeons, which are not known before the Tertiary. A branch from the Palaeoniscidae gave rise to the Platysomidae which existed in the Carboniferous and Permian periods; they are deep-bodied compressed fishes completely covered with ganoid scales, but with short jaws carrying blunt crushing teeth; they do not appear to have give rise to any later forms. The passage from the Palaeozoic into the *Mesozoic* era is not indicated by any marked changed as far as fishes are concerned. The more remarkable forms of the Trias are Sharklike fishes represented by ichthyodorulithes like *Nemacanthus*, *Liacanthus*, and *Hybodus*; and Cestracionsts represented by species of *Acrodus* and *Strophodus*. Of the Ganoid genera *Coelacanthus*, *Ambiypterus* (*Palaeoniscidae*, *Saurichthys*) persists from the Carboniferous epoch. *Ceratodus* appears for the first time.

The Hybodonts and Cestracionts continue in their fullest development Holocephales (*Ischyodus*), true Sharks (*Paloeoscellium*), Rays (*Squaloraja*, *Arthropterus*), and Sturgeons (*Chondrosteus*) make their first appearance; but they are sufficiently distinct from living types to be classed in separate genera, or even families. The Ganoids, especially Lepidosteoids, predominate over all the other fishes: *Lepidotus*, *Semionotus*, *Pholidophorus*, *Pachycormus*, *Eugnathus*, *Tetragonolepis*, are represented by numerous species; other remarkable genera are *Aspidorhynchus*, *Belonostomus*, *Saurostomus*, *Sauropsis*, *Thrissonotus*, *Conodus*, *Ptycholepis*, *Endactis*, *Centrolepis*, *Legnonotus*, *Oxygnathus*, *Heterolepidotus*, *Isocolum*, *Osteorhachis*, *Mesodon*. These genera offer evidence of great change since the preceding period, the majority not being represented in older strata, whilst, on the other hand, many are continued into the succeeding oolithic formations.

The homocercal termination of the vertebral column commences to supersede the heterocercal, and many of the genera have well ossified and distinctly segmented spinal columns. Also the cycloid form of scales becomes more common: one genus (*Leptolepis*) being, with regard to the preserved hard portions of its organisation, so similar to the Teleosteous type that some Palaeontologists refer it (with much reason) to that sub-class. The Cretaceous group offers clear evidence of the further advance towards the existing fauna. Teeth of Sharks of existing genera *Carcharias* (*Corax*), *Scyllium Notidanus*, and *Galeocerdo*, are common is some of the marine strata, whilst Hybodonts and Cestracionts are represented by a small number of species only; of the latter one new genus, *Ptychodus*, appears and disappears.

A very characteristic Ganoid genus, *Macropoma*, comprises homocercal fishes with rounded ganoid scales sculptured externally and pierced by prominent mucous tubes. *Caturus* becomes extinct. Teeth and scales of *Lepidotus* (with *Sphoerodus* as sub-genus), clearly a fresh water fish, are widely distributed in the Wealden, and finally disappear in the chalk; its body was covered with large rhomboidal ganoid scales. *Gyrodus* and *Aspidorhynchus* occur in the beds of Variance, *Coelodus* and *Amiopsis* (allied to Amra), in those of Comen, in Istria. But the Palaeichthyes are now in the minority; undoubted Teleosteans have appeared, for the first time, on the stage of life in numerous genera, many of which are identical with still existing fishes. Assuming that the ancestral fishes were Elasmobranchs, as we are bound to do from the fact that the flat bones of the head and of the pectoral girdle of bony fishes are derived from the dermal denticles of the Elasmobranch, the earliest bony fishes must have had a Crossopterygian or fringed structure in their lateral fins.

From the Crossopterygian fringe finned type was derived the ray-finned type by the reduction of the internal radials, and the lung-fish (Dipnoan) in another direction by the development of an elongated axis to form the biserial or pinnate type of fin. The Crossopterygian is therefore a central type and its fossil forms should be among the oldest. The more primitive and earlier forms retain the ganoid scaling, while some of the later have cycloid scales and perfectly homocercal tails. They first appear in the Permian, but were only abundant in the Trias and still more so in the Jurassic, while in the Cretaceous the Teleostei become the dominant type.

The family Eugnathidae, fishes much like a tarpon of large herring in shape and fins, appear first in the Trias and are common in the

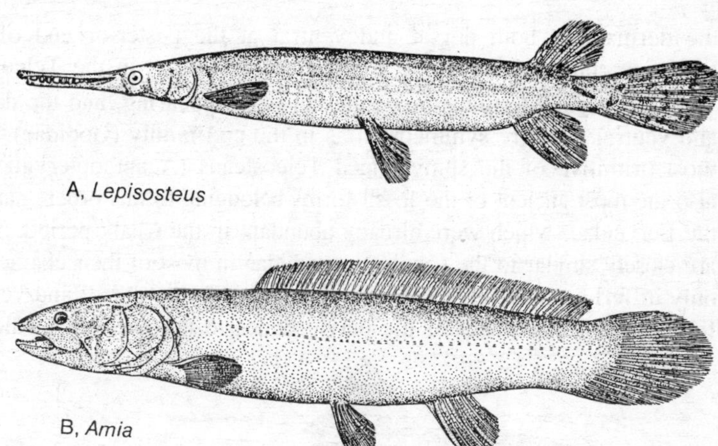

Fig. 5.8. Holosteans. A—The gar pike; B—the bowfin.

Jurassic; some of them have the Ganoid scaling, while in others the scales are thin, cycloid, and over lapping. The American Bowfin, *Amia*, is a surviving member of this group, retaining the incomplete vertebrae in the tail, and a bony plate under the throat, which are primitive features. *Caturus* and *Exrycormus* of the Jurassic lead directly to the soft-finned Teleostei (Malacopterygii) such as Tarpons (Elopidae) and Herrings (Clupeidae). The latter family were already abundant in Cretaceous times and are sometimes fossilized in dense shoals. The Pycnodonts, deep-bodied ganoids with crushing teeth, seem to have became extinct in the Lower Eocene without leaving any descendants.

Lepidosteus, the bony Pike of North America, is a pike-like, predaceous modification of the earlier Holostei, but in the fossil condition it is only known as far back as the Eocene, when it lived in Europe as well as in America. The Ancestry of the several divisions of Teleostei has not yet been worked out, but it is probable that they have descended separately from the fossil ganoids, and not all from the most primitive Malacopterygii above mentioned. It has been suggested that the Cat-fishes (Siluridae) were descended directly from the Chondrostei, and are therefore more allied to the Surgeons than to the Malacopterygii, but it is a question whether the equality of dermal rays and radials could have been evolved twice independently. There seems greater probability that the cod-like fishes (Anacanthini) may have been derived directly from the extinct Coelacanthidae; although the latter are placed among the Crossopterygians it is remarkable that the dermal ray both dorsal and ventral at the posterior end of the body are equal in number to the internal radials as in the Teleostei.

The original termination of the tail is disappearing, and the dorsal and ventral rays are symmetrical as in the cod family (Godidae). The most primitive of the spiny-finned Teleosteans (Acanthopterygii) are also the most ancient of the fossil forms belonging to this order, namely the Berycidae, which were already abundant in the Chalk period. They are closely similar to the ganoid Eugnathidae in most of their characters, only differing in the presence of a few spines in the dorsal and ventral fins and the attachment of the pelvic bones to the pectoral girdle.

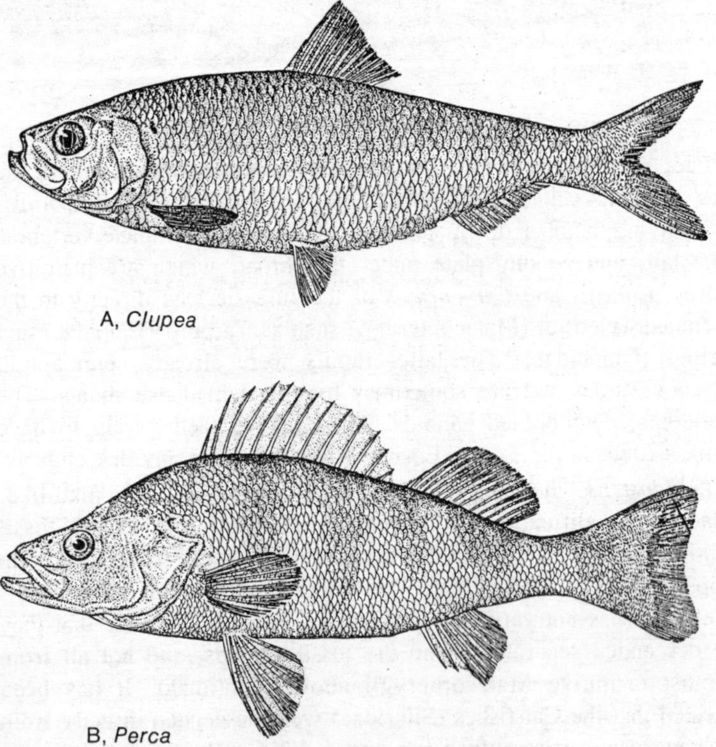

Fig. 5.9. Teleosts. A—Primitive type, the herring; B—an advanced, spiny teleost, the yellow perch.

In *Beryx* and *Holocentrum* the air-bladder even retains its communication with the digestive tube. There can be little doubt that these early examples of the spiny-finned type were evolved in the sea, and that the comparatively few existing fresh-water forms, such as the

Perch, have ascended from the sea at some later period. The above may seem a somewhat technical and condensed discussion, beyond the easy comprehension of the average reader. It is however impossible to state the most important facts concerning fossil fishes, and the relative age of the geological strata in which they occur, without the use of some technical terms. We will endeavours to give in simple language the chief conclusions which our present knowledge enables us to draw concerning the evolution of fishes.

In the broadest sense it may be said that there are two principal types of structure in fishes, the shark type and the bony type. At the present time fishes of these two types exist in large numbers and in great variety side by side in the sea. We cannot maintain that the differences of structure between sharks and bony fishes correspond to equally marked differences mode of life: the shark and the funny are both voracious monsters living as far as we can see much in the same way, both powerful swimmers both feeding on other fishes, both ranging through the surface water of the open ocean. We cannot perceive that the bony fish was derived from the shark-like fish by adaptation to a different mode of life in the sea, as the Amphibian was derived from the fish by adaptation to a life on dry land in its adult condition. What then is the explanation offered by the doctrine of evolution for the existence of these two types? When we turn to the study of fossil fishes, we find that in the most ancient stratified rocks both were represented, but the fossil forms of the bony belonged exclusively to certain groups of which at the present time very few species exist, and these only in fresh water, namely the Fringe-finned Ganoids, represented now by the African *Polypterus*, and the Lung fishes which occur now in tropical rivers or swamps. Both these kinds of fishes have lungs, or open air-bladders actually used for breathing air.

The Lung-fishes evidently resemble the ancestors of the Amphibia, while from the Fringefinned Ganoids we have a series of diverging forms leading to the great variety of existing bony fishes. The conclusion is that the bony fishes of the sea at the present day are descended from fresh-water fishes of ancient times, which were adapted to breathe air in order to supplement the original respiration by gill. In this way we can understand the origin of the airbladder which is wanting in the shark type. The air or gas-bladder is evolved from lungs not from air-bladder. At the same time the skeletal structures both within the body and in the skin underwent a change to the bony type with lungs the change of habitat from the sea to fresh water.

From the Fringe-finned Ganoids descended numerous diverging forms which populated the various fresh waters of the globe, and ultimately reached the sea again and established themselves there along with the fishes of the shark type. The first important change in the transition from the fringe-finned type to the modern fishes was the reduction of the internal skeleton of the paired fins, so that the basal fleshy lobe disappeared and the fin became fan-like. It is difficult to say with certainty, which of the extinct fan-finned Ganoids were first to reach the sea: it is probable enough that many of the groups lived along the sea coast, but we know that, excepting the Sturgeons, the only surviving forms *Lepidosteus* and *Amia* of N. America, live in fresh water, and the sea is inhabited by the latest results of evolution in fishes, the Teleostei, together with Elasmobranchs or fishes of the shark type.

6

Primitive Jawless

The lampreys and hagfishes, usually termed the "cyclostomes," are outstanding among living vertebrates in the presence of seemingly primitive characters. These peculiar forms of eel-like appearance are quite devoid of jaws such as characterize all other living vertebrates and are limbless, whereas all other living groups of fishes have well-developed paired fins. Turning to fossil types, we find that the most ancient of known vertebrates were forms usually grouped as the ostracoderms. These fishes were almost always incased in a heavy armor of bone or other hard material and thus seem quite unlike the soft-skinned modern cyclostomes in which the skeleton is entirely cartilaginous. But there, too jaws appear to have been totally absent and limbs, at the most, poorly developed. It is probable that the living cyclostomes and the fossil ostracoderms are members of a common stock of primitive ancestral vertebrates which we may term the class Agnatha, jawless vertebrates.

Cyclostomes

The living agnathous forms include the lampreys, such as *Petromyzon*, and the hagfishes (*Myxine*, *Bdellostoma*). They are semiparasitic in habit, attaching themselves to fish and living on their flesh. The body is elongate, eel-shaped; there are no scales or denticles in the tough and slimy skin. Paired fins are absent, although median fins are developed. From six to fourteen pairs of gills are present; these differ from the slitlike structures of other living vertebrates in being spherical pockets connected with the inside and the surface by small tubes. In the lampreys the margins of the round, jawless mouth form a sucking disk for attachment to the prey; in the hagfishes the

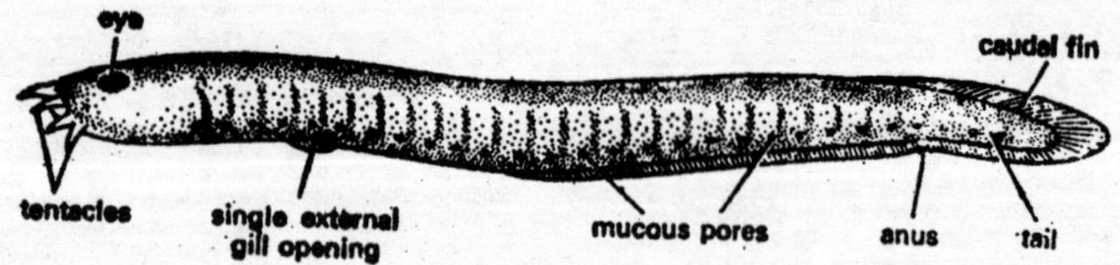

Fig. 6.1. Myxine glutinosa.

mouth is surrounded by tentacles. Jaws are functionally replaced by a long protrusile "tongue," armed with a patch of horny structures resembling teeth; this forms an efficient organ for rasping away the flesh of the lamprey's prey. There are paired eyes and a median pineal eye (well developed in lampreys, vestigial in hagfishes). Instead of the three semicircular ear canals of most vertebrates, lampreys have but two and hagfishes one. The olfactory organs are still more peculiar, for, instead of the double nostril found in all other living vertebrates, there is but a single median nasal sac.

The skeleton consists of uncalcified cartilage. The braincase is a rather specialized and complicated structure. An elaborate system of branchial arches is present in the lamprey; but these arches, instead of being separate elements, are fused into a peculiar basket inclosing the gills, while part of the branchial skeleton is modified into a support for the "tongue." The notochord is large and unrestricted. All forms have cartilaginous supports for the median fins, while *Petromyzon*, although lacking vertebral centra, has a row of small neural arches.

A striking difference between the lampreys and the hagfishes is shown in the position of the single nostril. In most vertebrates the nasal openings are near the front of the head, or even slightly on the under side. In the hagfishes the nostril opens at the tip of the "snout," but in the lampreys a large "upper lip" grows out from the front of the roof of the mouth to form the tip of the body; the originally ventral nostril (together with a second pocket-shaped structure, the hypophysis) is forced around forward and upward until in the adult it is found opening downward high on the top of the head just in front of the eyes—a situation unparalleled in other living vertebrates.

Possessing no bones, the true teeth, or other hard parts capable of preservation, the cyclostomes are unknown as fossils. Certain small Paleozoic tooth-like structures—conodonts—have been thought to be cyclostome "teeth"; but conodonts are apparently of invertebrate origin.

The lampreys and hagfishes are obviously lower in their plane of organization than any other living vertebrates. But, before concluding

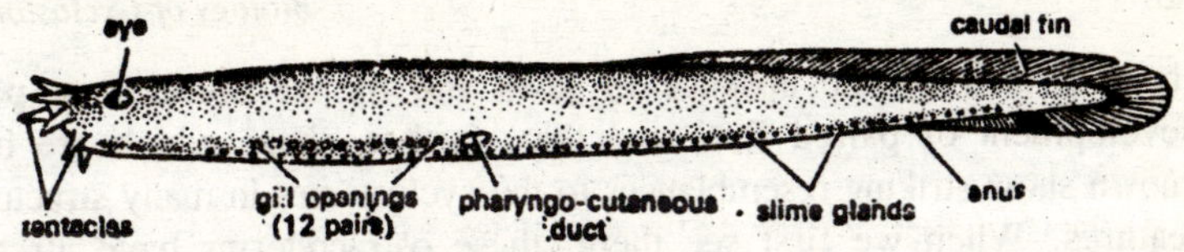

Fig. 6.2. Bdellostoma.

that they are really primitive types, we must take into account the fact that these forms are semiparasitic and that parasites are generally specialized and degenerate rather than truly primitive forms. Specialized they undoubtedly are in such respects as the peculiar "tongue." May it not be that many of the features of simplicity in their structure are really due to losses associated with parasitism rather than to true primitiveness? We may seek an answer in the study of related fossil types. As will be seen, many cyclostome features were present in the oldest-known fishes.

Ostracoderms

The oldest certain remains of vertebrates are fragments contained in Ordovician rocks from Colorado and other western states. The nature of the sediments suggests that these remains are the debris of forms living in inland waters, which, after death, were washed out into a shallow bay. This supports the theory that the vertebrates were of freshwater origin and helps to explain the almost utter absence of vertebrate remains in the dominantly marine sediments of the early Paleozoic. The fragments are too small to allow us to gain any idea of the appearance of these early types, but their microscopic structure shows that we have to do mainly with pieces of armor such as are found covering the bodies of vertebrates in the subsequent periods. It is of particular interest that some of these specimens show that bone had already developed in the earliest stage of known vertebrate history.

It is not until late Silurian and early Devonian times that we gain any adequate idea of these forms or, rather, their successors. Then, as the Silurian draws to an end, we find increasingly numerous remains of vertebrates quite unlike any existing forms in appearance. Almost all are covered with varied types of armor, a feature to which the name "ostracoderms" ("shell-skinned") is due. In some forms the internal skeleton of the head region is highly ossified; in others it is persistently cartilaginous; and in none have remains of the axial skeleton as yet been found. In bodily outlines and in the presence of hard skeletal parts they seem remote from the cyclostomes; but we find that all these forms, like living lampreys and hagfishes, are characterized by the absence of jaws and by the absence or poor development of paired fins; and that, further, those which are best known show striking resemblances to the cyclostomes in many structural features. When we first see them, these ostracoderms have already had a long history behind them and are divided into several distinct groups.

Osteostraci

Cephalaspis and its relatives constituting the order Osteostraci are known in more detail than any other ostracoderms, and hence merit first consideration. These types are abundant in late Silurian and Lower Devonian rocks, while a few forms lingered until the close of Devonian times.

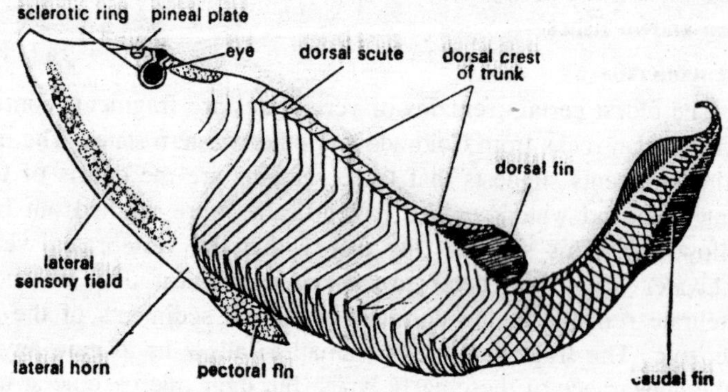

Fig. 6.3. Cephalaspis.

Few were of great size, the length ranging generally from a few inches to a foot or two. They were completely incased in an armor of plates and scales. When sectioned, these plates are seen to be composed of true bone, while on the surface was a tuberculated layer of dentine comparable to fused dermal denticles. The body scales were arranged in a series of vertical rows, each containing but a few elongate members. The body was rather fishlike in appearance, with one or two dorsal fins and a heterocercal caudal. The head was much flattened, the body higher, with a flat ventral surface and a triangular section. Structures are present which appear to be paired fins of a sort. At either lateral margin a continuous series of scales projects outward to form a fin fold somewhat comparable to a pelvic fin. Behind the head shield there are, in most genera, scale-covered flaps which are comparable to pectoral fins; their internal structure is unknown. The head—in fact, the whole area back to the shoulder region—was covered dorsally by a nearly solid shield of bone; in all but a few of the older and presumably more primitive genera there were prominent "horns" at the posterolateral corners. The eyes, directed upward as in many bottom-living types, were situated close together near the center of the shield; between them lay a median plate with an opening for the pineal eye. Anterior to this plate was a slit comparable to that which in lampreys contains the opening for the single nostril and hypophysis. In the center of the skull, behind the eyes, was an area filled by a series of small polygonal plates, and there were similar areas near each margin.

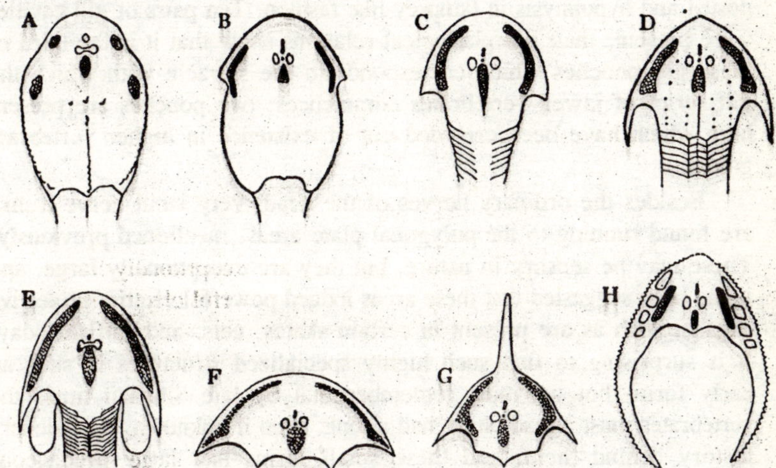

Fig. 6.4. Dorsal view of the head shield in various osteostracans. A—Trematаspis; B—Didymaspis; C—Kiaeraspis; D—Thyestes; E—Cephalaspis; F—Benneviaspis; G—Boreaspis; H—Scelerodus.

The shield folded under onto the margins of the flattened ventral surface. Behind its front edge was a small and obviously jawless mouth, while posterior to this a series of movable plates covered the throat region. At the edge of these plates were round external openings for the numerous gill pouches.

Recent work has disclosed much of the internal structure. It was long supposed that, while ostracoderms had superficial armor plates, the internal skeleton was purely cartilaginous. But in *Cephalaspis* an ossified cranial skeleton was present. Its composition was in contrast to that of most vertebrates; for, instead of a separate braincase and jointed branchial arches, there was a single unified structure underlying the entire dermal shield. This seems, at first sight, a highly specialized feature; but it is possible that it is really a primitive undifferentiated condition and that the establishment of independent units was a later development.

From the internal cavities most of the structure of the soft parts of the head can be made out. The brain and nerves and blood vessels are found to have been quite similar to those of lampreys and seem to show a very primitive vertebrate pattern. In the ear there were two semicircular canals, as in the lampreys; the dorsal slit referred to above is seen to lead into cavities which obviously lodged a single nostril and hypophysis in lamprey-like fashion. Ten pairs of gill cavities were present; their morphological relations show that it is the third of these gill pouches which corresponds to the spiracle with which the gill series of jawed vertebrates commences; two pouches are present here which have been crowded out of existence in higher vertebrate groups.

Besides the ordinary nerves of the head, very stout nerve trunks are found running to the polygonal plate areas, mentioned previously. These may be sensory in nature, but they are exceptionally large, and it has been suggested that these areas lodged powerful electric protective organs, such as are present in certain skates, eels, and catfish today. It is surprising to find such highly specialized structures in such an early form; but we must remember that by late Silurian times the vertebrates must already have had a long, even if unknown, evolutionary history behind them, and these small forms had large predaceous invertebrates as contemporaries.

The Osteostraci were a homogeneous group as regards their fundamental structures. There is, however, considerable variation in the contours of the head: there are great differences in the degree of development of the "horns"; the head may be broad or slender; a long rostral spine may be developed; the plate-areas may vary in size, and the lateral ones may be subdivided. *Tremataspis* is a Silurian genus which may be relatively primitive in nature. Here there are no "horns" or pectoral fins; instead, the head tapers smoothly back into the trunk, and a considerable portion of the trunk armor is fused with the head shield. Such an animal would have swum rather inefficiently, and with little steering ability, by a tadpole-like wriggling of the posterior part of the body. In *Cephalaspis* and other later genera a freeing of body segments from the shield would give greater mobility to the body, and the development of pectoral fin flaps would aid greatly in balance and control of direction. A similar development of paired-fin structures will be seen in other early vertebrates.

The cephalaspids were obviously, from their depressed shape and dorsally situated eyes, bottom-dwelling forms; their small mouths and expanded gill chambers suggest that they were forms which made their living by straining food particles from the mud of the stream bottoms.

It is obvious that these ancient types were fundamentally quite close to the modern lampreys in structure. Can the lampreys have descended from them? Many of the differences may be correlated with a change from a bottom-dwelling to a semiparasitic mode of existence. The most striking contrast lies between the well-ossified skeletal system of the ancient types and the purely cartilaginous skeleton of the lampreys. But similar degeneration is known to have occurred in other groups; and, despite their superficial dissimilarity, the osteostracans may possibly have been ancestral to the lampreys.

Anaspida

The order Anaspida includes a number of genera, such as *Birkenia* and *Lasanius*, which were widespread in the late Silurian; there were late survivors in the Upper Devonian of Canada. None was more than 10 inches in length, most about half that size. In the typical Silurian genera there was a complete covering of dermal armor, composed of a noncellular bony material. This appears to be the equivalent of the basal layer of the scales and plates of other early vertebrates and is suggestive of a stage in armor reduction. Scales arranged in regular rows covered the body; and the head region was protected by a complicated pattern of small plates, which appear to be more or less fused into a shield on the dorsal surface. Unlike any normal vertebrate of later times, the tail tilted downward, rather than upward, to give a

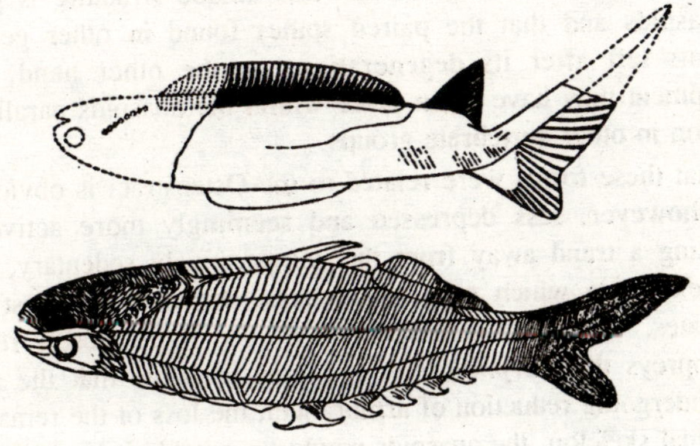

Fig. 6.5. Anaspids. Upper—Birkenia; Lower—Endeiolepis.

"reversed heterocercal" type of caudal fin, found among living animals only in the larval lamprey. So unexpected was this find that for many years these forms were restored bottom side up.

A series of spines lay along the dorsal side of the body. An anal fin was present. Of paired fins there were none in the typical Silurian genera, but there was a prominent, projecting spine in the position of a pectoral fin, and other spines more posteriorly placed. Behind the head a row of small, circular openings slanting back and down formed the exists from a series of gill pouches.

No trace of internal structure has been discovered; the skeleton was presumably cartilaginous. The dermal armor, however, reveals that the head structure was fundamentally similar to that of *Cephalaspis*. The two large orbits were somewhat farther apart and faced more laterally than in that form. Between them lay a plate pierced by an opening for the pineal eye; while the nostril, just as in the last order and in the lampreys, reached the surface through an opening high on the top of the head. The mouth was shaped more like that of higher vertebrates than in other agnathous types, but it is improbable that true jaw structures were connected with it.

Considerable variation from this typical structure is found in other anaspids. *Lasanius* of the Upper Silurian had lost most of its armor, except for the dorsal spines and a series of pectoral spines and their supports. Still more divergent is *Endeiolepis* of the Upper Devonian. The body was almost completely naked, and the dorsal spines were replaced by a long, soft dorsal fin. Vertically, a continuous fin fold, stiffened by scales, ran the length of the trunk on either side and formed a structure comparable to both pectoral and pelvic fins of higher vertebrates. It may be that this unique structure is primitive for anaspids and that the paired spines found in other genera are remnants left after its degeneration; on the other hand, this fin development may have taken place within the anaspids parallel to fin evolution in other vertebrate groups.

That these forms were related to the Osteostraci is obvious; they were, however, less depressed and seemingly more active types, suggesting a trend away from the comparatively sedentary, bottom-living existence which may have characterized the earliest jawless vertebrates. They may well have been even more closely related to the lampreys than *Cephalaspis*. It seems probable that the anaspids were undergoing reduction of armor; with the loss of the remainder of the dermal skeleton, the anaspids would be suitable lamprey ancestors.

Heterostraci

A very different group of jawless vertebrates is the order Heterostraci, whose members (like the cephalaspids) were abundant in the late Silurian and early Devonian but had disappeared completely by the end of the latter period. Perhaps the Heterostraci may be considered the oldest of all vertebrate orders, for some of the fragmentary dermal plates from the Ordovician appear to show a microscopic structure of heterostracan type. *Pteraspis* is the best-known member of the group; *Poraspis*, *Palaeaspis*, *Anglaspis*, and *Cyathaspis* are other characteristic forms. Usually there was a complex armor comparable to that of *Cephalaspis*, but lacking bone cells. In contrast with the cephalaspids and anaspids, the paired eyes were far apart on the sides of the head, the pineal eye often failed to pierce the top of the skull, and a striking difference lies in the fact that there was no opening for a dorsal nostril. The body was covered by scales, often diamond-shaped. In typical heterostracans the body was little flattened. The tail fin was of the reversed heterocercal type. There are no other median fins or paired fins, but there are dorsal and ventral spines on the trunk; frequently a prominent dorsal spine arises from the back of the head shield.

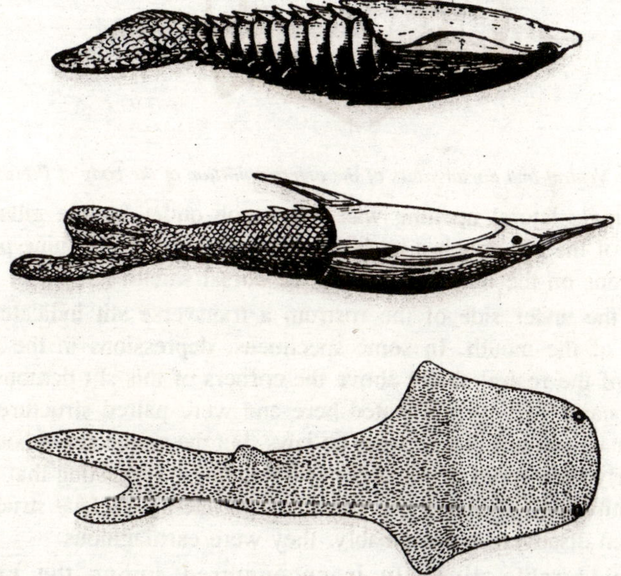

Fig. 6.6. Heterostraci and coelolepsids. Upper—Anglaspis; Center—Pteraspis; Lower—Thelodus.

The anterior part of the body, including the large gill chamber, was protected by stout armor. In the more primitive Silurian types this consisted of a single oval dorsal shield, a similar ventral element, and a pair of lateral gill plates. In later types there is a tendency for the armor to break up into a number of smaller elements. In *Pteraspis*, for example, there separates from the dorsal shield an anterior rostral element and smaller plates about the orbits.

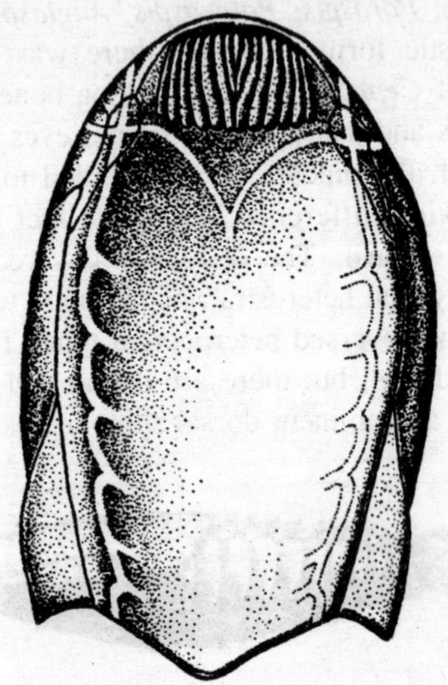

Fig. 6.7. Ventral and dorsal views of the anterior portion of the body of Pteraspis.

A single lateral opening was a common outlet for the gills; the position of the gill pouches is shown by a series of five to nine paired impressions on the inner surface of the dorsal shield.

On the under side of the rostrum a transverse slit indicates the position of the mouth. In some specimens, depressions in the inner surface of the rostral shield above the corners of this slit demonstrate that the nasal sacs were situated here and were paired structures, as in higher vertebrates. There were no jaws, but the mouth slit is bounded posteriorly by a series of parallel movable plates, suggesting that some type of nibbling movement was possible. No internal skeletal structures have been discovered; presumably, they were cartilaginous.

Considerable diversity is encountered among the known heterostracans. We have no knowledge of the appearance or structure of supposed Ordovician Heterostraci. The Upper Silurian genera, such as *Poraspis* and *Cyathaspis*, are more heavily armored than the Devonian forms and with the armor concentrated in a smaller number of plates. There is great variety, too, in the shape of the "head"; in some the rostrum is short and rounded, in others greatly elongated.

Flattened, bottom-dwelling relatives of the pteraspids are represented by *Drepanaspis* and other Devonian genera. The head and gill region, corresponding to the area included in the pteraspid shield, is very broad and flat but is covered by a series of larger and smaller plates generally comparable with those of *Pteraspis*. The eyes were quite small; the mouth apparently was similar to that of *Pteraspis*, but in the fossils it appears at the front end of the upper rather than the under surface.

The Heterostraci are obviously not closely related to the agnathous forms previously considered—cyclostomes, Osteostraci, Anaspida—all of which are peculiar in the possession of a single dorsal nostril. Possibly the Heterostraci are the remnants of a generalized primitive vertebrate group which might have given rise, by modification and specialization, to other jawless types. Still further, it is possible that the Heterostraci may be close to the line of ascent to jawed types, in which paired nostrils were retained. Known forms of the order, however, appear to be too late in time or too specialized on their own account to be considered as such ancestors.

Coelolepids

In the late Silurian and Lower Devonian are found a number of poorly known and tiny fishes which may be grouped in the order Coelolepida. *Coelolepis*, *Thelodus* and *Lanarkia* are representative genera. In many instances these forms are known only from patches of the scales which covered their bodies. These were minute structures, not overlapping but nevertheless forming a continuous covering for the body and hence to be considered as true scales rather than as dermal denticles of the shark type. In a few cases the complete body outline is known, but, even so, our knowledge of these fishes is slight. We know almost nothing of internal structures and very little even of superficial features. The tail is forked and apparently of the reversed heterocercal type; an anal fin may be present. The anterior part of the body (including the gill chamber) is broad and presumably flattened, as in most other ostracoderms. Flaps at the posterior end of this region in *Thelodus* may mark merely the region of the gill opening, as in Heterostraci, or may possibly be rudimentary pectoral fins, as in cephalaspids; in *Coelolepis*, however, these flaps are absent, and the head region tapers evenly into the trunk contours. In some instances pigment spots or openings show that the paired eyes were laterally placed; we have, however, no data on the pineal eye or nostrils. The mouth was small. Impressions of gill chambers have been seen in *Thelodus*, but the nature of the external opening is unknown.

They have frequently been allied with the Heterostraci because of the similarities in body contours, but there is no proof of such association. Their scales are comparable to the superficial portion of the plates and scales of other ostracoderms. It was originally suggested that the coelolepids demonstrated a stage in the development of armor and that typical bony plates and scales might arise by a fusion and deeper growth of small and superficial structures such as those present here. But it seems more probable today that we here have indications of a degenerative process—a reduction of armor, which, if carried a bit further, would lead to the condition of isolated dermal denticles, as seen in sharks. While it is customary to regard coelolepids as ostracoderms, even this is uncertain. It is not impossible that they may be forerunners of some group of the later jawed fish types.

Primitive Vertebrates

Classical theory maintained that cartilage was the older material. But the paleontological data strongly suggest that bone was a primitive adult skeletal material, cartilage an embryonic adaptation which appears in the adult as the result of degenerative processes.

The evidence from the Agnatha tends strongly to support the latter view. Modern jawless forms are boneless, those of the Silurian and Devonian all possess dermal bone, and in the cephalaspids there is, in addition, a well-developed internal bony skeleton. It may be argued that these ancient forms had still earlier cartilaginous forebears. If so, however, the time when bone was unknown must have been an exceedingly remote one, for bony tissues were present in the oldest vertebrate scraps from the Ordovician. If bone was a new invention among the older vertebrates, one would expect to find a progressive increase in ossification among the later representatives of the groups concerned. This, however, is the reverse of the true situation. In general, the later ostracoderms of the Devonian show a less substantial armor than that of Silurian forms; bone is regressive, not progressive, in its history among ostracoderms. As will be seen, a similar story of bone reduction is true of various other lower vertebrate groups.

It seems certain that in the ostracoderms we are dealing with a truly primitive series of vertebrate types. That the absence of jaws is a primitive feature need not be seriously questioned. This condition greatly limits the possible modes of life of primitive vertebrate types. The modern cyclostomes have been enabled to take up a predaceous mode of existence through the development of a rasping "tongue" as a substitute for jaws, and a type of nibbling may have been present in some ostracoderms. In general, however, the older vertebrates were limited, as to food supply, to tiny organisms and bottom detritus that could be strained through the gill apparatus, much as in lower chordates today. The gills, therefore, were not only a breathing, but also a feeding, device. In relation to this, we see that in most ostracoderms the major—anterior—part of the body was much expanded to form a large set of branchial chambers, surmounted by a brain and sense organs. Posterior to this, the trunk and tail appear as a relatively small locomotor appendage.

The modern cyclostomes lack paired limbs; among the ostracoderms we see various stages in the early development of paired-fin structures. Without such fins the locomotion of the oldest vertebrates must have been of the relatively ineffective and uncontrolled type seen in a frog tadpole. Paired flaps or rudders would aid greatly in preventing rolling and pitching and, if flexibility were attained, aid in steering. The finlike developments seen among the ostracoderms are not closely comparable to the "orthodox" fin structures of sharks or of higher bony fishes but presumably represent independent evolutionary developments. Similar essays in the establishment of paired-fin systems, often of curious types, will be seen among the placoderms. Paired fins, it seems probable, were developed in parallel fashion among various early vertebrates in relation to their needs for more efficient locomotion.

The ostracoderms are primitive vertebrates; but if we seek among the known forms for the ancestors of higher vertebrate groups, we meet with disappointment. The cephalaspids and anaspids show very definite characteristics (as in the development of nostril and hypophysis) which seem to ally them to the cyclostomes, and the anaspids especially may well be close to the cyclostome ancestry. It is, however, very improbable (although not impossible) that the ancestors of other vertebrate groups passed through a stage with the peculiar monorhine condition seen here.

The Heterostraci and coelolepids appear, in such features as their apparently diplorhine condition, to offer better prospects of relationship to higher vertebrates. We know, however, very little about the structure of the coelolepids, and the known Heterostraci are obviously too highly specialized in many ways to be considered in themselves as ancestors of any of the higher jawed vertebrates.

Our failure to find actual gnathostome ancestors among known ostracoderms is, of course, a result only to be expected from a broad consideration of the problem. The known forms are ancient vertebrates, it is true; but not the oldest of vertebrates. We know that the vertebrate stock was in existence a full period earlier, in the Ordovician; and it is possible that highly developed vertebrate types may have been in existence in the Cambrian, a hundred million years or more before the first adequately known types. Cephalaspids, anaspids, pteraspids, and coelolepids represent not the beginning but the end of a cycle of early vertebrate evolution—end-forms rather than generalized ancestral types.

The sea is generally assumed to have been the original home of life; lower chordates and lower living vertebrates are mainly marine types. It is, therefore, natural to believe that the vertebrates were originally dwellers in a salt-water environment.

This, however, is probably the reverse of the case. The active swimming characteristic of vertebrates suggests that their home lay in fresh waters where mobility was necessary to counteract the downward sweep of stream currents. Studies of kidney structures and function indicate that the primitive vertebrate kidney was one "invented" for use in a fresh-water environment and that this structure was variously modified by later marine types. The fossil record is in accord with these other lines of evidence. Nearly all records of Silurian and early Devonian vertebrates are of a type that suggests that the oldest fishes lived in inland waters. Most ostracoderms were apparently stream and pond dwellers; it is not until the later part of the Devonian that sharklike fishes and placoderms become numerous in the seas; few of the higher bony fishes appear to have been ocean dwellers before the Mesozoic.

Why should primitive vertebrates have been universally armored? Later armored types are usually protected against their own carnivorous relatives; but we believe that the earliest vertebrates were without biting jaws and incapable to ingesting other vertebrates as food. Obviously, the enemies must have lain among invertebrate types. The inland-dwelling vertebrate would not have come in contact with cephalopods or other marine forms which one might at first think of as possible enemies. We do, however, find that one group of predaceous invertebrates was present in the same beds with the early vertebrates—the eurypterids, aquatic, scorpion-like Paleozoic creatures of large size, with well-developed claws and biting mouth parts. It is probable that the early vertebrates furnished a food supply for the "water scorpions" and that vertebrate armor (and the electric organs of the cephalaspids) served as a defense against these carnivorous enemies. When, in the Devonian, faster-swimming fishes supplanted the comparatively sluggish ostracoderms, the eurypterids dwindled into insignificance and presently disappeared. The activity which characterizes the vertebrates may be related not merely to the taking-up of life in running streams but also to the necessity of escaping early eurypterid enemies.

7

NUTRITION

FEEDING MECHANISMS

In both groups of cyclostomes the teeth are entirely horny epidermal structures. The grasping teeth of the oral disc of the lamprey have no counterpart in myxinoids, although the single median, recurved palatine tooth may help the hagfish to maintain its purchase on its food during the tearing action of its jaws. The biting and rasping teeth of the lamprey are carried on tooth plates above and below the mouth, as well as on the apex of the piston cartilage, whereas the grasping jaws of the hagfish are borne on a cartilaginous dental plate, which is everted during feeding, at the same time closing the jaws laterally. Sliding in a groove on the upper surface of a fixed cartilaginous basal plate on the floor of the mouth, the U-shaped dental plate is moved forwards and backwards by the contraction of protractor and retractor muscles. As the dental plate is pulled over the edge of the basal plate the teeth are everted, and once the jaws come into contact with the food they are then rapidly retracted, snapping the teeth together so biting or tearing off pieces which are transferred to the mouth.

Although homologies with the hagfish feeding apparatus are obscure, the piston or tongue mechanism of the lamprey can be regarded as the functional equivalent of the dental plate. The long piston cartilage carries at its apex a transverse lingual lamina and above this a bilobed structure with longitudinal ridges bearing small denticles. With the retraction of the tongue these teeth are brought together, exerting a cutting action on the tissues of the prey. The main retractor muscle is the long cardio-apicalis running from the head of the tongue to the pericardial cartilage at the rear of the branchial basket. The protractors

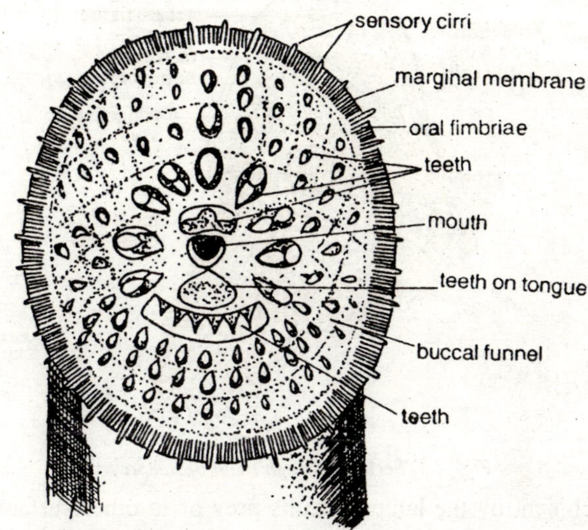

Fig. 7.1. Buccal funnel of Petromyzon.

arise mainly from the middle section of the piston cartilage, running forwards to attach to cartilages around the mouth, while shorter protractors from the head of the tongue are attached to the cornual cartilage. The complexity of the piston and pharyngeal pump mechanisms may be illustrated by the fact that these systems involve no fewer than twenty separate muscles. When feeding, the tongue rocks backwards and forwards, bringing into play the cutting actions of the bilobed head, as well as a rasping action of the teeth of the lingual lamina. Secretions of the salivary glands are poured out on to the wound, exerting cytolytic and anticoagulant effects on the tissues and body fluids of the host.

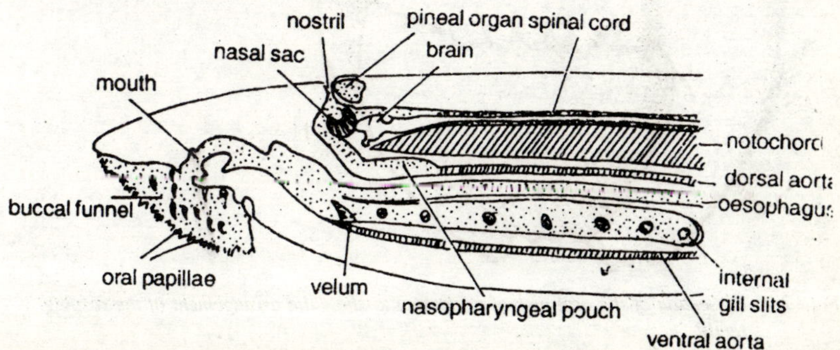

Fig. 7.2. Anterior region of Petromyzon.

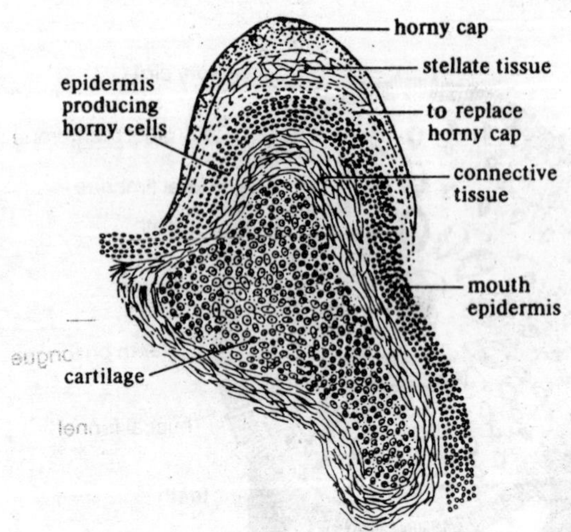

Fig. 7.3. Section through tooth of lamprey.

Attachment by the lamprey to its prey or to other surfaces, when it is not swimming, involve the development of reduced pressures within the buccal cavity. After the oral disc has been extended over the surface of attachment, its internal volume is reduced by contractions

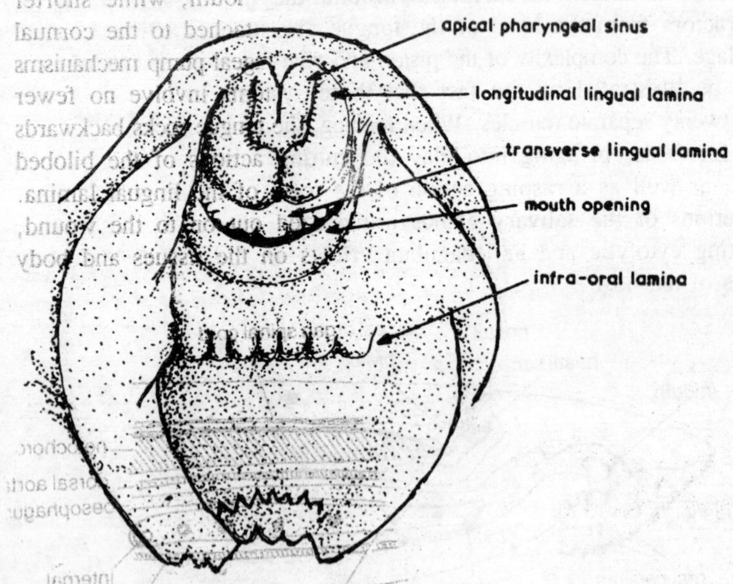

Fig. 7.4. Dissection of the oral disc of a lamprey to show the arrangement of the rasping teeth.

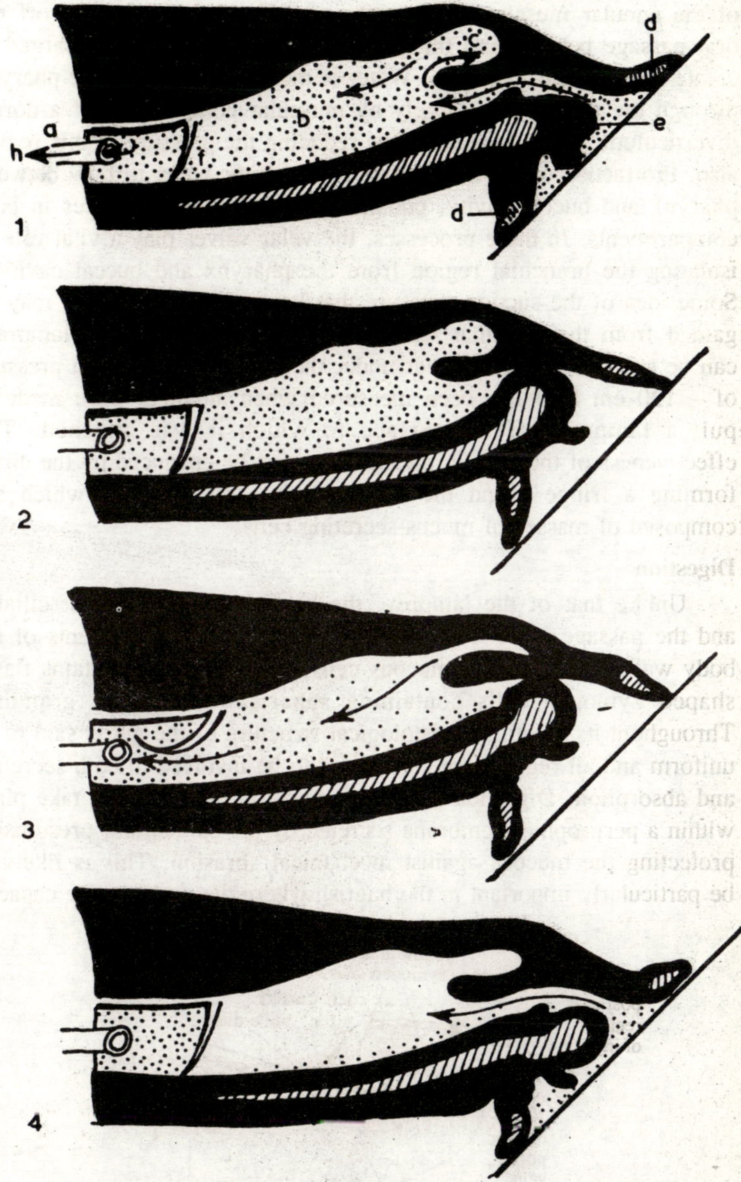

Fig. 7.5. The suction and feeding mechanism of the lamprey as seen in successive stages of its operation 1-4. a–oesophagus; b–pharynx; c–hydrosinus; d–annular muscle; e–tongue; f–velar tentacles; h–gill pouch.

of the annular muscles. The tongue is now retracted to seal off the oral passage behind, and the buccal cavity is once again enlarged to create the necessary vacuum. Water is then expelled from the pharynx through the water tube, largely by muscular compression of a dorsal diverticulum, the hydrosinus, thus reducing the pressure in this region also. Protraction of the tongue then allows some water to flow between pharynx and buccal cavity, equalizing the negative pressures in both compartments. In these processes, the velar valves play a vital role in isolating the branchial region from the pharynx and buccal cavities. Some idea of the suction pressure that lampreys can generate may be gained from the fact that when attached, even the largest lampreys can be suspended out of water under their own weight and a pressure of – 120 cm H_2O has been measured when attempts were made to pull a lamprey off the surface to which it was attached. The effectiveness of the suction mechanism must be enhanced by the cirrhi forming a fringe round the edges of the oral disc and which are composed of masses of mucus-secreting cells.

Digestion

Unlike that of the lamprey, the hagfish intestine is not ciliated and the passage of food is assisted by the general movements of the body wall. In addition to mucous cells, the epithelium contains flask-shaped zymogen cells containing spherical acidophilic granules. Throughout its length the histological structure of the gut is said to be uniform and all regions are considered to be involved in both secretion and absorption. Digestion and the evacuation of the faeces take place within a peritrophic membrane secreted by the epithelium, presumably protecting the mucosa against mechanical abrasion. This is likely to be particularly important in the hagfish where the proliferative capacity

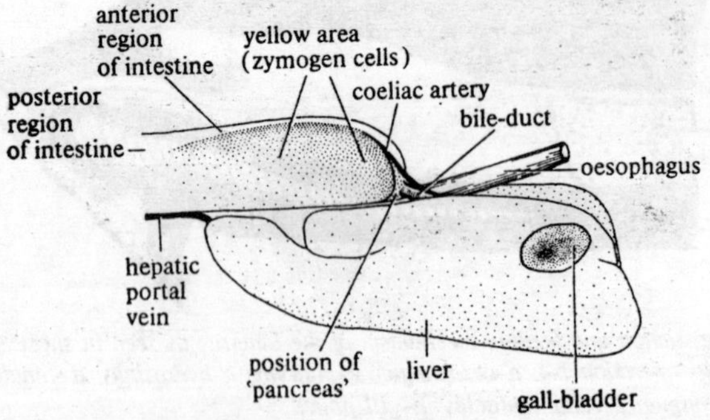

Fig. 7.6. Mid-gut of larval lamprey.

of the epithelium has been found to be exceptionally low in comparison with that of other vertebrates. The intestine is also remarkable for its high lipid content, representing about 13% of its wet weight.

Employing histochemical techniques, Adam (1963b) was able to detect an amylase with an optimum pH of 8-9 and also high lipase activity in the intestine of *Myxine*. In addition, a strong positive response was obtained with techniques used to detect leucoaminopeptidase activity. In common with those fish that are without a differentiated stomach, no HCl or peptic enzymes are produced in the digestive tract of cyclostomes. Nilsson and Fänge (1970) found some proteolytic activity at pH4 in intestinal extracts from *Myxine*, but this was believed to be due to intracellular cathepsins, suggesting that during the evolution of the vertebrate stomach, the extracellular peptic enzymes may have been developed from this type of tissue enzyme. Among the hagfish enzymes that are active in the alkaline range of pH, these authors identified trypsin-like and chymotrypsin-like activity. Both enzymes occurred throughout the length of the intestine, although they were less prominent in the posterior third. This may be contrasted with conditions in the larval lamprey where the zymogen cells are concentrated to a marked degree at the extreme anterior end of the intestine. In the hagfish, Nilsson and Fänge also identified a leucoaminopeptidase and dipeptidase as well as carboxypeptidase that acts together with chymotrypsin. In other vertebrates these enzymes occur on the brush border membranes or within the mucosal cells and are concerned in the final degradation of proteins to amino acids.

Most of our information on the lamprey digestive tract comes from studies carried out on the ammocoete or on the migratory stages of the adult and little is known of conditions in the adult feeding stages. In the ammocoete of *L. planeri*, zymogen cells with large acidophilic granules are confined to a short region at the anterior end of the intestine, adjacent to its junction with the oesophagus and similar cells are also found in the paired intestinal diverticula of the Southern hemisphere genus *Geotria*. This led Barrington (1972) to suggest that the concentration of zymogen cells in the anterior intestine and their aggregation in the diverticula of *Geotria* might parallel the early evolutionary history of the exocrine pancreas of the gnathostomes. Such a view would then place the lampreys closer to the higher vertebrates than the myxinoids, where these developments have not occurred. In the other Southern-genus, *Mordacia*, where there is only a single diverticulum in place of the paired structures of *Geotria*, no proteolytic activity has been detected in this region of the intestine, whereas in *L. planeri* Barrington was able to demonstrate a protease with an optimum pH of about 7.5, confined to the anterior region where the zymogen cells occurred. In *Mordacia*, proteolytic activity was present throughout the rest of the intestine but significantly, all the amylolytic and most of the lipolytic activity was restricted to the diverticulum.

In their ultrastructure, the secretory cells of the lamprey intestine resemble those of the hagfish and their high tryptophan content presumably reflects the presence of trypsin and chymotrypsin. It has been suggested that they may differ from pancreatic acinar cells in having both absorptive and secretory functions, perhaps in this respect representing an early stage in the evolution of the zymogen cell, before these two functions had become separated in distinct and differentiated cell types. During metamorphosis, the zymogen, mucous and ciliated cells of the anterior intestine are largely replaced by ion-transport cells. In contrast to the posterior regions, this part of the adult gut has been regarded as a storage rather than an absorptive area, perhaps foreshadowing the evolution of the vertebrate stomach.

Feeding lampreys are remarkable for the amounts of food that they consume, although it is possible that the results obtained in laboratory experiments may be biased by the fact that food fishes are ready to hand, whereas under natural conditions a considerable time may be spent in the location of prey. In one series of experiments, Farmer *et al.*, (1975) established that young feeding sea lampreys, *P. marinus*, when presented with trout whose blood cells had been labelled with radioactive chromium (^{51}Cr), fed at rates varying from 2.9-29.8% of their wet body weight a day. Such rates of feeding are greater than in most teleost species, perhaps reflecting the ease with which a blood diet can be assimilated. This, together with its high protein content must also contribute to the high conversion efficiency of the lamprey which, in these experiments, was found to vary from 10.6-56.9% compared with an upper limit of about 30% measured in teleosts. As might be expected, faecal losses were very low, representing only about 3% of the energy content of the food, as against figures of about 10% reported in some fish species.

The contrast between the high efficiency rates of the adult lamprey and the low rates recorded in the ammocoete has been linked with the changes that take place at metamorphosis in the histology of the gut and its circulatory system. These include an increase in the absorptive surface through the development of longitudinal folds, improvements in the arterial supply, the development of vascular couples in which the venous and arterial blood flows in opposite directions through the intestinal wall, the disappearance of muscle fibres separating the blood vessels of the lamina propria from the epithelium and finally, the development of a more efficient venous return through the hepatic portal system.

Respiration

One of the more interesting aspects of cyclostome physiology is the existence within the group of three quite distinct methods of gill ventilation. That of the ammocoete has been regarded as the most primitive and, as in the filter feeding protochordates, the pharynx is undivided and the unidirectional water current serves both for feeding and respiration. With the development of macrophagous feeding in the agnathans, it became necessary to separate these functions, but this division has been achieved in quite different ways in the hagfishes and the lampreys; the myxinoids making use of the nasohypophysial tract to convey the water current to the posteriorly placed gills; the adult lamprey developing a separate water tube connected to the internal gill openings, making it possible to use pumping, tidal mechanisms of gill ventilation that can be operated even when the animal is engaged by its sucker.

Respiratory Current

In the hagfish, the respiratory current is produced by the pulsations of the complex scroll-like velar folds, operated by extrinsic muscles attached to the neurocranium and axial skeleton and inserted on the cartilages that support them. These folds are housed in a velar chamber developed from the nasohypophysial tract. When at rest, the folds are rolled upwards towards the dorsal side of the chamber, subsequently unrolling ventrally and laterally. This has the effect of scooping up the water and deflecting it backwards towards the pharynx, while at the same time, the reduction in pressure within the velar chamber draws in fresh water through the nostril.

Although most authors believe that the velum of lampreys and myxinoids are homologous, this view is not without its difficulties and although a comparable structure may have been present in the fossil agnathans, we have no direct evidence to support this belief. The ammocoete velum consists of two simple flap-like paddles, but these are arranged vertically rather than horizontally as in the hagfish. Here its movements are produced by intrinsic muscles and it serves not only as a respiratory pump, but also acts as a valve preventing the reflux of water from the pharynx into the mouth during the expiratory phase of the respiratory cycle. Neither in the hagfish nor in the ammocoete does the velum act alone in producing the respiratory current. In the hagfish, the activity of the velum is assisted by peristaltic contractions of the muscular gill pouches and their afferent and efferent ducts. In the ammocoete, the current is augmented by alternate contractions and expansions of the branchial basket; the inhalatory phase being assisted by the passive elastic recoil of the cartilaginous framework of the pharyngeal wall.

Given the engagement of the lamprey mouth and sucker during feeding and the forward position of the gills, the tidal method of ventilation represents the only possible alternative to the solution adopted by the myxinoids. In the lamprey, the velum has completely lost its water-pumping functions and has been reduced to a small flap-like valve, preventing the entrance of food into the water tube and playing a role in the maintenance of suction. As in the ammocoete, the water current is produced by alternate changes in the volume of the branchial basket; contraction of the branchial constrictors and the diagonal muscles reducing its volume and elastic recoil bringing about its subsequent enlargement. The direction of the water flow is controlled by two sets of valves in the external gill openings; namely an outer ectal valve and an inner pair of ental valves. Water flow from the water tube into the pharynx can be controlled by the velar valves. Even when lampreys are swimming, the tidal pumping mechanism is thought to be maintained, although it has been shown that water flow through the mouth and out of the gills is also possible under certain circumstances.

Morphological and Functional Aspects of Cyclostome Gills

In spite of the evolutionary distance separating the cyclostomes from the teleosts and the fundamental differences in the mechanics of gill ventilation in the two groups, there are some remarkable parallels in the gross and microscopic structure of the gills and their functional correlates. These resemblances have been shown to extend to the relation between filament lengths and numbers and body weight, the spacing and areas of the secondary lamellae and the total gill areas.

Amongst the factors that influence the rate of gaseous exchange, gill areas are of obvious importance, although it should be remembered that since the gills are exposed to osmotic or ionic stresses, gill areas may represent a compromise between respiratory requirements and osmotic limitations. In the ammocoete of *L. fluviatilis*, gill areas have been estimated at between 1462-2717 mm^2 g^{-1} and similar values have been recorded for the adult (1402-2337 mm^2 g^{-1}). Both of these values come well within the upper end of the range for teleosts and are comparable with those of the most active species.

In the fine structure of the gills and the water-blood pathway there are further parallels with gnathostome fishes. Except in the marginal and basal regions the surface of the gill lamella is composed of two cell layers an outer layer of platelet cells whose surfaces are covered with microvilli and below them, the basal cells. These rest on a basement membrane containing collagen fibres, in places interdigitating into the plasma membranes of the pillar cells to form the collagen columns that support the lamellae and maintain the arterial blood pressure. The pillar cells have cytoplasmic extensions - pillar cell flanges which meet those of adjacent cells to enclose the blood spaces. Estimates of the thickness of the water-blood barrier have given values of 4.57 μm and 1.21 μm in an ammocoete and adult of *L. planeri* of similar body weight.

As might be expected from the close parallels in their gill morphology, the lamprey gill appears to be quite as efficient as that of a gnathostome fish as a medium for gaseous exchange. This efficiency has been measured by Lewis (1976) from the diffusing capacity D_t calculated from the expression D_t = K.*A*/*t* where:

K is Krogh's (1919) permeation constant (0.00015 ml cm^2 mm Hg^{-1} min^{-1});

A the total gill area; and

t the thickness of the water-blood barrier.

When expressed in terms of body weight, values of D_t for adult *L. planeri* and *P. marinus* were 1.84 kg^{-1} and 0.51 kg^{-1}; quite similar to those of teleosts.

In larval lampreys, the ventilation of the gills is adjusted to varying oxygen demands by alterations in the frequency of pumping or in the volume of water passing over the gills at each contraction (stroke volume). At 20°C the mean ventilation rate (volume of water passing over the gills in unit time) was found to be 0.60 ml min^{-1} g^{-1} and the maximum rate for any animal was 3.3 ml min^{-1} g^{-1} Various stimuli were effective in increasing ventilation volumes, but after exercise, mechanical obstruction of the water current, low oxygen tensions or increased carbon dioxide concentrations, the stroke volumes were raised to a greater extent than the ventilation frequencies. Stroke volumes varied from 4-25 μl g^{-1} and these, like the ventilation volumes, lie within the normal ranges for fishes i.e. 2-31 μl g^{-1} and 0.09-13 ml g^{-1} min^{-1}. For an adult lamprey, *L. (Entosphenus) tridentata*, ventilation volumes varied from 0.25-10.5 ml g^{-1}I min^{-1}; values which again are very similar to those of teleosts. These comparisons indicate that, in spite of the vast differences in the ventilatory mechanisms of ammocoetes, adult lampreys and fishes, each appears to be equally effective in delivering water to the gills.

In teleosts, the rate of gaseous exchange between water and blood can be regulated by changes in the flow pattern of the blood in the secondary lamellae of the gills and, since microfilaments are present within the pillar cells of lamprey gills a similar form of control may also exist in these animals.

The amount of oxygen that passes into the blood in the gills will depend on the surface area of the gills, the diffusion distance and the oxygen gradient between blood and water. The efficiency with which the animal is able to utilize the oxygen contained in the respiratory water can be assessed by calculating the extraction rates, represented by the percentage difference in the oxygen content of the inspired and the exhaled water. For larval stages of *P. marinus* this has been found to be about 43% compared with values of 10-28.4 % in the adult lamprey, *L. (Entosphenus) tridentata*. These apparently higher extraction rates in the ammocoete are in line with the general picture in fish, where low extraction rates are characteristic of active species living in running water, while more sluggish species living in still water tend to show higher rates.

Because of its habit of feeding with its head and branchial region frequently burrowed deeply into the carcasses of fish, it has been suggested that cutaneous respiration may be a significant factor in the oxygen uptake of the hagfish. This is borne out by the fact that *Myxine* can survive for a considerable period after its nasal opening has been obstructed. In the case of the lamprey, *L. fluviatilis*, it has been estimated that up to 18% of its oxygen uptake may take place through the skin and these animals are certainly capable of surviving for several days out of water at temperatures below 10°C. On the other hand, survival is reduced to a matter of only a few hours if the gills are covered, indicating that when the animal is out of water, respiration may still be taking place through the moist gill surfaces. Cutaneous respiration may be significant in small ammocoetes which, once they have settled down under laboratory conditions at low temperatures and relatively high oxygen tensions, may cease to show either velar or branchial contractions.

Oxygen Consumption

Both larval and adult lampreys lend themselves to measurements of standard oxygen consumption, defined as the nearest practicable approach to the basal metabolic rate. Suitable conditions can be provided for measurements on ammocoetes by employing an artificial substrate of glass beads into which they will burrow and settle down quietly. Adults can be maintained within glass tubes to which they will attach themselves and remain motionless for long periods.

The importance of allowing the ammocoete to burrow has been demonstrated by comparing the oxygen consumption of burrowed and unburrowed animals. The unburrowed ammocoetes took longer to reach a constant level of oxygen consumption than the burrowed larvae and this level was always higher than for animals that had been provided with a substrate. Measurements of the oxygen consumption of burrowed ammocoetes show little variation between species and by vertebrate standards the rates are low, although greater than the rates of oxygen consumption of the protochordate *Amphioxus*, whose burrowing habits are not very dissimilar to those of an ammocoete.

The contrast in the metabolism of larval and adult lampreys may be illustrated by following the changes in oxygen consumption that take place during metamorphosis in the closely related parasitic and non-parasitic species, *L. fluviatilis* and *L. planeri*. Before metamorphosis, the mean larval rates of oxygen consumption at 10°C were 30 $\mu g\ g^{-1}h^{-1}$ for *planeri* and 43 $\mu g\ g^{-1}h^{-1}$ for *fluviatilis* rising at the completion of transformation to 74 $\mu g\ g^{-1}h^{-1}$ and 89 $\mu g\ g^{-1}h^{-1}$ in the young adults (macrophthalmia), representing a doubling of the rate over that of the ammocoete stage. In addition, with the eruption of the eyes, a circadian rhythm of oxygen consumption develops, involving an increase in the rate during the dark phase. In the case of the non-parasitic *planeri*, which almost immediately enters the phase of sexual maturation, sex differences in oxygen consumption begin to appear, similar to those observed in the sexually mature *fluviatilis*. Curiously, the rates of oxygen consumption (expressed in terms of body weight) are remarkably similar in adults of the two species, in spite of the vast differences in their body weights. Thus, for adult males and females of *fluviatilis* with a mean body weight of 45 g, the rates were 90 and 50 $\mu g\ g^{-1}h^{-1}$ compared to 108 and 65 $\mu g\ g^{-1}h^{-1}$ for male and female *planeri*, whose mean body weight was only 2.5 g. For a wide variety of animal groups, the relationship between metabolic rates and body weights can be expressed by $M = a W^b$, where M is the metabolic rate, W the body weight, a a constant of proportionality and b the exponent. For most species that have been studied the value of b lies between 0.7-0.8, but although this applies to the ammocoete, the value for the adult lamprey is close to unity.

The limited information on the metabolic rates of hag fishes is interesting in so far as it suggests that there may be pronounced species differences in oxygen consumption related to their ecology and more especially to the depths of water they normally frequent. Laboratory experiments on the comparatively shallow water species *E. stouti* indicated low oxygen consumption rates of 12-15 $\mu g\ g^{-1}h^{-1}$ for animals acclimated at temperatures of 4° and 10°C, resting quietly on the bottom. More recently, a remote-controlled underwater vehicle (monitored by television cameras), has been used in conjunction with a respirometer, to measure oxygen consumption in a number of benthic and abyssal animals, including the hagfish, *E. deani*, on the sea bed under natural conditions. Measurements on this hagfish at a depth of 1230 m and a temperature of 3.5°C in the San Diego Trough gave an average value of only 3.1 $\mu g\ g^{-1}h^{-1}$ over a period of 13 h and during this time, no significant changes in the rate of oxygen uptake were observed. In the same area, a very similar respiration rate was also recorded in a deep sea benthic teleost, the rattail, *Coryphaenoides aerolepis*. Moreover, later measurements on an abyssopelagic rattail, *C. armatus* at depths of 2700-3600 m and a temperature of 3°C in the North West Atlantic again produced very low values of 4.9-5.3 $\mu g\ g^{-1}h^{-1}$. Such low metabolic rates are now believed to be characteristic for deep water species and have been related to an environment of low temperatures and high pressures, where opportunities for feeding are infrequent and availability of food is low.

Even making allowances for the rather low temperatures at which these measurements were made, the metabolic rates of the hagfishes are in striking contrast to lampreys of a similar size range. They may be compared with oxygen consumption rates at 5°C of 15 $\mu g\ g^{-1}h^{-1}$ for adult river lampreys (*L. fluviatilis*) in winter rising to about 50 $\mu g\ g^{-1}h^{-1}$ at spawning time and a figure of 53 $\mu g\ g^{-1}h^{-1}$ for the larger sea lamprey, *P. marinus* at the same temperature.

The extent to which the rates of oxygen consumption of the lampreys can be raised during activity may be illustrated by the experiments carried out by Beamish (1973) on the effects of swimming activity in the sea lamprey, *P. marinus*. For example, in the case of feeding stage adults with mean body weights of 52 g, he recorded mean oxygen consumption rates of 475 μg $g^{-1}h^{-1}$ in animals swimming for 20-60 minutes at speeds of 30-40 cm s^{-1} at 10°C. Taking the upper values for oxygen uptake of swimming lampreys, these represented active rates some seven times higher than the resting rates; increases that are of the same order of magnitude as those recorded in the highly active trout, whose oxygen uptake during exercise may be from four to eight times higher than its resting values.

Responses to Oxygen Depletion

To a far greater extent than the myxinoids, both larval and adult lampreys must be able to cope with changes in their external environment, particularly changes in temperature and in the oxygen content of the water. Larval lampreys do not normally live in stagnant water, but the currents over the burrows may be very slow and both oxygen tensions and water temperatures will show quite large seasonal variations. Naturally, the degree of oxygen depletion that the ammocoete is able to tolerate will be dependent on temperature. For example, at 5°C ammocoetes of *I. hubbsi* were able to survive for at least 4 days at oxygen tensions of only 7-10 mm Hg, whereas at a temperature of 22.5°C, survival for a similar period was only possible at much higher oxygen tensions of 19-21 mm Hg.

When exposed to low oxygen or high carbon dioxide concentrations, the ammocoete displays a number of behavioural adaptations, tending either to minimize the oxygen deficiency or to move the animal into a more favourable environment. Physiological adaptations include increases in ventilatory frequency or stroke volume. The increase in the volume of water passing over the gills must be capable of compensating for the decreased oxygen content of the water as well as for the additional energy expenditure involved in the greater respiratory effort. Under conditions of severe oxygen deficiency, ammocoetes may show pumping rates up to 200 beats min^{-1}, but at tensions approaching the lethal levels, the rate of pumping again begins to decline.

Adult lampreys show a remarkable ability to maintain, or even increase their oxygen uptake with decreased oxygen concentrations in the water. Starting from air saturation values of 100%, *L. fluviatilis* increased its rate of oxygen uptake at 9.5°C from 38 μg $g^{-1}h^{-1}$ to 52 μg $g^{-1}h^{-1}$ at 30% air saturation.

In adult lampreys, both ventilatory frequencies and heart rates increase as oxygen tensions are reduced and in the case of *L. fluviatilis* the former may reach levels some three times higher than those of resting animals. At air saturations of less than 20%, hyperventilation becomes much more marked and at this stage the animals begin to show restless, continuous swimming activity, often leaping clear of the water. Similar avoidance responses are observed in fish when oxygen concentrations approach the lethal levels. Under conditions of hypoxia, teleosts and elasmobranchs develop a slowing of the heart beat (bradycardia) attributed to increased vagal tone and in trout this is known to be accompanied by an increased stroke volume, thus maintaining a relatively constant cardiac output. This is in marked contrast to the condition in the lamprey, where the heart rate increases slightly under reduced oxygen tensions and in this connection it should be noted that vagal stimulation of the lamprey heart or the administration of acetylcholine, results in an acceleration of the heart rate.

The increased oxygen uptake of the lamprey under hypoxic conditions is almost certainly associated with the additional energy expenditure involved in increased gill ventilation. In fish generally, the metabolic cost of ventilation has been put at about 30% of the standard rates, but this will almost certainly vary widely with the habits and environments of different species. For example, this energy expenditure will tend to be higher for fish living in still water, where more active pumping is required, and in these cases, ventilation volumes will tend to be low and oxygen extraction rates high. For species living in fast-flowing water, the energy costs will tend to be lower, but larger volumes of water will be passed over the gills, from which comparatively less oxygen will be extracted. This would be advantageous in so far as it would keep oxygen tensions high over the entire length of the secondary lamellae. It has usually been assumed that the tidal respiration of lampreys would be more costly in terms of energy than the mechanisms employed by fish, but this has been questioned on the grounds that the passive elastic recoil during the expiratory phase of the lamprey cycle might well compensate for any general inefficiency in tidal breathing.

8

Blood Vascular System

In its basic plan, the circulatory system of the cyclostome conforms to the characteristic craniate pattern. In the pharyngeal region, the ventral aorta gives rise to a series of aortic arches supplying the gills, of which there are eight in the lamprey; the first – the hyoid – serving the anterior hemibranch of the most anterior gill pouch. The remaining afferent arteries divide to supply the anterior and posterior hemibranchs of adjacent gill pouches, and a similar pattern is seen in the efferent arteries, joining the dorsal aorta below the notochord. In myxinoids, the number of arterial arches varies with the number of gill pouches and unlike those of lampreys or other craniates, each arch supplies the hemibranchs of the same pouch. This also applies to the efferent vessels, which join two lateral aortae above the pharynx. These unite behind the branchial region to form the median dorsal aorta, which is continued forwards between the lateral vessels. The homologies of these vessels are by no means certain, but Holmgren (1946) considers that in its embryonic development, the dorsal aorta of *Myxine* is closer to the gnathostome pattern than that of the lamprey.

The venous system shows the typical craniate pattern of paired anterior and posterior cardinal veins, the latter formed by the bifurcation of the caudal vein. The anterior cardinals of the hagfish are asymmetrical, the left emptying into the sinus venosus independently, while the right cardinal joins a median inferior jugular draining the posterior tongue muscles. The proximal part of the right anterior cardinal constitutes an anterior portal vessel, derived from the right Cuvierian duct and opening into the contractile portal heart. In the lamprey both anterior cardinals enter the single right duct of Cuvier, independently

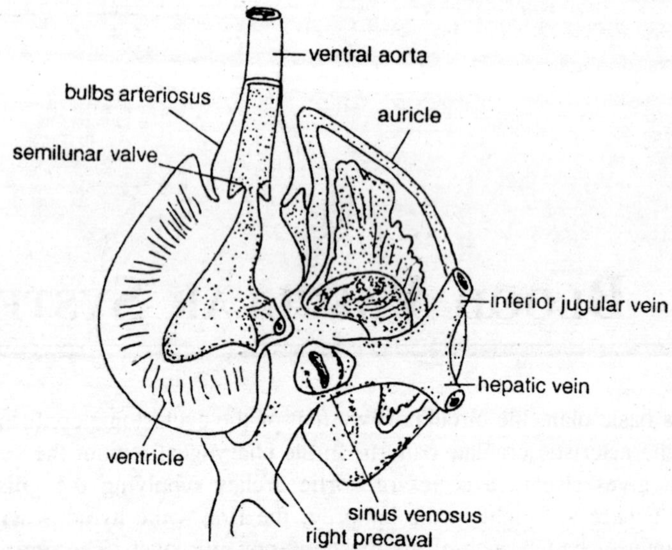

Fig. 8.1. Heart of lamprey.

of the inferior jugular. In both groups an hepatic portal system is present, but neither possess a renal portal system. In the myxinoids, the portal heart receives blood from the branchial region through the anterior portal vessel and from the gut through the intestinal vein, transmitting it to the liver through a common portal vein. In the lamprey, the portal system is concerned solely with blood from the gut conveyed to the liver via the mesenteric vein.

A characteristic of the cyclostome circulation is the presence of an extensive system of sinuses or blood lacunae, communicating on one side with the arterial system and on the other with the veins. This open type of circulation is particularly well developed in the hagfishes, where the blood pressures are low and the maintenance of the circulation is aided by 'accessory hearts', supplementing the pumping activity of the branchial or systemic heart. In addition, the general movements of the body of the hagfish make a significant contribution to the circulation by exerting pressure on the sinuses, thus forcing blood into the veins. In the lamprey, where the sinuses are well developed in the branchial region, they play an important part in the mechanics of respiratory movements. In the myxinoids, the sinus system is particularly important in the head regions, where apart from the skin and brain, there are no true veins and these lacunae replace the capillary bed.

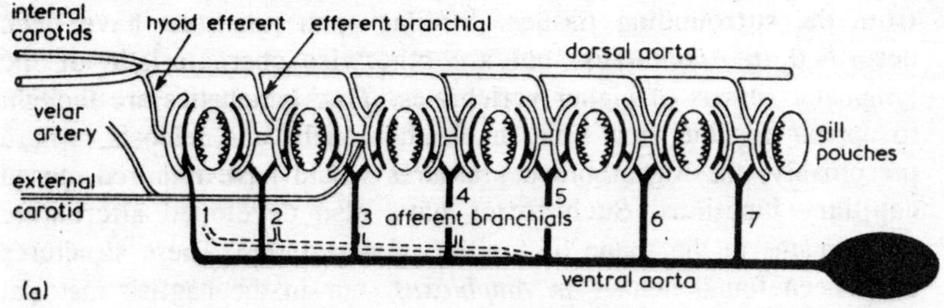

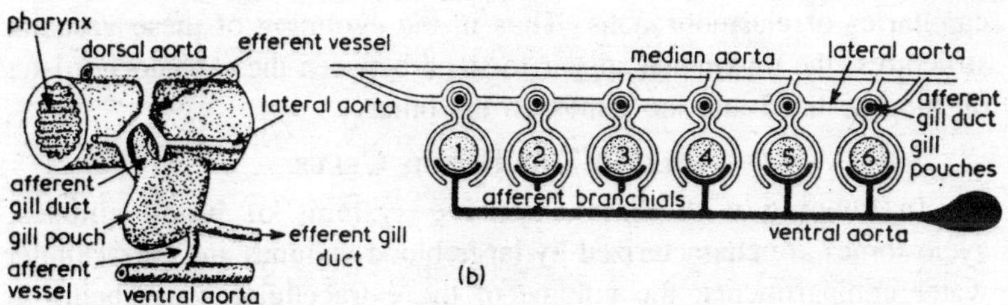

Fig. 8.2. The branchial circulation of a lamprey (a) and a hagfish (b). Details of the gill pouch and its vascular relations are shown in (c).

These conditions have been compared to the primitive type of circulation seen in many invertebrates and in the cephalochordates, where peristalsis of the blood vessels and contractions of the body wall musculature are important factors in circulatory dynamics. Further primitive features have also been described in the fine structure of the capillaries of the neurohypophysis and thyroid of the hagfish, where open connections have been described between the endothelial cells. These openings, measuring from 20–50 μm (even in some cases up to

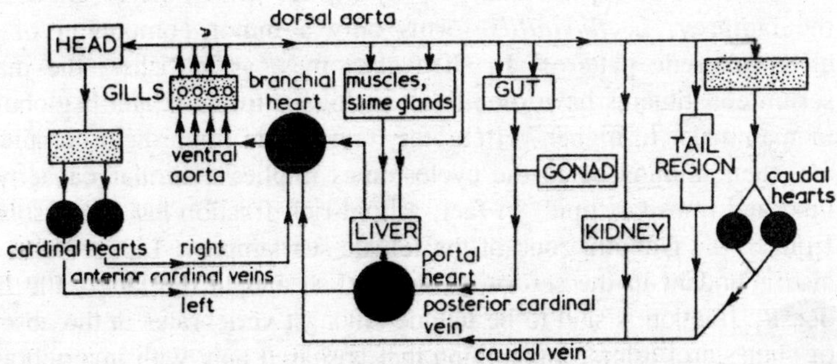

Fig. 8.3. Diagrammatic representation of the circulatory system of a hagfish.

1.0 mm), would allow macromolecules to be taken up into the blood from the surrounding tissues. Similar open junctions have been described in *Amphioxus*, but are otherwise characteristic of the lymphatic vessels of higher vertebrates. True lymphatics are thought to appear for the first time in elasmobranchs and teleosts, where presumably, the higher blood pressures would have required closed capillary junctions. Such forms have also developed alternative mechanisms in the shape of capillary fenestrations. These structures have been found neither in *Amphioxus*, nor in the hagfish material examined by Casley-Smith, but they are known to occur in the capillaries of elasmobranchs. Thus in the evolution of these vascular structures, the myxinoids appear to stand between the cephalochordates on the one hand and the fishes on the other.

Blood and Blood Cells

In keeping with their extensive systems of blood sinuses, cyclostomes are characterized by large blood volumes and extracellular water compartments; the volume of the extracellular water being of the same order as in the elasmobranchs and distinctly greater than in teleosts. Correlated with their more extensive lacunar system, blood volumes in hagfishes are larger than in lampreys.

Both in hagfish and lamprey, the protein concentrations of the plasma are higher than in elasmobranchs or teleosts. In these fishes concentrations are said to range from 2.5-3.5%, but they may reach over 7% in the sea lamprey during its parasitic phase, falling in the non-feeding migrant stage to reach minimum values (2.1-3.8%) in the sexually mature animal. In contrast to mammalian sera where the dominant component is the albumin fraction, an analogous fraction is either absent or represented only by a trace in *Myxine* serum and in the lamprey, *L. fluviatilis* forms only a minor component of the electrophoretic patterns. In all cyclostomes, as in fishes, the major serum constituents have mobilities analogous to the α and β globulins of mammals. In higher vertebrates, these form lipoprotein complexes and their dominance in the cyclostomes implies a similar capacity to bind and transport lipid. In fact, a lipid-rich fraction has been isolated from β_2 the globulin zone of the female sea lamprey. Lipoproteins are also abundant in the serum of the hagfish and in the latter, the high density fraction is said to be unique amongst vertebrates in the absence of cholesterol esters; a condition that is shared only with invertebrates. Lampreys differ from gnathostomes in that vitamin A is transported in association with these plasma lipoproteins, rather than being bound to specific retinol-binding proteins. Serum fractions with mobilities analogous with mammalian gamma globulins are also present in cyclostome sera, although these do not show immunological activity. These same serum fractions, however contain transferrins; single-chained proteins with molecular weights of 75-80000, able to bind iron at two binding sites, suggesting that they may have been produced as a result of gene duplication. At least in the higher vertebrates, transferrins may show bactericidal activity, but this has not been demonstrated in the cyclostomes.

The blood cells of lampreys fall into the same broad categories as those of higher vertebrates, although some doubt still remains on the functional and structural correspondence of certain elements of the leucocyte series. Among these, granulocytes, eosinophils and neutrophils can be distinguished and macrophages are present, perhaps corresponding to the monocytes described by earlier workers and apparently derived from large lymphocytes. The cells of the granulocyte and erythrocyte series originate in small lymphocytes, which these authors regard as stem cells; a view that differs radically from that proposed by several previous authors who had traced the origins of the haemopoietic stem cells to fixed reticular cells in the haemopoietic tissues.

Initially, at the time when the prolarva is hatching, the formation of blood cells occurs in the blood islands around the remaining yolk mass, but as the yolk is absorbed and the intestine opens, haemopoiesis is taken over by the typhlosole. By the time the ammocoete has reached lengths of about 20 mm, haemopoiesis begins in the intertubular and adipose tissues of the kidney fold and together with the intestine, these remain the active sites throughout the remainder of larval life. With the approach of metamorphosis, these larval blood-forming tissues undergo involution and their functions are taken over by the fat column above the spinal cord, which remains the active haemopoietic tissues of the adult lamprey until the end of its spawning run, when it too undergoes massive degeneration. This change in haemopoietic sites at metamorphosis coincides with a switch from larval to adult haemoglobins, although it is not known if these are contained in distinct erythrocyte populations or are solely produced in separate haemopoietic sites. The nucleated red cells of cyclostomes are exceptionally large and those of the lamprey have a volume some eight times that of the human red cell, containing about six times as much haemoglobin.

Less is known of the developmental history of the blood cells of myxinoids, but it now appears that the main areas of blood cell formation are the intestinal submucosa and the central mass of the pronephros; the first of these tissues making up as much as 10% of the whole intestinal volume. The implication of these sites in haemopoiesis is supported by the intense activity in these tissues of delta-aminolevulinic acid dehydrase (ALAD); a key enzyme in porphyrin synthesis. Large numbers of mononucleate, lymphocyte-like cells are present in the blood of *Myxine*, but these are said to be difficult to distinguish from spindle cells, thrombocytes, monocytes or haemocytoblasts described by other authors. The mitotic activity of erythroblasts in the circulating blood has been confirmed by their incorporation of H^3-labelled thymidine, but strong RNA synthesis is indicated in spindle cells and some lymphocyte-like blast cells by their intense uptake of H^3-labelled uridine. The identification of cells functionally equivalent to the lymphocytes of higher vertebrates is made more difficult by the fact that these, unlike the lymphocytes of the lamprey, are relatively insensitive to radiation treatment. Macrophages are present in the peritoneal fluid and have been activated by thioglycollate treatment. Ultrastructural studies on the granulocytes of the intestinal wall of *Myxine* suggests that these animals, unlike the lamprey, may possess only a single type of granulocyte with heterophilic granules and equivalent to the neutrophil of the higher vertebrates.

The oxygen capacity of cyclostome blood appears to be lower than that of active teleosts and for the lamprey, *L. (Entosphenus) tridentata* this has been put at 9.15 vol.%. Haematocrits, often used as an indication of oxygen capacity, vary considerably throughout the life cycle of the lamprey and show a dramatic fall during and after spawning, when the haemopoietic activity of the fat column is declining. For the ammocoete, values of 24.7 and 28.7% have been recorded, but in adult lampreys records vary from 21.5-64%, coming within the range encountered in active teleosts and higher than the haematocrits of lungfishes or elasmobranchs. In spite of this being a time of intense activity, the spawning period is characterized by a massive destruction of blood cells in the liver and kidney, accompanying the final involution of the haemopoietic tissue of the fat column. As a result, haematocrits decline sharply and the blood of some spawning or spent lampreys may become virtually colourless. Meanwhile, bile pigments derived from the breakdown of the haemoglobin may be responsible for the change in liver colour from orange to green that occurs in some animals. In the sea lamprey, *P. marinus* the skin of the sexually mature animal changes from a greyish metallic blue to orange and this also has been attributed to the accumulation of biliverdin. Haemoglobin concentration in the lamprey (5.8-13.0 g 100 ml^{-1}) are comparable with the range in active teleosts (7-17%), but reports for several species of hagfish suggest that these animals may have distinctly lower haematocrits (20-25%) and lower haemoglobin concentrations (4.6 g 100 ml^{-1}) than adult lampreys.

Cyclostome Haemoglobins

Physiological Properties

In the structure and properties of their haemoglobins, the cyclostomes have been regarded as the most primitive of vertebrates. Whereas the haemoglobins of the higher vertebrates are composed of four chains α_2 β_2 with four haem oxygen binding groups and molecular weights of about 68 000, the haemoglobins of lampreys and hagfishes are monomeric, with only one haem group to each molecule and their molecular weights are only about one quarter those of the higher vertebrates. Not surprisingly, in view of their monomeric character, cyclostome haemoglobins, and more especially those of hagfishes, show at best only slight indications of the allosteric effects that have developed in the multi-chained haemoglobins of the higher vertebrates, including cooperativity between the haem binding sites and the modifying effects of pH and CO_2 (Bohr effects).

In relation to the oxygen transport functions of the blood and the conditions under which oxygen is taken up by the respiratory organs and unloaded to the tissues, the important factors are defined by the oxygen dissociation curves. Among these are the shape of the curve for percentage oxygen saturation (*y*) plotted against partial oxygen pressure (*p*), the oxygen affinity of the haemoglobin (partial oxygen pressure at 50% saturation - P_{50}) the effects of pH and CO_2, and temperature dependence and allosteric effects, involving substances such as the organic phosphates which may modify the oxygen equilibrium. When plotted as log $y/100 - y$ against log p, the data fall approximately on a straight line (except at extremes of high and low saturations) and the slope of the line *n* or sigmoid coefficient, expresses the degree of interaction between haem binding sites or the degree of co-operativity between the sub-units, such as occurs in tetrameric vertebrate haemoglobins, where oxygenation of one haem group favours the oxygenation of another. At values for *n* close to unity, as in the monomeric haemoglobin or myoglobin, the dissociation curve has the form of a rectangular hyperbola, but above values of about 1.6 the curve becomes recognizably sigmoidal in shape. This has important consequences in the greater efficiency with which the haemoglobin

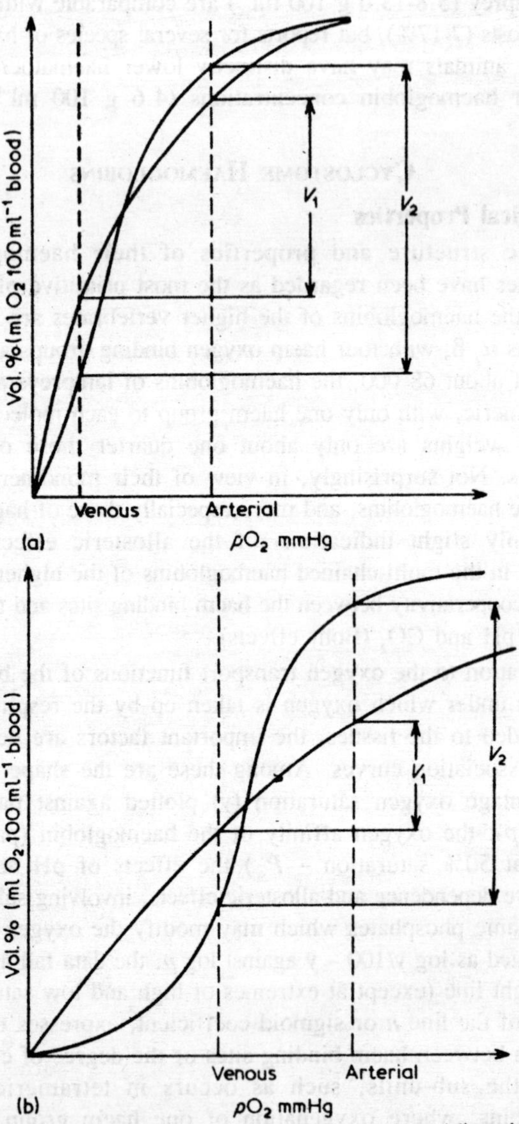

Fig. 8.4. The influence of co-operativity on the volumes of oxygen delivered to the tissues in haemoglobins of high (a) and low (b) oxygen affinity. V_1—volume of oxygen delivered without co-operativity. V_2—volume delivered by a haemoglobin showing co-operativity.

takes up or unloads oxygen to the tissues. Thus a shift of the sigmoid curve to the right and a decrease in oxygen affinity, means that at a given pO_2 it will deliver more oxygen to the tissues than in the case of the hyperbolic curve of the monomer and is better adapted to unloading its oxygen over a wider and higher range of oxygen tensions. At the same time, it has been emphasized that from the standpoint of oxygen transport, it is not the shape of the curve in itself that is significant, but rather the influence of this shape on the actual amounts of oxygen that are transported. The higher the range of oxygen tensions over which unloading occurs, the greater will be the gradient between blood and tissues and the more efficient the delivery. Thus, with or without co-operativity, the further the curve is shifted to the right, the lower the oxygen affinity and the greater the volumes of oxygen that can be delivered.

Although they share the monomeric character of invertebrate globins, the physiological properties of the cyclostome haemoglobins (and more especially those of the lampreys) are unique in showing a significant degree of cooperativity ($n>1.0$) and distinct Bohr effects. The latter is of great importance in coupling oxygen and CO_2 transport; the CO_2 produced by tissue metabolism favouring the delivery of oxygen, while oxygenation in the respiratory organs promotes the removal of the CO_2. Compared to the monomeric invertebrate globins or the myoglobin of vertebrates which have $n=1.0$ and high oxygen affinities, the cyclostomes may be regarded as transitional to the heterotetrameric haemoglobins of higher vertebrates, where n is usually greater than 2.5 and oxygen affinities are low. Although their possession of a monomeric haemoglobin might be thought to exclude haem interactions, the existence in lampreys (and to a lesser extent in hagfishes) of haemoglobins which show values of n in excess of unity, especially in concentrated solutions or at lower pH, indicates some degree of co-operativity. These properties are explained by the tendency for the monomeric globin to aggregate to form homodimers and temporary tetramers. The equilibrium between these aggregated sub-units and the monomers is dependent on oxygen tension; deoxygenation or reduced pH favouring aggregation, whereas oxygenation promotes the dissociation of the sub-units. In the higher vertebrates, the oxygen affinity of the haemoglobin is controlled by organic phosphates whose binding is dependent on pH, thus increasing the Bohr effect. Neither in lampreys nor in hagfishes do these compounds affect the oxygen equilibria and this form of regulation must therefore have been evolved after the divergence of agnathans and gnathostomes.

Information now exists on the oxygen equilibria of a variety of lamprey species, both adult and ammocoete and on the haemoglobins of three species of myxinoids. In drawing conclusions from data of this type, considerable care needs to be exercised because of the variety of techniques that have been used and the differences in experimental conditions, particularly in regard to temperature, haemoglobin concentrations or pH. All these factors may significantly modify the oxygen equilibria. For example, in solutions oxygen affinities are usually higher, Bohr effects reduced or abolished and haem interaction tends to be increased. The physiological properties of the haemoglobin are also modified by the intracellular environment of the erythrocyte, so that interpretations in terms of the whole animal should preferably be based on whole blood rather than on haemolysates.

Among the more significant points to emerge from Table 6.2 are the following:

1. In general hagfish haemoglobins have higher oxygen affinities than the haemoglobins of adult lampreys.
2. Amongst lamprey haemoglobins, those of the ammocoete always show considerably higher oxygen affinities than those of the adult.
3. In lampreys there appears to be a broad correlation between the adult oxygen affinities and the body size and migratory habits of the different species. This may be illustrated by the series *P. marinus*, *L. tridentata*, *L. fluviatilis*, *I. unicuspis* and the non-parasitic *L. lamottenii*. Here the P_{50} ranges from about 17-20 mm Hg in the large, wide-ranging *P. marinus* or *L. tridentata* to 2.0-8.0 mm Hg in the dwarf and non-migratory brook lamprey.
4. The haemoglobins of the three hagfish species show distinct levels of oxygen affinity. The lowest values are those of *E. burgeri*, which vary from 6-25 mm Hg depending on pH and concentration. In *M. glutinosa* two independent reports have given values of 4.2 and 9.0 mm Hg, but the highest oxygen affinity is almost certainly that of *E. stouti* (2.4 mm Hg).
5. As judged by the values for *n*, haem interactions are much more marked in lampreys than in hagfish haemoglobins.
6. Bohr effects in the hagfishes are slight or absent, but appear to reach significant levels in lampreys.

In an aquatic vertebrate the haemoglobin must be adapted to combining with oxygen at the gills over the range of partial pressures that exist in the water of its environment, while at the same time being able to unload its oxygen at tissue pressures compatible with their metabolic rates. Hence, the more active species tend to have haemoglobins with low oxygen affinities and large Bohr effects, while the converse tends to apply in more sluggish species living in water relatively depleted of oxygen. Sigmoidal oxygen dissociation curves, which tend to be more pronounced in the active species, can be regarded as a compromise between the high oxygen affinity needed for loading in the gills and the lower affinity that would favour the delivery of oxygen to the tissues. For example, the carp, a more sluggish teleost living in still water with lower oxygen tensions, has a high affinity haemoglobin with a P_{50} of 4.0 mm Hg, whereas the active trout has a low affinity haemoglobin with a P_{50} of 38 mm Hg. It has also been maintained that Bohr effects tend to be smaller in species that live in environments where CO_2 tensions are variable.

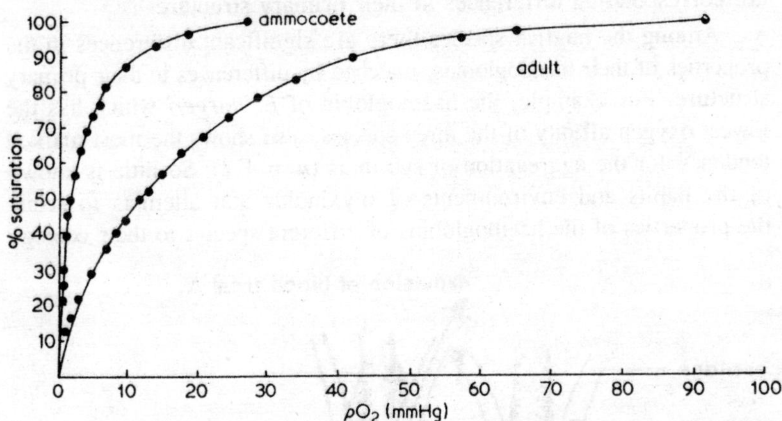

Fig. 8.5. Oxygen dissociation curves for the haemoglobins of ammocoetes and adults of the non-parasitic lamprey, L. planeri.

Within the cyclostomes there are similar ecological parallels between the low metabolic rates, burrowing habits and environment of the hagfishes and larval lampreys, which are reflected in the comparatively high oxygen affinities of their haemoglobins. In the case of the myxinoids, Manwell (1963) linked their slight or non-existent Bohr effects with what he termed their 'saprophytic' mode of feeding. This may involve the hagfish burrowing into the carcasses of animals, where they may be confronted by high or variable CO_2 tensions. Much more significant Bohr effects have been recorded in the haemoglobins of larval lampreys, but although these animals spend most of their time beneath the mud surface, the tip of the oral hood usually projects from the mouth of the burrow and the respiratory current is therefore drawn direct from the water flowing above it.

In keeping with their much higher rates of oxygen consumption and highly active life, the haemoglobins of adult lampreys have much lower oxygen affinities, pronounced Bohr effects and values of n well in excess of unity. Indeed, the substitution of adult for larval haemoglobins at metamorphosis is a graphic illustration of the modification of the properties of haemoglobins in response to changes in ecological requirements. Similar changes occur in other vertebrates, for example the change from foetal to adult haemoglobins in mammals or the transition from the larval haemoglobins of the tadpole to those of the adult frog. In lampreys, the larval and adult haemoglobins differ in their electrophoretic patterns, but it is not yet known whether there are corresponding differences in their primary structure.

Among the hagfish species there are significant differences in the properties of their haemoglobins, matched by differences in their primary structure. For example, the haemoglobin of *E. burgeri* which has the lowest oxygen affinity of the three species, also shows the most marked tendency for the aggregation of sub-units (n = 1.2). So little is known of the habits and environments of myxinoids that attempts to relate the properties of the haemoglobins of different species to their ecology

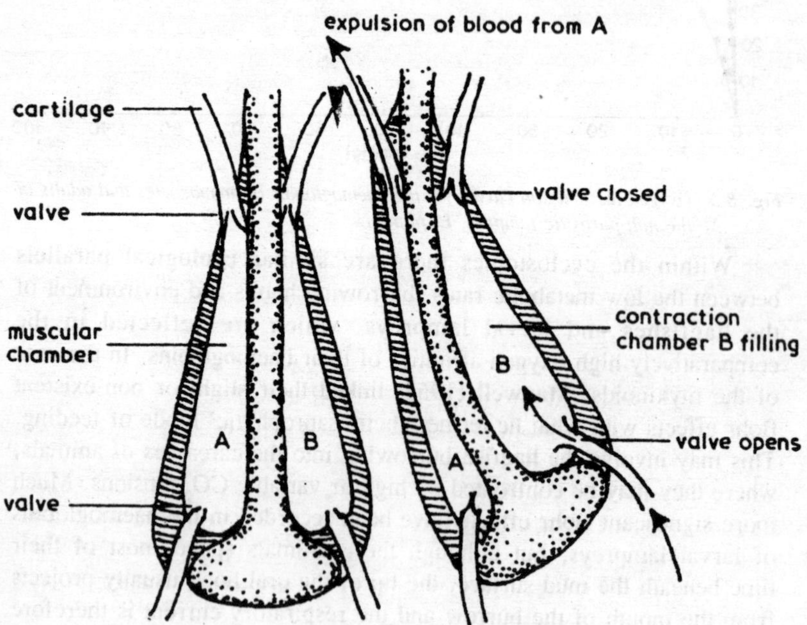

Fig. 8.6. The caudal heart of the hagfish.

must be highly speculative. Nevertheless, it may be recalled that *E. burgeri* is a species that inhabits shallow water and undergoes a seasonal migration; habits that could well be related to the distinctive characteristics of its haemoglobins.

In relation to the conditions under which blood gas transport occurs, the little information that we have on internal oxygen tensions points to an interesting contrast between the lampreys and hagfishes. In *Eptatretus*, Manwell found that the arterial blood was never more than 50% saturated, while the venous blood was almost completely deoxygenated. From the oxygen dissociation curve it was clear therefore, that blood gas transport would take place at very low oxygen tensions, normally less than 10 mm Hg. This would locate oxygen transfer on the steepest part of the hyperbolic curve. On the other hand, in *L. tridentata*, arterial blood saturations range from 94—100 % and arterial tensions from 58-77 mm Hg. Venous blood had saturation values of 75% and oxygen tensions of 20-40 mm Hg. Such differences must obviously have important consequences for the relative volumes of oxygen transferred to the tissue in the two groups of cyclostomes.

Multiple Haemoglobins

In common with many invertebrates, fish, amphibia and reptiles, the blood of cyclostomes usually contains a mixture of several haemoglobin species, differing in physico-chemical characteristics and often in amino acid composition. The properties of the blood may therefore be a result of the interaction of the various components and many of the more recent biochemical studies on cyclostome haemoglobins have been concerned with the characteristics of these isolated haemoglobin species.

In the hagfish, *E. burgeri* four main components F^1-F^4 have been isolated and yet a further component was identified in some animals. F^1 and F^2 are present in proportions of 17% and 13% and are virtually identical in their amino acid composition. In this respect, they differ from F^3 and F^4 which together make up about 35% of the haemoglobin. F^1 and F^2 have similar oxygen affinities, which are lower than those of F^3 and F^4. None of the isolated components showed a significant degree of haem interaction and when the various components were combined these interactions were only observed between F^3 and F^4. This combination produced higher oxygen affinities than any other component alone and had an *n* value of 1.31. From the sedimentation data it was inferred that this combination formed tetramers, which would therefore be hybrid molecules. Only F^3 showed any tendency to aggregate by itself. Unlike the haemoglobins of lampreys, which hardly aggregate at all in the oxygenated state, this did occur in *E. burgeri* at high concentrations, although in the deoxygenated state it occurs more readily at lower concentrations. Because of the aggregation behaviour of F^3 and F^4, it was suggested by the authors that these components might be comparable with the α and β chains of higher vertebrates, but we should be very cautious in drawing far-reaching conclusions from this type of experiment, where the conditions may be very far removed from those that exist *in vivo*.

In the hagfish, *E. stouti*, electrophoretic patterns have disclosed the presence of 2-3 major and at least two minor haemoglobin components each differing in amino acid composition. As in *E. burgeri*, the sedimentation data indicate some aggregation in the oxygenated state. All these components differ in electrophoretic mobility, but have identical oxygen affinities.

The three main haemoglobin components of *Myxine* have distinct oxygen affinities, but the value of *n* is close to unity and does not suggest the existence of haem interactions. In whole haemolysates, the addition of CO_2 increases the small Bohr effect and also the value of *n*, indicating some aggregation of sub-units which is absent in the separate components.

Among the lampreys, no less than six haemoglobin species have been recognized in *P. marinus* differing in sedimentation properties, isoelectric points, oxygen affinities, *n* values and in primary structure. Of these, fractions I and IV differ in oxygen equilibria from III, V and VI. In the deoxygenated state, IV, V and VI aggregate to form homo- and heterotetramers, but II does not do so and I and III are intermediate in character. In *L. fluviatilis* on the other hand, only two major components are present and these are identical in their primary structure, except that in one the N-terminal residue is blocked, while in the other it is free. In the closely related *L. planeri* also, two components can be recognized in electrophoretic patterns and it seems quite likely that these may have the same kind of relationship to one another as the components of *L. fluviatilis*. Two components are present in *L. (Entosphenus) japonicus* in proportions of 92.3% and 7-8% with different electrophoretic properties, but their amino acid composition seems to be identical to the haemoglobins of *L. fluviatilis*.

These examples of haemoglobin heterogeneity in the cyclostomes raise the much debated problem of its possible biological significance. While it may be natural to assume that the existence of these multiple haemoglobins in the lower vertebrates has some respiratory significance, perhaps as an adaptation to variations in ambient conditions, this would be alien to the strictly neutralist view, which maintains that a majority of duplicate or redundant gene copies have no particular adaptive significance, but permit the accumulation of random mutations, free from the rigorous restraints of natural selection. However, the fact that haemoglobin heterogeneity is so widespread in aquatic poikilotherms does seem to indicate that it may be related to the greater environmental variability that confronts these animals, particularly in relation to oxygen tensions and temperature. On the other hand, it has been pointed out that Antarctic fish, living in a relatively stable environment, also exhibit multiple haemoglobins. Referring to the situation in the hagfish *Myxine glutinosa*, Bauer *et al.*, (1975) suggest that a rise in ambient temperature might favour the increased synthesis of Hb II, whose higher oxygen affinity could then counteract the impaired oxygen uptake due to the high heat of oxygenation. In the case of *E. stouti*, where the oxygen equilibria of the major components are almost identical and therefore would be neutral to selection, it has been emphasized that each has a different isoelectric point. Because of the Donnan equilibrium, the pH within the red cell would therefore vary with the changes in the proportions of the haemoglobins and, since a small Bohr effect is present, these could result in changes in oxygen affinity that might be of some adaptive value. This could only apply however, if the various haemoglobin species are present within the same erythrocyte.

Cyclostome Hearts and Vascular Physiology

Relative to their body weight the cyclostomes have remarkably large hearts. The heart ratio for the lamprey is greater than for any other vertebrate group, with the exception of mammals, and even the somewhat smaller heart ratio of *Myxine* is conspicuously greater than that of the teleosts. As a general rule, vertebrate heart ratios are broadly correlated with the activity and metabolic rates of the various species and this also applies within the life cycle of the lamprey, where in keeping with its lower metabolic rate and more sedentary habits, the heart ratio of the ammocoete is only about a half that of an adult lamprey. Although the absence in the cyclostome heart of a coronary circulation might be expected to limit their capacity for sustained work at heavy loads, the characteristically spongy texture of the myocardium probably assists in the interchange between blood and muscle.

A noticeable feature of cyclostome hearts is the presence of large numbers of chromaffin cells and the high concentrations of catecholamines. In lampreys, the sinus venosus has high concentrations of both adrenalin and noradrenalin, whereas the atrium and ventricle contain mostly adrenalin. On the other hand, in *Myxine*, adrenalin predominates in the ventricle, but noradrenalin in the atrium and portal heart. The hagfish heart is the only vertebrate heart to be completely aneural, whereas the lamprey heart is innervated by branches from the vagus. No evidence has been found for the innervation of the chromaffin cell system and although cholinergic fibres have been traced to neuromuscular endings on the muscle fibres of the sinus venosus, none have been identified either in the atrium or the ventricle.

In addition to the main systemic or branchial heart, hagfishes have several accessory contractile structures (cardinal, portal and caudal hearts) which tend to work independently of the systemic heart and which have different inherent rhythms of contraction. Unlike the other accessory hearts, the portal heart is a true heart with muscular tissue of the same type as that of the systemic heart. It has been suggested that one of the main functions of this organ may be to assist in overcoming the capillary resistance of the liver. Its contractions begin in the proximal part of the supraintestinal vein, which has therefore been regarded as a pacemaker region. At temperatures of about 20°C, its frequency is said to average 47 beats min^{-1} falling to 24 beats min^{-1} at 10°C. The pressures that it generates are said to vary from 3-6 mm Hg and its output has been estimated at 0.1-0.3 ml min^{-1}. The caudal heart is a quite unique structure, located below the notochord and pumping blood from the large subcutaneous sinuses of the tail region into the caudal vein. It consists of a median cartilage with two chambers, one on each side, communicating separately with the caudal vein and provided with valves, preventing reflux from the vein. Valves also occur at the openings from the heart chambers into the sinuses. The filling and emptying of the heart chambers is brought about by extrinsic muscles, attached at one end to the median cartilage and at the other to a posterior cartilaginous knob. By alternate contractions of the muscles of either side, the chambers are alternately compressed and dilated. Histologically, the muscles are of the skeletal type and are innervated from the spinal cord.

Circulatory Dynamics

What fragmentary information we have on the intravascular pressures of the cyclostomes, indicates that these are much lower in the hagfishes than in lampreys. This helps to explain the necessity for accessory hearts in the myxinoids. Under resting conditions systolic pressures in the systemic heart of the hagfish vary from about 1.0-8 mm Hg falling to zero at diastole. Nevertheless, at maximum load with increased venous pressures, the heart is capable of systolic pressures of 20-30 mm Hg. When the aortic valves are open, the pressure drop in the ventral aorta is negligible, but the gill capillaries impose a considerable resistance between the dorsal and ventral aortae. As a result, pressures measured in the dorsal aorta were only about 3-5 mm Hg and the pulse pressure was of the order of 2 mm Hg. The average pressure drop in the gill capillaries is said to be about 2-4 mm Hg. In *Myxine*, Johansen detected a distinct pulse in the dorsal aortic pressures, which he attributed to the rhythmic contractions of the gill pouches, although this was not observed by Chapman *et al.* (1963) in *Eptatretus*. No measurements of intracardial pressures have been made on lampreys, but pressures of 18-34 mm Hg have been recorded in the dorsal aorta. Making allowance for a pressure drop in the gill capillaries of the same order as in teleosts (40-50%), this would suggest that the pressures in the ventral aorta and ventricle may be within a range of 30-60 mm Hg at systole. The recorded pressure in the dorsal aorta is similar to that of teleosts and very much higher than in the hagfish. Furthermore, pulse pressures in the lamprey are said to be about 9 mm compared to only 2 mm in the myxinoid. Although there are no records of vascular pressures in the ammocoete, the relatively small size of its heart suggests that these would be lower than in the adult lamprey. In addition, the extensive changes that have been described in the larval circulation at the time of metamorphosis, including improvements in arterial supply and venous return would almost certainly lead to a more efficient circulation.

In elasmobranchs and lungfishes where the heart is contained within a non-compliant pericardium, venous pressures around the heart are usually negative, favouring the aspiration of the venous blood on its return to the heart. Negative pressures of −0.7 to −3.0 mm Hg have also been recorded in the lamprey, which also possesses a rigid closed pericardium.

Heart rates in the systemic heart of the hagfish are said to vary from 18 beats min^{-1} at 11°C to 26 at 18°C, whereas in the isolated heart the rates may be restricted to 5 min^{-1} at 8°C rising to 12 beats min^{-1} at 15°C and to 24 at 20°C. Frequencies tend to be higher in the portal heart and in intact animals, rates of 24 min^{-1} at 11°C and 30 beats min^{-1} at 18°C have been recorded. The caudal heart tends to beat intermittently and irregularly, but when it is contracting it may show a rate as high as 88 beats min^{-1} at 16°C. Being no larger than a pinhead, this structure can make only a local and minor impact on the circulation and no obvious ill effects are said to follow its removal.

In line with their higher metabolic rates, the heart rates of lampreys are considerably higher than those of the hagfish. Frequencies of 40-60 min^{-1} have been recorded for the isolated heart of *L. fluviatilis* and in anaesthetized or decapitated specimens of *L. tridentata*, rates rose from 25 min^{-1} at 5°C to 110-130 beats min^{-1} at 25°C. In *L. fluviatilis*, the heart rates of intact animals showed a steady increase throughout the migratory period and at a constant temperature of 16°C increased from 33 beats min^{-1} in December to nearly 50 beats min^{-1} in March. Parallel with these increases in heart rate there were similar increases in ventilatory frequency and standard oxygen consumption. Judging from the limited information obtained on *L. tridentata*, the cardiac output of the lamprey heart in resting animals (32 ml kg^{-1} min^{-1}) comes well within the range recorded in teleosts (5-100 ml kg^{-1} min^{-1}) or elasmobranchs (21-25 ml kg^{-1} min^{-1}).

Circulatory Control

In keeping with the habits of an animal that may spend most of its life in almost total quiescence, punctuated by short bursts of vigorous activity, the systemic heart of the hagfish shows exceptionally low systolic pressures. In higher vertebrates, cholinergic and adrenergic cardiac nerves regulate the heart in response to varying demands, either through changes in the heart rate (chronotropic effects) or by alterations in the force of its contractions (inotropic effects). Except for embryonic hearts, the systemic heart of the myxinoids is unique in the absence of these cardiac nerves, but it is still able to adjust its activity to meet environmental demands or sudden increases in its locomotory activity.

In the resting condition or when the heart is isolated, it is purely myogenic with its own inherent rhythmic activity. This is attributable to the spontaneous firing of pacemaker cells, widely dispersed throughout the cardiac tissue. If a ligature is tied between atrium and ventricle so as to effectively separate the two chambers, the ventricle shows a decreased rate of contraction. Thus, each chamber has its own pacemaker activity, but while this is of similar magnitude in the atrium and sinus venosus, it appears to be weaker in the ventricle. Electrical stimulation of this chamber has no apparent effect on its rhythm, but the atrium and sinus venosus show a considerable increase in the frequency of their contractions. Characteristic pacemaker potentials have been recorded by Arlock (1975) from cells in the atrium of the systemic heart of *Myxine*, but not from the ventricle. These show a depolarization phase of about 25 mV, following a slow depolarization, and during subsequent repolarization the potentials reached a higher value of about 30 mV.

The immediate stimulus that is responsible for raising the output of the hagfish heart is an increase in the backflow of venous blood. The resulting distention causes an acceleration in heart rate and an increase in the force of its contractions, obeying Starling's Law of the Heart. This ability of cardiac pacemakers to alter their rate of spontaneous firing in response to changes in tension or pressure is widespread in vertebrates and has been referred to as 'intrinsic rate regulation'. This could be regarded as a primitive condition, phylogenetically predating the development of neural control systems.

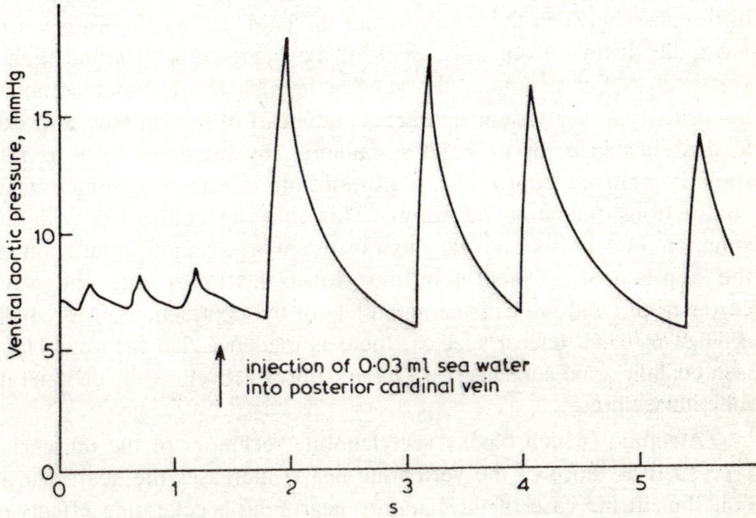

Fig. 8.7. The response of the hagfish heart to increased venous inflow.

Although the lamprey heart has a vagal nerve supply, a similar type of intrinsic rate regulation is still present and the heart rate is sensitive to moderate distention of the ventricle. The quiescent heart of *P. marinus* responds to an increase in the intraventricular pressure from 0-15 mm Hg by raising its rate of contraction from 24-44 beats min^{-1}, subsequently returning towards normal as the pressure decreases.

In response to stimulation, the portal heart of the hagfish behaves in a similar way to that of the systemic heart, showing acceleration with weak electrical stimulation and inhibition when the stimulation is more intense. Like the systemic heart, it also responds to increases in venous inflow or to slight mechanical stimulation by an increase in the rate of contraction and in stroke volume. Pacemaker potentials have also been observed in the portal heart and the pacemaker region is apparently located in the proximal part of the portal vein where this enters the heart. The caudal heart on the other hand, is under reflex control from the spinal cord and its contractions disappear when the latter is destroyed. It is also said to stop beating after intense mechanical stimulation of the skin surface or during vigorous movements of the body. Responses to postural changes have been observed as when the tail of the hagfish is raised above the level of its head, resulting in increased rates of contraction.

Pharmacology of Cyclostome Hearts

In the response of the heart to drugs, not only are there significant differences between the lamprey and the hagfish, but these responses are quite distinct from those seen in the hearts of fishes and higher vertebrates. For example, the aneural hagfish heart shows complete insensitivity to acetylcholine, whereas the heart of the lamprey responds to this substance, or to vagal stimulation, by increases in heart rate (positive chronotropy) and a diminution in the amplitude of its contractions (negative inotropy). This may be contrasted with the situation in teleosts and elasmobranchs, where vagal stimulation or the application of acetylcholine slows the heart rate (negative chronotropy) and reduces the amplitude of the contractions. Moreover, at least in some teleost species, there is evidence that the vagus may also contain some adrenergic excitatory fibres as well as the cholinergic inhibitory fibres.

Atropine (which blocks acetylcholine receptors of the muscarine type such as those of the vertebrate heart) increases the heart rate in fish, but in the case of the lamprey heart, the accelerating effects of acetylcholine are blocked by curare and not by atropine. This indicates that the acetylcholine receptors are of the nicotinic type. In fact, the application of nicotine mimics the positive chronotropic and negative inotropic effects of acetylcholine itself, whereas muscarine or pilocarpine are without effect. Thus the lamprey heart resembles that of a fish in its cholinergic nerve supply, but differs in the type of acetylcholine receptors.

The high concentrations of catecholamines and the large numbers of chromaffin cells in the cyclostome heart have been referred to earlier. The lamprey heart is particularly sensitive to reserpine, which depletes catecholamine stores and limits the uptake of these compounds. After treatment with reserpine, the heart stops in diastole, although its contractions can be restored by the application of acetylcholine. Large doses of noradrenalin exert positive inotropic and chronotropic effects on the heart of the lamprey, acting *via* β-adrenergic receptors, but these effects are never as pronounced as those observed with acetylcholine. However, even in high concentrations, catecholamines are without any effect on the hagfish heart. At the same time, reserpine treatment may reduce the normal response to increased venous inflow and this response can to a certain extent be restored with the subsequent application of catecholamines. As is the case in the lamprey, large doses of reserpine stop the hagfish heart in diastole, but in this animal, contractions can then be restarted by catecholamines and not, as in the lamprey by acetylcholine. It has been suggested that acetylcholine may act indirectly on the lamprey heart through liberating catecholamines from the chromaffin cells, but the relative insensitivity of the heart to these compounds presents some difficulties for this interpretation. On the other hand in the isolated ventricle of the ammocoete, colchicine (known to inhibit the release of hormones and neurotransmitters, including catecholamines) has been found to inhibit the chronotropic effects of acetylcholine.

From the hagfish heart, Jensen (1963) isolated a highly active cardioactive agent which he called eptatretin. This is apparently peculiar to the hagfish and is believed to be an unstable amine, but not a catecholamine. Although its role in cardioregulation is unknown, Jensen suggested that it might facilitate the repolarization of the pacemaker potentials to the threshold levels required for their spontaneous discharge.

Cardiovascular Responses of Lampreys during Activity and Under Conditions of Oxygen Depletion

In the lamprey, *L. tridentata*, Johansen *et al.* (1973) have been able to analyse some of the cardiovascular changes that occur during exercise or in hypoxia. With decreasing oxygen concentrations, both arterial and venous oxygen tensions declined and the differences between them were reduced, as was the blood/water oxygen gradient. This reduction in blood oxygen tensions and particularly the venous tension, means that these are now positioned on the steepest part of the oxygen dissociation curve and oxygen utilization (expressed as the percentage difference in arterial and venous oxygen tensions) increased some three- or four-fold. A very significant factor in the resistance to hypoxia is the high degree of venous saturation. This is said to act as a reservoir compensating for the reduced availability of oxygen in the water.

During activity, the venous oxygen tensions decrease to a greater extent than the arterial tensions, but because of their positions relative to the oxygen dissociation curve and the relatively high oxygen affinity of the haemoglobin, the reduction in arterial tension has little effect on arterial saturation. As a result, oxygen utilization is approximately doubled.

Nothing in our fragmentary knowledge of respiratory or cardiovascular physiology in the lampreys seems to explain their relatively poor swimming performance and lack of staying ability. Structurally, the gills are very similar to those of teleosts and their areas are comparable with those of the most active fishes. The tidal respiration of the adult lamprey might be more energy costly, but as yet we have no evidence to support this view. Neither does the relative size of the heart or its functional capacities encourage the belief that cardiovascular factors are likely to limit the activities of these animals. Perhaps, as Beamish (1974) has suggested, the answer may lie more in the direction of hydrodynamic factors or even in some biochemical or physiological characteristics of the parietal muscles. Attempts to compare the energy costs of swimming in lampreys and teleosts have been complicated by uncertainty over the distances actually travelled by the migrating sea lamprey, but assuming that they follow a direct route, the energy cost has been calculated as 1.31-1.48 cal g^{-1} km^{-1}. This would be much higher than the figure of 0.33-0.42 cal g^{-1} km^{-1} reported for the eel, using rather similar swimming mechanics.

9

ENDOSKELETON

CRANIAL SKELETON

Neurocranium

In comparison with the difficulties that we face when attempting to relate the splanchnocranium of the cyclostomes to that of the gnathostomes, the neurocranium presents relatively few problems of interpretation and, in its broad outlines, this structure conforms to a pattern that is readily recognizable in the skull of higher vertebrates, consisting essentially of the sense capsules, parachordals and trabeculae. In its simplest form, in the ammocoete, the brain is supported ventrally by a cartilaginous framework attached to the cranial extension of the notochord, with paired parachordals continuing forwards as the 'trabeculae'. The latter diverge again in the midline to form a hypophysial fenestra at the site of the pituitary and where the trabeculae

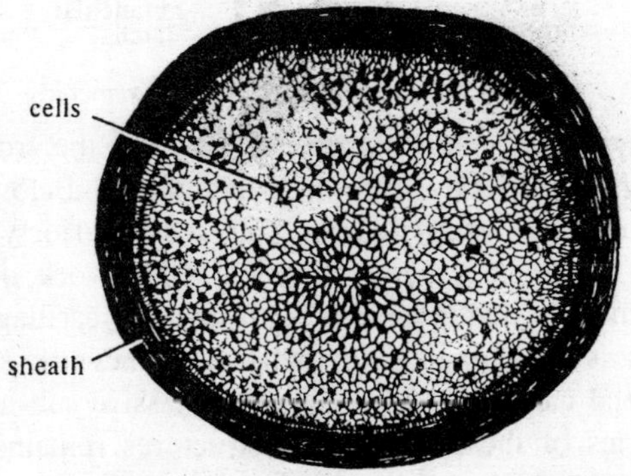

Fig. 9.1. Transverse section through notochord of lamprey.

and parachordals meet, there is a small, basitrabecular process. The ear capsule is united to the parachordals, but the small olfactory capsule is connected to the membranes covering the brain only by fibrous tissue. In the course of metamorphosis, the trabeculae extend to form lateral cranial walls and the auditory capsules are united by a cartilaginous bridge (tectum synoticum). The anterior part of the hypophysial fenestra is reduced by the development of an intertrabecular plate and the developing eye becomes supported by a massive sub-ocular arch formed from an extension of the basitrabecular process.

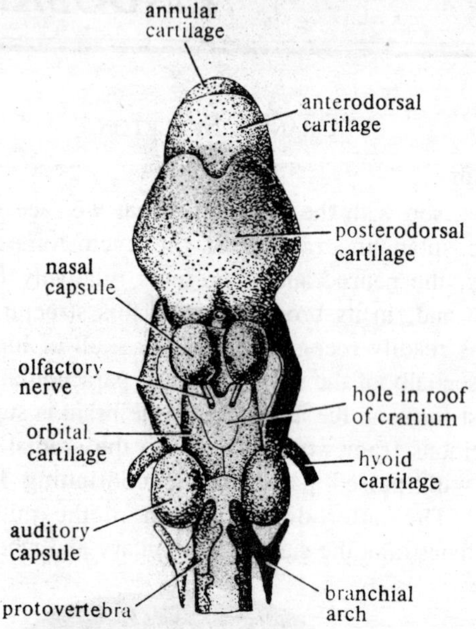

Fig. 9.2. Dorsal view of skull of Petromyzon.

In the myxinoids, the trabeculae arise from the front of the otic capsule rather than directly from the parachordals. Below the pituitary there is a subhypophysial plate, connected anteriorly to the nasal capsule. The latter is it hemicylindrical framework of longitudinal rods, joined in front and behind by half hoops of cartilage. Peculiar to the hagfishes is an elongated cylindrical framework supporting the nasohypophysial canal and below this is a massive sub-nasal cartilage. The homologies of these specialized structures remains obscure and their inclusion in the neurocranium is quite arbitrary.

The precise homologies of the cyclostome trabeculae have been disputed, but it now seems probable that at least a part of these

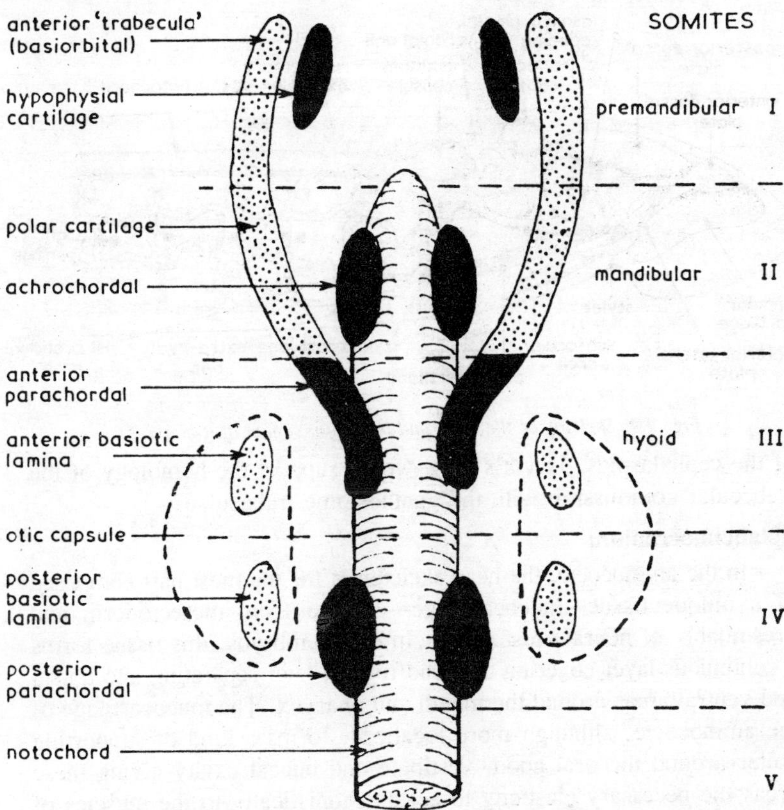

Fig. 9.3. Mesodermal elements of the lamprey cranium and their embryonic sources.

structures may correspond to the gnathostome polar cartilages, lying between the cranial end of the parachordals and the trabeculae. An alternative view is that the true trabeculae of the lamprey may be represented by the transverse commissure linking the cranial ends of the 'trabeculae' and which are also present in the myxinoid embryo. This interpretation is strengthened by the claim that at least in the lamprey, this commissure is ectomesenchymal in origin. Jarvik (1964) believes that as in the gnathostomes, the cyclostome visceral arches were originally made up of three separate elements, which in the cephalaspids had fused to one another and to the cranium to form a continuous unit. The medial part of the mandibular arch (infrapharyngeal) still lay in its original position and would correspond to the trabeculae. True trabeculae, in the sense of paired longitudinal rods would therefore have been absent. If the condition in the cyclostomes has followed that

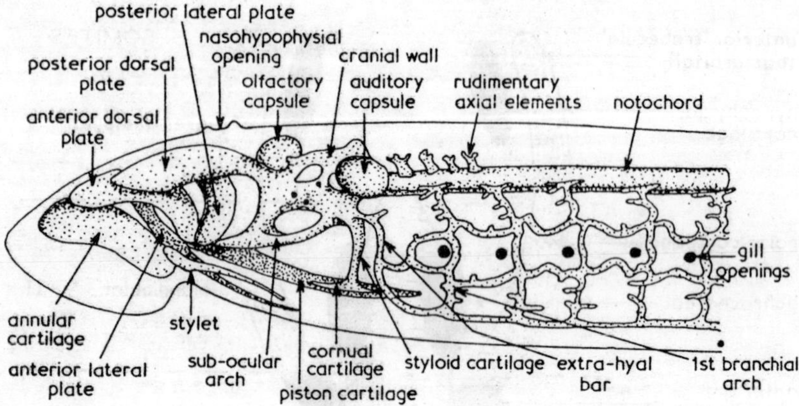

Fig. 9.4. Skeleton of the head and branchial region of a lamprey.

of the cephalaspids, Jarvik's view would support the homology of the trabecular commissure with the gnathostome trabeculae.

Splanchnocranium

In the ammocoete, the head skeleton is for the most part composed of a unique tissue—mucocartilage—derived from mesectoderm and presumably of neural crest origin. In early embryos, this tissue forms a continuous layer covering the head region, later separating into dorsal and ventral areas around the mouth and pharynx. The mucocartilage of the ammocoete, although more localized, forms a kind of supporting collar around the oral hood, vestibule and buccal cavity giving these areas the necessary elasticity to act antagonistically to the muscles of these regions.

During metamorphosis, the cartilaginous skeleton of the adult lamprey develops in several ways. In some areas, the mucocartilage of the ammocoete breaks down and is replaced, sometimes quite independently, by hyaline cartilage. In other cases, larval mucocartilage may be transformed directly into adult cartilage. Finally, some adult cartilages are neoformations developed from embryonic blastemas.

During metamorphosis, the major modifications take place in the buccal region involving the elongation of the prenasal region and the transformation of the oral hood into the suctorial funnel of the adult. The latter is supported by an annular cartilage, carrying slender stylets serving as attachments for muscles that are able to alter the position of the funnel relative to the body axis. The roof of the funnel is supported by an anterior dorsal plate which can be moved vertically to assist in creating the necessary suction pressure for attachment.

Behind this are anterior lateral cartilages, whose compression occludes the passage between the oral funnel and the buccal cavity. A similar function can also be attributed to the large posterior dorsal cartilage, joined behind to the trabecular commissure and forming the roof of the mouth cavity. The movements of these dorsal plates relative to one another are also involved in the changes in the volume of the hydro sinus that plays such an important part in the suction mechanism. A further pair of posterior lateral cartilages, attached by elastic tissue to the subocular arch, are concerned in lateral compression of the buccal cavity.

The skeleton of the branchial region remains basically similar to that of the ammocoete. The most important changes are the anchorage of the whole structure to the cranial skeleton through the union between the dorsal longitudinal bars and the parachordals. At its anterior end, the branchial skeleton now shows an additional arch, passing forwards and upwards from the first branchial arch to unite with the cranium at the point where the velar arch (styloid cartilages) join the sub-ocular arch. With a few exceptions, all the cartilages of the adult myxinoid skull are fused to a form a continuous structures. From the front of the 'trabeculae' extends the dorsal longitudinal palatine bar, joined to the auditory capsule behind, and separated from it by a fenestra through which the facial nerve (VII) emerges. At their cranial ends these longitudinal plates are joined by the palatine commissure, carrying the palatine tooth. This palatine bar may correspond to the subocular arch of the lamprey, where it forms the floor of the orbit, and the subocular fenestra of the latter would then have its counterpart in the fenestra between the palatine bar and 'trabeculae' of the hagfish, both of which carry the main roots of the trigeminal (V). The complex 'tongue'

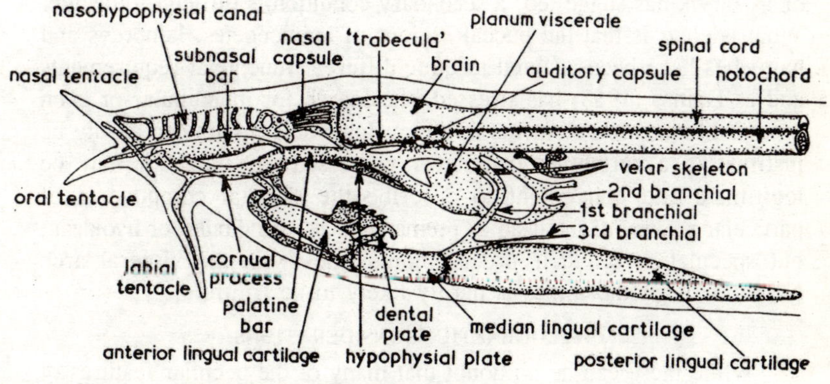

Fig. 9.5. Head cartilages of a hagfish.

skeleton is attached to the first branchial arch through the posterior end of the median lingual cartilage and in front of the latter are two pairs of anterior lingual cartilages, one medial and two lateral. These have been compared with the medial, apical and lateral apical cartilages of the lamprey tongue skeleton, while the paired middle and single posterior segment of the hagfish apparatus have been likened to the anterior 'head' and the main body of the lamprey piston cartilage. Jarvik (1965) on the other hand has discounted a strict homology between the two structures, although he allows the possibility that both may have been derived from some basal elements of the visceral arch system of the ancestral vertebrate.

In the hagfish the posterior and functional gill sacs have no branchial skeleton, other than cartilaginous rings around their external openings. The first so-called branchial arch attached above to the otic capsule and below to the median lingual cartilage has been compared to the first visceral arch of the lamprey, both lying outside the corresponding nerve and therefore referred to as extrabranchials. The second arch of the hagfish is united to the first and joins the posterior angle of the velar plate (planum viscerale) supporting the walls of the buccal cavity in front of the velar folds. The third rudimentary arch emerges from the posterior end of the median lingual plate.

Interpretations of the cyclostome skull have been based mainly on extrapolation from conditions in the gnathostomes and this has led many authors to search for the same visceral arch elements that are found in the skulls of the jawed vertebrates. The rationale for this approach is at least open to question. Whether the fused and continuous branchial skeleton of the cyclostome or the cephalaspid is a primitive or as Jarvik has suggested, a secondary condition is difficult to decide. What is clear is that the buccal regions of ammocoetes, lampreys and hagfishes have been adapted to quite different functional requirements and as Damas (1958) has stressed, the search for mandibular or even premandibular arch elements in the cyclostome cranium can hardly be justified in a vertebrate that never possessed true jaws. It may be legitimate and convenient to describe the skeletal components of particular regions of the head as premandibular, mandibular or hyoidean, but speculation on their precise homologies with the visceral arch elements of gnathostomes is hardly likely to be fruitful.

Phylogenetic Considerations

While there can be no doubt that many of the peculiar features of the cyclostome splanchnocranium have arisen as adaptations to their highly specialized mode of feeding, some of the difficulties in detecting a common organization plan in the cyclostome cranium may be due to the extensive evolutionary regression that has probably occurred. In lampreys, the presence of the rudiments of an axial skeleton in the form of vestigial cartilages around the notochord, seems to offer clear evidence of such secondary reduction. Skeletal regression is a well-established phenomenon in the evolutionary history of several major vertebrate groups and Orvig (1968) believes that the Devonian cephalaspids also show evidence of a reduction in the bony exoskeleton. This trend is believed to have involved a reduction in the middle vascular layer of the corium in the trunk region behind the cephalic shield. During its metamorphosis, the corium of the lamprey also exhibits three zones; the middle of which is a vascular layer. Behind the head this layer is absent, suggesting that the naked condition of the petromyzonids may represent the final stage in an evolutionary reduction that had already begun in its middle Devonian cephalaspidomorph ancestors.

In its distribution and development, the unique mucocartilage of the ammocoete has sometimes been regarded as a relic of a continuous endoskeleton of the cephalaspid type. Such comparisons are really more appropriate to the mesectoderm that precedes ontogenetically the differentiation of the mucocartilage, and which is presumably derived from the neural crest. As Damas has shown, this tissue does in fact, form a continuous envelope, covering at its maximum extent, the cephalic and branchial regions. In later stages, its distribution becomes more restricted through the increased development of other structures, including the musculature. The relationship between the mucocartilage of the ammocoete and the adult cartilages are still far from clear. Earlier work had suggested that mucocartilage was entirely converted into true cartilage, but other authors have maintained that cartilage cells never arise directly from the tissue of the mucocartilage. The truth perhaps may lie between these two extremes. Certainly, the basal cartilages of the skull and branchial regions are formed before the embryonic mesenchyme has differentiated into mucocartilage. What is more, adult cartilage appears during metamorphosis, in places where no mucocartilage exists and at some sites, cartilage may apparently be formed within the degenerating larval muscles. Johnels (1948) considered mucocartilage to be a specialized larval tissue, developed secondarily to cartilage in an evolutionary sense and thus should not be regarded as an ontogenetic stage in the formation of true cartilage. Phylogenetically therefore, mucocartilage would be of more recent origin than cartilage and in this respect the adult lamprey would be more primitive than its larval stage. Like the rudiments of the paired eyes, the larval mucocartilage could thus be thought of as a kind of persistent embryonic primordium, whose further development has been suspended. This view is re-inforced by the relatively late histological differentiation of the mucocartilage, which for a long period, retains the characteristics of an embryonic mesenchyme. It is also consistent with the histochemical development of the tissue. Throughout the whole of the larval period, the intercellular matrix is characterized by the presence of hyaluronic acid typical of a loose connective tissue. In older ammocoetes, chondroitin sulphates, characteristic of cartilage matrix begin to appear in the tissue, although this is not yet reflected by changes in its morphological character. Significantly, as the animals approach metamorphosis, there is an increase in the synthesis of chondroitin sulphates, showing that the cells are beginning to differentiate in the direction of cartilage.

Apart from the question of the distribution of mesectoderm or mucocartilage, there are a number of other indications that the cartilaginous skeleton of the present day cyclostomes may be a relic of what was at one time, a more extensive exoskeleton. The great variability that is seen in the detailed morphology of the prebranchial arches in myxinoids, or in the branchial basket and vertebral rudiments of lampreys, is consistent with the variability that is characteristic of many vestigial structures no longer subject to selective pressures. In lampreys, particularly during metamorphosis, there are tendencies towards aberrant cartilage development, involving in some cases, reduction of existing cartilages and in other cases, their transformation into connective tissues. In addition, at some sites, blastemas, otherwise identical to those which produce cartilage, may fail to do so and subsequently regress.

The existence in the petromyzonid ancestry of a massive bony cranium of the cephalaspid type has been questioned because of its implications for the musculature of the head. As in the gnathostomes, the presence of an immoveable occipital region of the skull might be expected to have led to the disappearance of the myotomes of the post-otic region. However, in the head of the lamprey the myotomal regions of the 4th and 5th head segments remain functional, projecting forwards over the dorsal surface of the head, and it is partly from these post-otic myotomes that the unique corneal accommodatory eye muscles are developed. If, as Janvier (1975) believes, these muscles were absent in the cephalaspid eye, this would tend to support the view that the corresponding myotomes had already disappeared from the occipital region. Thus, as Damas (1944) has pointed out, the derivation of petromyzonids from cephalaspid-like ancestors would appear to be contrary to Dollo's law, since it would involve the reappearance in phylogeny of structures that had previously been suppressed. On the other hand, because of the entirely different position of the branchial region and the cranium in lampreys and cephalaspids, the relationships of the myotomes and the skull may also have been quite different. The absence of the forward extension of the epibranchial parts of the myotomes in the cephalaspids need not necessarily imply that the corresponding myotomes had disappeared and their present projection over the head of the lamprey could have been a secondary development.

In the pteraspids, the existence of a phylogenetic trend towards skeletal regression is even more problematical and the naked condition of the myxinoids may well be a primitive character, derived from an ancestral form (which may or may not have been related to the pteraspidorphs) which had not, at that evolutionary stage, acquired a rigid exoskeleton. In relation to the internal factors that may govern calcification, Moss (1968) has referred to the view that there may be a critical level of ionic strength in the body fluids, beyond which calcification does not occur. Living agnathans (presumably referring more specifically to the hagfishes) are said to have higher ionic strengths than either teleosts or elasmobranchs, which might have something to do with the absence of hard tissues in myxinoids, although in this case it would be necessary to assume lower levels of ionic strength in the ancestral and fossil agnathans with their calcified tissues.

Attempts have been made to identify in myxinoid embryos, a structure resembling the head shield of the pteraspids, but these are far from convincing. On the other hand, it does seem likely that the cartilaginous endoskeleton of the hagfish has undergone extensive modification and regression. This is obviously true of the visceral arches. Holmgren (1946) who accepted the relationship of the myxinoids to the pteraspidomorphs, explained the caudal position of the functional gill sacs as a result of the intercalation of myotomes between the anterior end of the branchial region and the skull, rather than to their backward migration. Such an extension of this 'neck region' could not have occurred in a pteraspid-like ancestor, where this part of the body was covered by a bony head shield. The backward movement of the gills and the extension of the prebranchial zone could therefore only have taken place in such ancestral forms if the dermal skeleton had already regressed.

10

MUSCULAR SYSTEM

The somatic musculature of the cyclostomes shows a primitive and complete segmentation, extending in a continuous series of myotomes from the tip of the tail to the head. The parietal muscles of the lamprey extend over the dorsal surface of the head as far forwards as the posterior dorsal plates and, in the hagfish, to the area of the nasal capsules. In lampreys, where the branchial region is placed immediately behind the head, the continuity of the myotomes is interrupted by the gill ports, which split up the myotomes into epibranchial and hypobranchial sections. This is not the case in myxinoids, even in those species with separate gill openings, and the gill ports are placed below the ventral margins of the parietal muscles. The intervening region is occupied by the complex longitudinal musculature of the tongue mechanism. In the hypobranchial zone of the lamprey the segmentation no longer corresponds with that of the epibranchial region above. In front of the first gill pore and under the eye, these muscles are connected indirectly to the annular cartilage through a sub-ocular muscle, controlling the side to side and vertical movements of the head. The small corneal accommodatory muscle attached to the outer sclerotic coat of the eyeball is also derived in part from the epibranchial sections of post-otic myotomes.

Unlike the parietal muscles of the lamprey that meet in the mid-ventral line, those of the hagfish end ventro-laterally at the level of the slime glands and the ventral surface is covered by a layer of unsegmented oblique muscles and segmental rectus muscles; an arrangement that helps to explain their ability to coil and contort their bodies, as well as the extreme extensibility demanded by their voracious feeding habits.

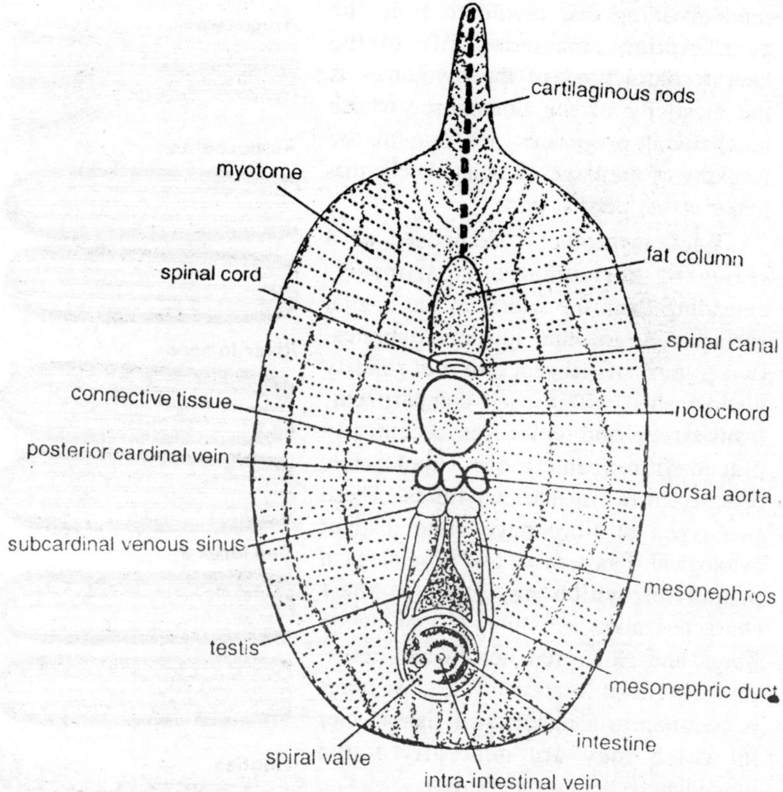

Fig. 10.1. T.S. trunk region of lamprey.

Contrasting with the lamprey, all the head muscles of the hagfish belong to the visceral division, originating from branchial mesoderm and not from the myotomes. Accordingly they are innervated by cranial nerves V-X. In both groups, the muscles that operate the complex feeding mechanisms are controlled by the trigeminal complex. Other important visceral muscles are those of the velum and branchial regions.

Organization of the Myotomes

Cyclostome myotomes may be described as ≶-shaped with one forward and two backwardly directed flexures. In addition, each myotome is directed obliquely backwards from the medial to the lateral surface, so that successive units are telescoped into one another and a single transverse section will pass through a number of successive myotomes. Within each myotome, the muscle fibres run parallel to the long axis of the body and are attached at either end to the connective tissue septa dividing one myotome from the next. Acting antagonistically to the lateral contractions of the myotomes is the elasticity of the notochord, whose mechanical properties are due to the turgidity of the large vacuolated cells that make up its central core.

Each myotome is composed of a series of horizontal compartments, extending from the lateral to the medial surfaces. As in other lower vertebrates, two types of fibres can be readily distinguished within each compartment; transparent and colourless, centrally placed fibres and more superficial, opaque, brownish parietal fibres. These two types of fibre have quite distinct cytological, biochemical and physiological properties which have led to their characterization as white, fast or twitch fibres and slow, red non-twitch fibres. The superficial location of the slow fibres is common to a majority of fish species in which they are generally found immediately below the skin.

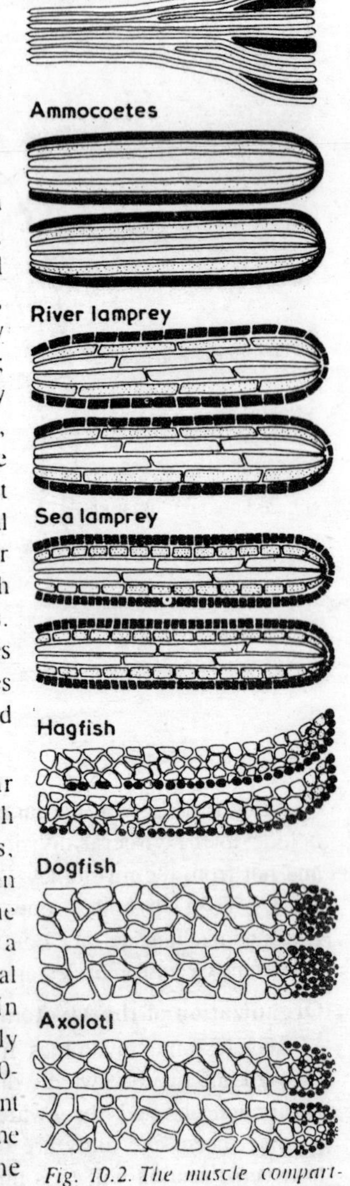

Fig. 10.2. The muscle compartments of Amphioxus and lower vertebrates.

The approximately rectangular muscle compartments of the hagfish consist of from 2-4 layers of fast fibres, stretching horizontally from the skin surface to the medial border of the myotome; the slow fibres forming a single layer over the ventral and lateral surfaces of each compartment. In lampreys, each compartment is usually made up of four layers of fast fibres, 50-60 μm in diameter, devoid of a basement membrane and each consisting of some 30-40 individual fibres. Unlike the hagfish, the fast fibres are surrounded on all except the medial side by a layer of slow fibres with a diameter of about 65 μm and a common basement membrane. The larger diameters of the slow fibres is a general feature and seems to be even more marked in hagfish muscles.

Within the chordates, Flood (1973) has described a phyletic trend in the organization of the axial musculature. In the cephalochordates, the myotomes are made up of a continuous system of muscle plates with no compartmentalization, although Flood suggests that differences in mitochondrial content of the myofibrils may be an indication of incipient differentiation towards the muscle types of higher forms. In the larval lamprey, each lamella in the compartment is regarded as a cellular unit and division into separate fibres only occurs in the adult lamprey. Compartmentalization is also seen in fishes and urodeles, but in these groups the slow fibres are more restricted in their distribution and tend to be confined to the lateral surfaces.

Compared to conditions in parasitic lampreys, the muscle compartments of the non-parasitic *L. lamottenii* show some interesting features. Here, the muscle compartments contain only three, rather than the usual 4-5 layers of central fibres that are characteristic of the parasitic lampreys. Moreover, the middle member of these central fibres extends as a continuous sheet from medial to skin surface, recalling the undivided muscle plates of the ammocoete. This failure

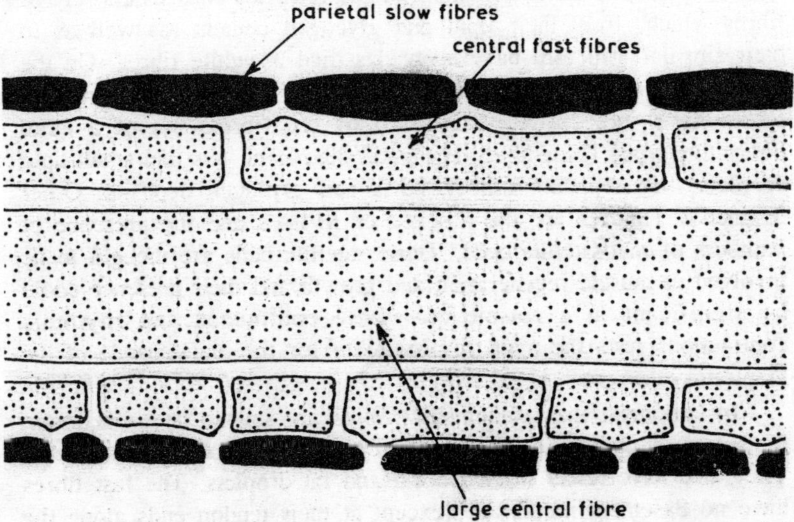

Fig. 10.3. Diagrammatic transverse section through the parietal muscle of a non-parasitic lamprey, L. lamottenii.

on the part of an adult brook lamprey fully to develop characteristics normally associated with the adult form, has been noted in a number of other morphological and physiological features and has been interpreted as evidence of paedomorphic trends.

Cytological Differentiation of the Muscle Fibres

In the parietal muscles of hagfishes and lampreys, the cytological differentiation of fast and slow fibres follows a similar pattern. In *Myxine*, the fast white fibres, apart from their larger diameters, have sparse mitochondria and a low content of lipid and oxidative enzymes. The arrangement of the myofibrils is said to conform to the type described as 'Feldstruktur' in which the fibrils are continuously and evenly distributed within the fibres. The myofibrils are separated by straight Z discs and distinct M bands are present. The smaller slow fibres have a much higher content of lipid and mitochondria. The myofibrils branch and anastomose to a greater extent than in the fast fibres and are surrounded by a conspicuous capillary network. Among cyclostome parietal fibres, a third and intermediate type has been described. These have diameters between those of the typical slow and fast fibres, but with greater numbers of mitochondria than the latter and with less densely packed myofibrils. However, these do not branch and anastomose like those of the typical slow fibres. Of the visceral muscles, the longitudinal tongue retractor consists entirely of fibres which, from their lipid and glycogen content, as well as in their fibrillar structure have been classified as white fibres. On the other hand, two other visceral muscles—the craniovelaris and spinovelaris—are entirely composed of red fibres, but of a quite distinctive type. These have very small diameters, few and small lipid droplets, but abundant mitochondria and glycogen deposits. Other distinctive features are the absence of M-lines and the presence of numbers of myosatellite cells. These satellite cells are thought to be involved in muscle regeneration and growth and their presence could be significant in a muscle on whose continuous and rhythmic contractions rests the main responsibility for the maintenance of the respiratory current.

In lampreys, the smaller slow fibres are said to have maximum diameters of about 70 μm compared to 164 μm for the fast fibres. They also have fewer mitochondria and fat droplets. The fast fibres have no basement membrane, except at their tendon ends along the medial surface. As in *Myxine*, an intermediate type of fibre has been distinguished in the parietal muscles, lying immediately adjacent to the single layer of slow fibres. Among the visceral muscles of lampreys, a distinctive type of 'myotube' fibre occurs in the muscles of the velum and in the branchial constrictors. In this type of muscle, the myofibrils are concentrated around the periphery, surrounding a central core of sarcoplasm containing the nucleus. These fibres have been compared to the nodal tissue of the mammalian myocardium and although their physiological characteristics have not been investigated in detail, their peculiar structural features could well be related to the rhythmical activity involved in respiratory movements. Another visceral muscle, the tongue retractor of the adult lamprey is considered to be transitional between typical fast and slow fibres in its diameter, arrangement of myofibrils, lipid content and mode of innervation. In addition to their distinctive morphological and biochemical characteristics, vertebrate fast and slow fibres are now known to be distinguished by their relative content of fast or slow myosin isoenzymes. Differentiation of the two types of muscle apparently develops ontogenetically through changes in the biosynthesis of these myosins, probably under the influence of their distinctive motor innervation.

Innervation

The fast and slow fibres of the cyclostomes differ fundamentally in their modes of innervation. Fast, central fibres are innervated at their ends by the plate-like terminals of motor nerves and are activated by propagated action potentials. The slow, parietal fibres on the other hand, have a distributed innervation by motor nerves which pass along the length of the fibre, branching to form 'bouton-like' terminals fitting into deep invaginations on the surface of the muscle and activating it through junction potentials. At least in the case of the lamprey, it has now been established that the axons innervating the two types of somatic muscle come from distinct fast and slow motoneurones in the spinal cord.

The fast fibres of lampreys have nerve endings near the myotendinous junctions at both septal ends of the fibre. At the same time only those fibres closest to the abdominal surface have a direct motor innervation—the rest, like vertebrate smooth muscle, being electrically coupled to innervated fibres. With the exception of cardiac muscle, this indirect stimulation is a unique feature for a striated muscle. Its morphological basis may be the absence of a common basement membrane, the close packing of the individual fibres and the presence of the desmosome-like junctions described by Jasper (1967), which may enable the entire muscle lamella to function as a single unit.

In place of the double-ended innervation of the lamprey central fibres, the fast muscle fibres of the hagfish are innervated at only one of their septal ends by motor nerves ending in a typical motor end plate. In both hagfish and lamprey, the same motor axon may innervate corresponding fast fibres in neighbouring myotomes, thus co-ordinating contractions over a wider area. This is also true of the slow fibres of the hagfish, although in this case they receive a motor nerve supply from both ends.

Of the other muscle types that have been distinguished, the tongue retractor of the lamprey (regarded as intermediate in structure and lipid content between fast and slow fibres) has no multiple innervation like that of a typical slow fibre. However, in spite of its monosynaptic innervation and activation by conducted action potentials, sensitivity to acetylcholine has been demonstrated over the entire length of the fibre, in this respect resembling the slow fibre and suggesting that this represents a primitive and transitional condition from which the differentiated innervation patterns of fast and slow fibres might have evolved. Like those of the lamprey heart, these cholinoreceptors are of the nicotinic rather than the muscarinic type.

In the parietal muscles of the hagfish, the intermediate type of fibre that has been distinguished by morphological criteria, resembles the typical slow fibres in being innervated at only one end by plate-like terminals. Of the visceral muscles that have been investigated, the longitudinal lingual muscle consists solely of fibres of the fast type with single axon innervation, whereas the spinovelaris and craniovelaris fibres are all of the slow type with distributed synaptic sites. According to Nicolaysen (1966) these are about 10 μm apart, whereas in the slow parietal muscles there are said to be about 28 synaptic sites to each fibre and these were estimated as being about 100 μm apart. Korneliussen (1973) found some evidence for monoaminergic innervation of the red muscle of the craniovelaris, although it is possible that the catecholamines are involved in controlling the metabolism of the lipid and glycogen stores.

Functional Comparisons of Fast and Slow Muscles

As implied in the descriptive terminology applied to them, the most obvious physiological characteristics of the two main muscle types are the speed and duration of their contractions. These differences in mechanical response can be demonstrated by indirect stimulation of hagfish muscle tissue containing both types of fibre. Under these conditions, a weak stimulus elicits a twitch response of short duration,

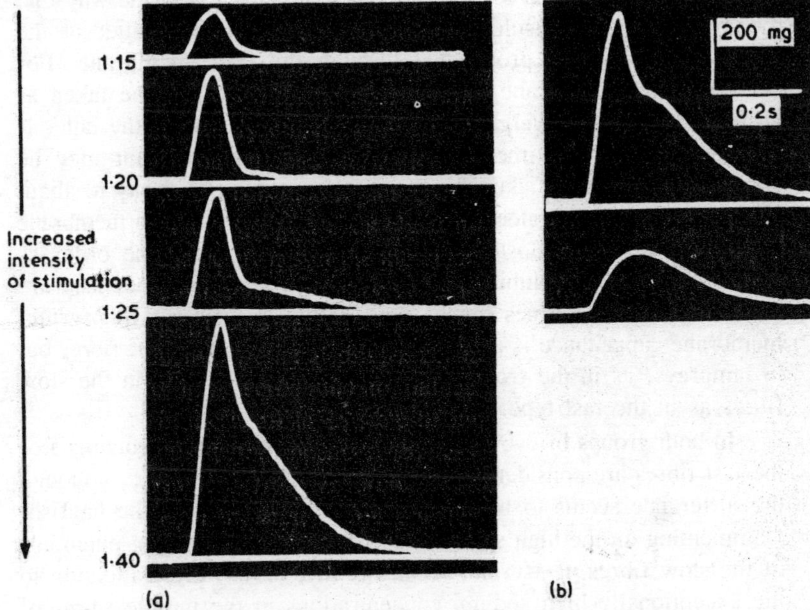

Fig. 10.4. Contractions of fast and slow fibres of the hagfish.

but as the intensity of stimulation is increased, a slow component appears, increasing in amplitude as more and more fibres are brought into play. With repeated stimulation, the fast component is eventually blocked and the slow component isolated. This has a peak contraction time of about 120 ms and a duration of about 500 ms, compared to a peak contraction time of 60-80 ms and a duration of 150 ms for the fast fibres. In the visceral muscles of the hagfish, peak contraction times in the longitudinal lingual muscle (consisting solely of fast fibres) are about 60 ms, while in the slow velar muscles the corresponding figure is about 125 ms.

When stimulated, the fast fibres respond in an all-or-none fashion, with the appearance of an overshooting action potential and the minimum duration of stimulus required to produce this response is proportional to the time constant. In the slow fibres, on the other hand, there is a graded response, varying with the extent of the membrane depolarization. However, in *Myxine*, the summation of junction potentials is not very effective in raising the amplitude of the response and the ceiling of depolarization is reached quite early in the train of impulses. The fast fibres require a higher frequency of stimulation for tetanic fusion and similar differences have been observed in the visceral muscles of *Myxine*.

Electrical constants for the two types of fibre reflect the way that membrane changes resulting from the release of transmitter at the motor terminals, are propagated through the fibre membrane. For example, high membrane resistance or capacitance may be taken as an indication of relatively low ionic permeabilities, while the latter is also responsible for time effects. Thus, the time constant may be expressed as the time taken for the electronic pulse to decay to about 37% of its maximum value. In their absolute values, the high membrane resistances of the cyclostome slow fibres are of the same order as those observed in amphibians and reptiles and in cyclostomes they are some two to three times higher than in the fast fibres. In *Myxine*, membrane capacitance is apparently similar in both types of fibre, but in lampreys, as in the frog they are about twice as high in the slow fibres as in the fast type.

In both groups of cyclostomes, the resting membrane potentials of the fast fibres are considerably larger than in the slow fibres, although the difference seems to be more marked in the case of the hagfish. Commenting on the high specific resistance and low resting potentials of the slow fibres of *Myxine*, Alnaes *et al.*, (1964) drew attention to the exceptionally high sodium concentrations in the muscle fibres of the hagfish, pointing out that if this reflected an unusually high sodium and a relatively low potassium conductance, these factors might explain their electrical characteristics. On the other hand, it may be noted that these conditions are not present in the lamprey, where the membrane resistance of the slow fibres may be even higher than it is in the hagfish. In an examination of sodium and potassium concentrations in muscles of various types in a number of marine vertebrates and invertebrates, Nesterov (1972) has claimed that the more rapidly contracting and active muscle fibres contain lower sodium and higher potassium concentrations. For example, in three teleost species, the potassium concentrations and the Na/K ratio were always much higher in the deeper white muscle than in the more superficially placed and presumably red fibres. Unfortunately, nothing is at present known of ionic distributions in the central and parietal muscles of the cyclostome.

Electrophysiological studies have revealed the existence in fast and slow fibres of miniature potentials, believed to represent the release of quanta of transmitter at the motor terminals. In fast fibres of the hagfish these potentials are observed only close to one end of the fibre and never at both ends. Moreover, with increasing distances from the end of the fibre, their amplitude decreases in accordance with the electrical properties of the membrane. On the other hand, the miniature

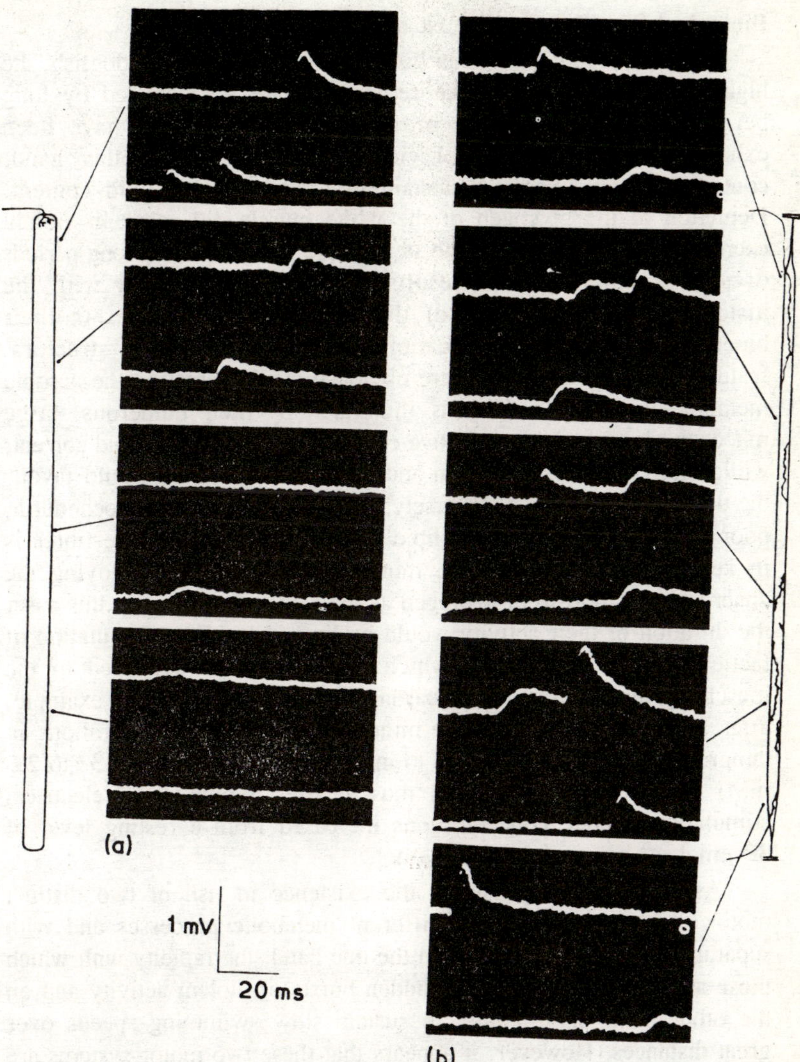

Fig. 10.5. Distribution of miniature potentials in fast fibres (a) and slow fibres (b) of Myxine.

junction potentials of the slow fibres are observed along the entire length of the fibre and show a varying time course. This picture is of course, consistent with the innervation of the slow fibres by multiple, distributed endings and with the presence in the fast fibres of a single motor end plate at one end of the fibre.

Biological Significance of Fast and Slow Muscles

As Bone (1966) showed in his experiments with spinal dogfish, the higher glycogen stores of the red muscle are not depleted by long periods of slow swimming or even after the animals have been exhausted by short bursts of violent activity. On the other hand, continuous slow exercise substantially reduced their lipid content. Depletion of the glycogen of the white muscle did occur in violent exercise, but these carbohydrate stores were not affected by long periods of slow swimming. The results are clearly consistent with the histological characteristics of the two types of muscle. In their biochemical constitution, location in the myotomes and structural features, the slow red fibres are obviously well adapted to the aerobic metabolism of lipids. This is illustrated by their numerous, large mitochondria (containing oxidative enzymes) and their high lipid content, while their superficial position and abundant capillaries would favour the delivery of oxygen. Conversely, the small and scanty mitochondria, poor vascularization and low lipid content of the fast, white fibres is in keeping with their role in more violent activity, employing the anaerobic metabolism of glycogen as their energy source. In this case, the duration of their activity would be limited by the accumulation of lactic acid and the rate at which oxygen can be delivered to the muscle by the respiratory and cardiovascular systems. As an example, after forced swimming for 20 minutes, lactic acid concentrations in lamprey blood have been found to increase sevenfold (from 0.32 to 2.3 mM) and in strips of parietal muscle after 5 minutes of electrical stimulation, lactate concentrations increased from a resting level of 30 μmol g^{-1} wet weight to 58 μmol.

As Bone has emphasized, the existence in fish of two distinct motor systems, powered by different metabolic processes and with separate innervation, explains on the one hand, the rapidity with which these animals are fatigued by sudden bursts of violent activity and on the other hand, their ability to sustain slow swimming speeds over great distances. However, it appears that these two motor systems are not always completely separate and that in some of the higher teleosts, the fast fibres are also active during continuous swimming at speeds well below the maximum the fish is able to sustain for long periods. These differences appear to be related to the type of innervation of the fast fibres. Whereas, in the dogfish, the herring and the cyclostomes these fibres are focally innervated at one end, in the gadoids and in the goldfish or carp which appear to use their fast fibres at cruising speeds, they are multiply innervated.

As outlined earlier, the lampreys are notable for their relatively poor performance in slow sustained swimming and are able to maintain their maximum speeds only over short distances. Attempts to correlate the locomotory abilities of different fish species with the distribution of the two types of muscle have not been entirely convincing and, at present, a satisfactory histological or physiological characterization of 'sprinters' or 'stayers' is hardly practicable. This is not surprising when we consider the complexity of the factors that are involved in swimming performance. In addition to the characteristics of the muscular system, these would include the cardiovascular and respiratory systems and hydrodynamic factors. Among the latter, mucus secretions play an important part in aquatic animals by reducing frictional drag and, in some teleosts, a 25% dilution of mucus has been found to reduce friction by 50-60%. On the other hand, in the hagfish, in spite of its prodigious slime producing capabilities, this factor appears to be less important and it has been reported that mucus from *E. stouti* only produced a 12% reduction in friction when compared to a sea water standard.

Reference has already been made to the sharp decline in the swimming endurance of the sea lamprey at sexual maturity. A possible clue to this inability to sustain the maximum swimming speeds of which its white musculature are capable, may lie in a reduction in the capacity of the liver to synthesize glycogen and glucose from the lactic acid, produced as a result of sustained muscular effort. During the upstream migration, degenerative changes occur in the liver, which are associated with starvation rather than sexual maturation. These changes are manifested by an alteration in liver colour from brown to green, as a result of the accumulation of bile pigments, resulting from haemoglobin degradation. Significantly, the green liver of the river lamprey, *L. fluviatilis* has been found to possess only half the gluconeogenic capacity of the brown liver as gauged from the relative activities of two enzymes (phosphoenolpyruvate carboxylkinase and pyruvate carboxylase) involved in the early stages of gluconeogenic pathways from lactate.

11

OUTGO SYSTEM

The entirely marine hagfishes, living in water of full salinity and often at considerable depths, maintain their body fluids at concentrations close to, if not completely identical with those of their environment. Thus, like the marine invertebrates and protochordates they are described as isosmotic and osmotic conformers. While they are in freshwater, lampreys on the other hand must be able to sustain their body fluids at concentrations far above those of the environment (hyperosmotic regulation), whereas in their marine phase they maintain internal concentrations little more than a third of those of the sea water (hyposmotic regulation).

OSMOTIC AND IONIC CONCENTRATIONS

Body Fluids

In the concentration of their body fluids, the two groups of cyclostomes represent almost the extreme poles of the vertebrate series. The hagfishes have the highest blood concentrations of any vertebrate (1060 mOsm) except for the elasmobranchs, whereas the ammocoete of *L. planeri* has a concentration of only 205 mOsm, parallelled only by polypteroids and lungfishes. The high concentrations of hagfish blood are almost entirely due to inorganic ions, which account for about 98% of the osmolar concentration, the small deficit being made up mainly by urea and trimethylamine oxide. Yet, in spite of this osmotic equilibrium between the blood and the sea water there are some significant differences in the distribution of various ions, indicating some active regulation of the internal ionic composition.

In all species of lamprey that have been studied, the lowest blood concentrations are found in the larval stages and serum osmolarity begins to increase during and after metamorphosis coinciding with the change from the larval to adult kidney. Comparisons of the freshwater serum concentrations in the three species, *L. planeri*, *L. fluviatilis* and *P. marinus* suggests that values tend to increase with body size, although in this respect, the dwarf freshwater race of the sea lamprey, *P. marinus* may be exceptional in showing somewhat higher osmolarities than the larger anadromous form of this species. However, the freshwater values for the latter represent a somewhat artificial situation and their lower serum osmolarity may reflect a failure of these animals to develop completely the mechanisms of hyperosmotic regulation at a period in the life cycle when these would normally be superseded by adaptations to life in a marine environment. In both forms of the sea lamprey, the blood concentrations are high when the animals first enter the rivers on their spawning migration, decreasing with sexual maturity and falling precipitously after the completion of spawning.

Ionic Composition of the Tissues

Indirect estimates have been made of the osmotic composition of hagfish muscle, based on the volume and composition of the extracellular fluid. These show the usual differences in the ionic composition of cell contents and extracellular fluids. Thus, sodium and chloride are in higher concentrations in the fluids bathing the surfaces of the cells, whereas concentrations of potassium and phosphate are higher within the cell than outside. A striking feature is the fact that inorganic and organic ions together account for rather less than half the intracellular osmotic activity; the remainder being due mainly to organic nitrogenous compounds, notably amino acids, trimethylamine oxide and betaine. This is precisely the kind of situation that exists amongst marine invertebrates, such as the decapod crustaceans, where a high proportion of the osmotic activity of their muscle tissue is also attributable to these organic compounds.

Osmotic Relations with the Environment

Myxinoids

In their natural environment, hagfishes are likely to encounter only minimal changes in the concentration of the sea water and to this extent, experiments involving the dilution or concentration of the ambient water could be regarded as unphysiological. Nevertheless, these laboratory experiments have provided some clues to normal physiological processes or properties, including the permeability of the body surfaces to water or ions and the animal's potential for regulating its body volume. It should also be borne in mind that when the hagfish is feeding, the ingestion of fish tissues and body fluids might well create a temporary, although slight, osmotic load.

In spite of their tendency to suffer from abrupt changes in salinity, myxinoids have shown a surprising capacity to tolerate slow and gradual acclimation to dilution or concentration of the sea water. Both *Myxine glutinosa* and *Eptatretus (Paramyxine) atami* die quite quickly if they are transferred directly from normal sea water to dilutions of 20-25‰. On the other hand, *Myxine* may be acclimated successfully to slowly increasing or decreasing salinities and has been maintained for several weeks in a range of salinities from 600-1500 mOsm. In general, it appears that at least some members of the genus *Eptatretus* tend to be more tolerant of salinity changes than *Myxine*. The Pacific hagfish, *E. stouti* has tolerated direct transfer to 80% or 120% sea water for at least 7 days, after which they were returned to normal sea water. In 80% sea water the body swelled rapidly and during the first day, their weight increased by about 10%. This was followed by a slow return to their original weight over a period of several days, but when replaced in normal sea water, the animals died. In 120% sea water the animals initially lost about 25% of their body weight, subsequently remaining at this level. When finally returned to full strength sea water, they recovered or even slightly overshot their initial weight.

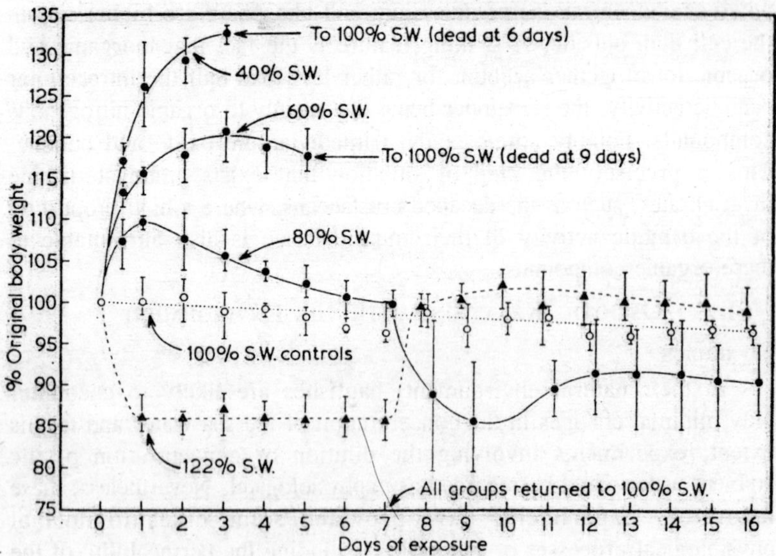

Fig. 11.1. Changes in the body weight of the hagfish, Eptatretus stoutii during exposure to concentrated or dilute sea water.

All the experiments carried out on hagfishes in varying concentrations of the external medium agree that the animals remain virtually isosmotic throughout, behaving as almost perfect osmoconformers. The rapidity of the weight changes in dilute or concentrated sea water indicate a high degree of permeability to water of the body surfaces, including the gills. This has been confirmed by measurements of the water flux using tritiated water. In *Eptatretus*, an exchange rate of 2287 ml kg^{-1} h^{-1} has been observed. This is some 5-10 times higher than the rates for freshwater teleosts measured by the same techniques and 20-50 higher than those recorded in marine teleosts. In these respects, a parallel has been drawn between the conditions in the myxinoids and in the decapod crustaceans. In the latter group, the freshwater forms such as the crayfish show much lower water permeability than the isosmotic marine crabs, whose high

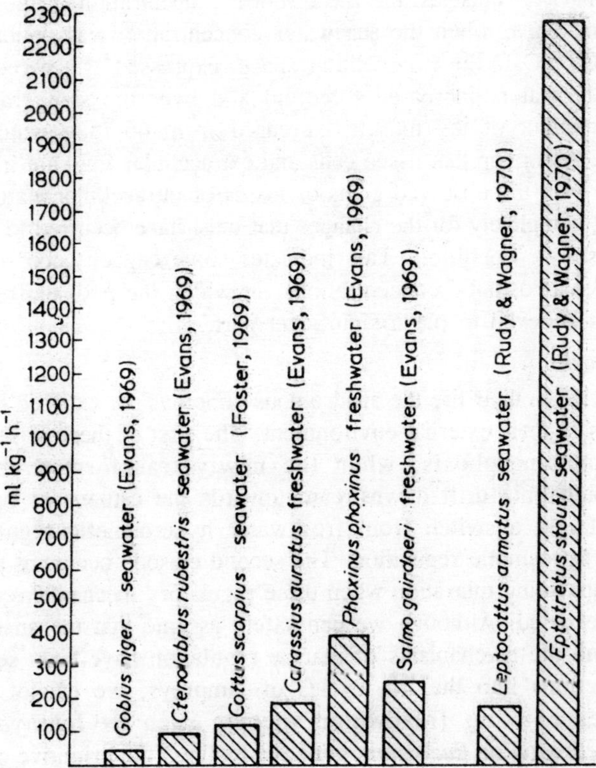

Fig. 11.2. Water influx in the hagfish, Eptatretus stouti compared with that of marine and freshwater teleosts.

exchange rates are similar to those of the hagfish. High water permeability of this order could only be tolerated by animals which would not normally experience serious osmotic gradients and where there would therefore be little osmotic movement of water through the body surfaces, to and from the external medium.

Experiments involving the subjection of hag fishes to varying salinities have also demonstrated how osmotic stresses are transmitted to the tissues of an isosmotic animal through the medium of the extracellular fluids. These stresses are expressed by the transfer of water from cell to extracellular fluid with increasing concentration of the external medium and a movement in the reverse direction when the external medium is diluted. When *Myxine* was subjected to concentrations of sea water from 600-1500 mOsm, a linear relationship was observed between the volume of the extracellular fluid compartment of the parietal muscles and the osmotic concentration of the external medium. Thus, when the sea water concentration was doubled from 700-1400 mOsm the extracellular space, expressed as a percentage of the body water, increased threefold and over the same range, the water content of the muscle increased from 66-78 g water 100 g tissue^{-1}. Assuming that tissue cells and extracellular fluid are in osmotic equilibrium, the observed gains or losses of intracellular water do not account completely for the changes that must have occurred to maintain the isosmotic condition. This indicates some capacity for regulating intracellular osmotic concentrations, in which the pool of free amino acids is believed to playa significant part.

Lampreys

Twice in their life the anadromous lampreys are exposed to drastic changes in their external environment. The first of these crises occurs after metamorphosis, when the newly transformed adults or macrophthalmia drift downstream towards the estuary or open sea, necessitating a switch from freshwater hyperosmotic regulation to marine hyposmotic regulation. The second episode occurs at the onset of the spawning migration when these regulatory mechanisms are once more reversed. Although we can safely assume that the anadromous habit and the mechanisms of marine regulation have been secondary introductions into the life cycles of lampreys, we cannot be sure whether existing freshwater species such as *Ichthyomyzon*, *Tetrapleurodon* or *Eudontomyzon* have retained the primitive condition or, like some of the non-parasitic lampreys, have descended from anadromous forms. In the latter case, it might be expected that these freshwater species would retain at least some limited capacity for regulation in saline media, but this has not yet been tested experimentally. In this connection, comparisons of the hyposmotic abilities of the closely related freshwater and anadromous forms of the sea lamprey, *P. marinus* are of particular interest in showing that a measurable change in osmotic performance may occur within a relatively short period of time.

At the larval stage, lampreys are quite incapable of hyposmotic regulation and, in ammocoetes of both anadromous and freshwater sea lampreys, serum osmolarity rises abruptly in water of more than 10 ‰ salinity (about 256 mOsm). In the anadromous form, the ammocoete values in freshwater (225 mOsm) or in 8‰ sea water (233 mOsm) are not significantly different, but in 10‰ sea water they reached a serum osmolarity of 299 mOsm after 8 days. During this gradual acclimation to increasing concentrations, body weight decreases with the osmotic loss of water to the external medium, blood volumes are drastically reduced and the concentration of the blood, although remaining hyperosmotic, rises with that of the water.

The transition from freshwater to marine regulation is dependent on morphological and physiological changes that occur during metamorphosis, particularly those affecting the gills, kidneys, gut and general body surfaces. Only when these changes have been completed are the animals able to move towards saline environments. At this time the macrophthalmia of *L. fluviatilis* or *P. marinus* can usually be transferred direct from freshwater to full strength sea water. However, even then the switch from freshwater to marine regulation is not irreversible and downstream migrant stages are able to survive a return from salt water to fresh.

In their studies on the landlocked sea lamprey, Mathers and Beamish (1974) have shown that marine regulation is first developed after metamorphosis, when the macrophthalmia, unlike the ammocoete, are able to regulate successfully in 10‰ sea water, maintaining levels of serum osmolarity similar to those of the adults in freshwater. On the other hand, in water of this concentration the blood concentration of the ammocoete increased to about 10% above freshwater values. Young adult lampreys, that had just begun to feed, were able to regulate at least for short periods in higher salinities up to normal sea water, but only the larger individuals were able to cope for long periods with salinities above 10‰. At a salinity of 26‰ animals with lengths of 127-188 mm suffered a 50% mortality within a 10 day period and only individuals of 280 mm or more were able to survive in full-strength sea water. This increase in regulatory capacity with age and body size is no doubt partly attributable to a decrease in the relative body surface, effectively reducing the area exposed to ionic and osmotic gradients.

In parallel experiments on the anadromous sea lamprey, small feeding adults only recently transformed and with body lengths of 135-140 mm were subjected to a series of increasing salinities from freshwater up to normal sea water. Until salinities exceeded 16‰ serum osmolarities showed no increases over the freshwater values and even beyond this point, increases in blood concentration were only slight. In 34‰ sea water, serum osmolarity averaged 263 mOsm, representing a concentration less than a third that of the external medium. In a smaller series of large feeding adults with body lengths of 260-310 mm, serum concentrations rose from 200 mOsm in fresh water to about 210 mOsm at 8‰ and about 250 mOsm in 26‰. In upstream migrants nearing sexual maturity, the capacity for marine regulation is clearly lost and serum osmolarities increased sharply in 16‰ sea water. This rise continued in 26‰ sea water and was accompanied by heavy mortality.

In these experiments, the serum osmolarities for the smaller sea lampreys are lower than those suggested for large and mature specimens of the same species, for which determinations have been made soon after the animals first entered the rivers. When measured in freshwater, these have given values varying from 220-291 mOsm. For sea lampreys actually taken in sea water we have only two determinations; the first (when recalculated from the freezing point) gives a value of 315 mOsm and the second suggests a concentration of about 346 mOsm, although in the latter case there is an element of uncertainty since haemolysis had already occurred. While it is difficult to form a clear picture from this data of conditions during the marine phase, the information seems to suggest that the values obtained by Beamish *et al.*, for recently transformed animals under laboratory conditions may represent rather low values for the blood concentrations of larger animals in the sea and in this connection it may be noted that the 34‰ sea water used in these experiments gave an osmolar concentration of only 919 mOsm kg^{-1}.

Nevertheless, in spite of these uncertainties, comparisons with the landlocked sea lamprey, show quite clearly that this dwarf form has lost some of its capacity for hyposmotic regulation and at higher salinities its performance is markedly inferior to that of the ancestral anadromous form. Once the salinity of the external medium has risen beyond the isosmotic point, serum osmolarity for the anadromous lamprey was always well below the levels observed in the landlocked form. It has been widely assumed that the origin of the freshwater race can be traced back to the recession of the Wisconsin ice sheet, which allowed the anadromous parent stock to invade Lake Ontario. If this assumption is valid, the quite pronounced changes in the osmoregulatory ability of the landlocked race, must have been accomplished within a period of separation of no more than about 8000 years. This interpretation would be in agreement with a suggestion that the differentiation of the freshwater forms has involved selection operating within the parent population in favour of smaller body size, shorter parasitic phase and lower potential fecundity, combined with some restriction in migratory habits and capacity for marine osmoregulation.

Physiological Mechanisms

Cyclostome Kidneys

Cyclostomes are characterized by their possession of a persistent pronephros, although in lampreys this is functional only in the late embryonic and prolarval stages and, in the hagfish, the pronephric tubules do not retain a connection with the main kidney duct and cannot therefore contribute to the formation of urine. The lamprey pronephros consists of 3-5 coiled tubules, served by a single glomus, receiving blood from the dorsal aorta. Each tubule opens by a ciliated funnel into the pericardial coelom and is connected at its distal end with the nephric duct. After hatching, the functions of the pronephros are gradually taken over by the continuous development of mesonephric tubules, some 5-6 segments behind its posterior end and in the adult lamprey, although the funnels remain, the tubules have disappeared.

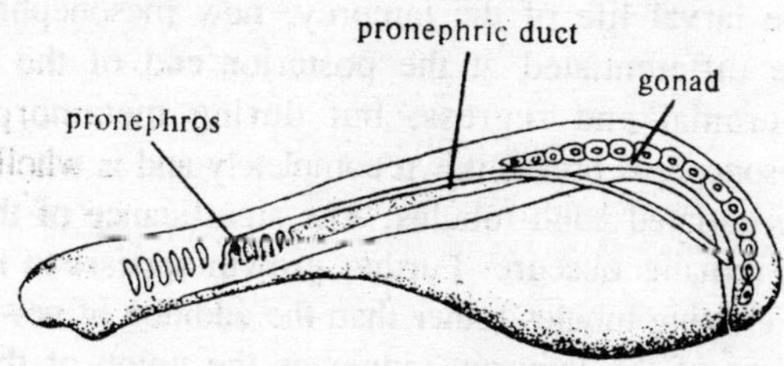

Fig. 11.3. Diagram to show arrangement of the pronephros in a freshly hatched lamprey.

In the hagfish, the pronephric tubules are initially segmental, but are then increased in number by branching until, in the adult, several hundred small tubules are present, opening by ciliated funnels into the pericardial coelom and communicating internally with the central mass of lympho-myeloid tissue. The single large glomus is thought to be concerned in the formation of the peritoneal fluid and there is some evidence that the central mass acts as a phagocytic filter between the peritoneal fluid and the blood.

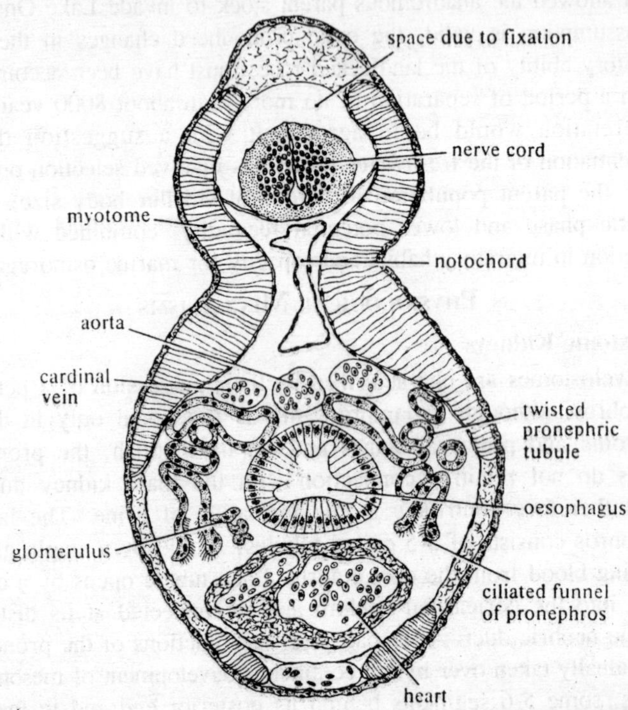

Fig. 11.4. Section through newly hatched larva of Lamptera binding the pharynx.

During the larval life of the lamprey, new mesonephric tubules continue to be differentiated at the posterior end of the kidney as those at the cranial end regress, but during metamorphosis the ammocoete mesonephros breaks down completely and is wholly replaced by newly differentiated adult tubules. The significance of this curious transformation remains obscure. Further growth appears to involve the lengthening of existing tubules rather than the addition of new nephrons. A unique feature of the lamprey kidney is the union of the separate glomeruli to form a continuous rope-like strand surrounded by the capsules into which open the ciliated funnels of a number of nephric tubules. In the adult, unlike the ammocoete, these capsules lack the usual double-layered structure, the visceral layer consisting only of podocytes covering the glomerular surface. Each nephron has a short narrow neck segment, followed by the long convoluted proximal tubule, lined by columnar cells bearing microvilli. The short distal segment consists of cells rich in mitochondria and this section, together with the terminal region of the proximal tubule is arranged in ascending and descending loops which have been compared to the loop of Henle of the mammalian or bird kidney. The final collecting ducts are formed by the distal segments of four adjacent nephrons.

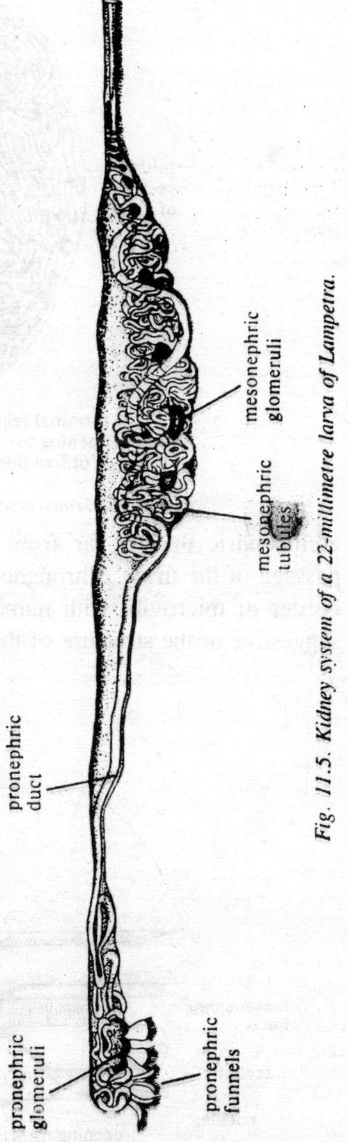

Fig. 11.5. Kidney system of a 22-millimetre larva of Lampetra.

The hagfish mesonephros is represented by little more than the longitudinal archinephric ducts and the segmentally arranged glomeruli. In the more caudal regions glomeruli may be absent, but conditions in the embryo suggest that this may be a secondary loss. Similarly, there may be a short cranial section where the ducts end blindly and are devoid of glomeruli, while isolated glomeruli unconnected with the ducts may be present in the area between pronephros and mesonephros. This variability, together with the branching and convolutions of the ducts in the embryo of *Eptatretus* have sometimes been adduced as evidence of phylogenetic regression in the myxinoid kidney, both as regards its extension and tubular development. However, in spite of its rudimentary tubule, the hagfish kidney appears to be able to carry out some of the functions associated with the highly differentiated tubule of lampreys or higher vertebrates and there are indications that the

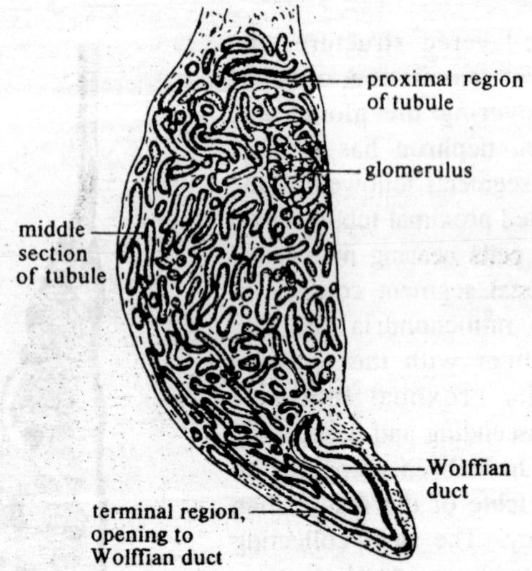

Fig. 11.6. Transverse section of kidney of Lampetra.

archinephric duct is far from being simply a passive route for the passage of the urine. Throughout its length the epithelium has a brush border of microvilli with numerous apical pits, tubules and vesicles, suggestive of the structure of the proximal tubule of higher vertebrates.

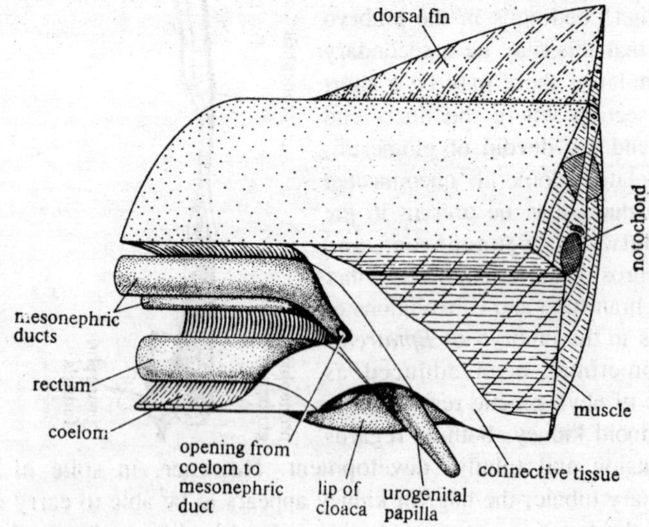

Fig. 11.7. Cloacal region of fully adult Lampetra.

Moreover, these structural indications of endocytosis have been confirmed by the experimental uptake of exogenous macromolecules.

Marine Regulation

Lampreys

In their marine osmoregulatory mechanisms, lampreys have developed methods similar to those evolved by marine and euryhaline teleosts and these are capable of maintaining rather similar levels of hyposmolarity. This involves the replacement of the considerable volumes of water lost osmotically through the body surfaces by swallowing and absorbing sea water. The rate at which water is swallowed is known only for river lampreys *L. fluviatilis* held in 50% sea water. Under these conditions, rates varies from 5-22% body weight day^{-1} of which about 75% is absorbed mainly in the anterior intestine. The mechanism of water absorption depends on the active uptake of sodium and chloride ions in the gut epithelium; the water then following passively along the osmotic gradient across the gut wall. Divalent ions—magnesium, calcium and sulphate—remain in the gut lumen and are apparently eliminated by this route, although it is possible that there may also be some active excretion of these ions into the luminal contents of the gut. Apart from the replacement of water losses, the lamprey in sea water will also be permeable to ions diffusing through the body surfaces. As in teleosts, the elimination of monovalent ions is achieved extra-renally, through the activity of specialized chloride cells, located in the interplatelet areas of the gills. These are characterized by their abundant mitochondria and prominent smooth endoplasmic reticulum. In freshwater, the majority of these ion-

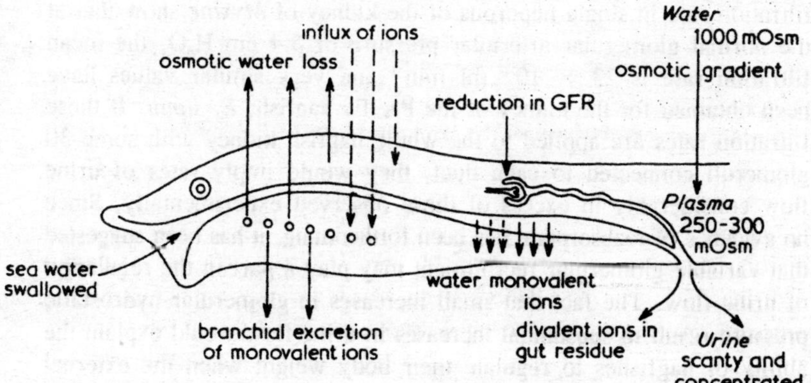

Fig. 11.8. Marine regulatory mechanisms of the lamprey.

transporting cells are covered by a layer of superficial cells, whereas in sea water their tips are exposed. In areas where teleost chloride cells occur, a sodium-potassium ATPase is thought to be involved in the extrusion of sodium ions and the inward movement of potassium. In lamprey gills, this enzyme has been detected in the gills of feeding adults of *P. marinus* and its activity apparently increased at higher salinities. On the other hand, in freshwater migrant stages this Na-K ATPase activity was observed only rarely.

In 50 % sea water the river lamprey (*L. fluviatilis*) swallows water and excretes chloride ions at a rate similar to that of the eel in normal sea water, but for the teleost, the osmotic gradient would be three times greater than for the lamprey. This indicates a much higher permeability to water and ions in the lamprey, which to this extent is less efficiently adapted to marine life. In half strength sea water only small volumes of urine are produced (0—6.2 ml kg^{-1} day^{-1}) and this is hyposmotic to the blood. Judging from the high content of divalent ions, it is believed that these may be actively secreted by the kidney tubule, whereas the monovalent ions are excreted mainly by the gills.

Myxinoids

In its natural habitat, the isosmotic condition of the hagfish would involve little or no osmotic influx or efflux of water. Urine volumes are therefore small and of a similar order to those of the lamprey, *L. fluviatilis* in 50% sea water. Thus, in *Myxine*, rates of urine flow of 1-11 ml kg^{-1} day^{-1} have been recorded and 4-10 ml kg^{-1} day^{-1} in *Eptatretus*. The rate of urine flow is thought to be dependent on the glomerular filtration rate, with little or no reabsorption of water in the rudimentary tubule or archinephric duct. Measurements of the filtration rate in single nephrons of the kidney of *Myxine* show that at the normal glomerular arteriolar pressure of 5.4 cm H_2O, the mean filtration rate is 23×10^{-6} ml min^{-1} and very similar values have been obtained for the kidney of the Pacific hagfish, *E. stouti*. If these filtration rates are applied to the whole hagfish kidney with some 30 glomeruli connected to each duct, they would imply rates of urine flow considerably in excess of those observed experimentally. Since no evidence of reabsorption has been forthcoming, it has been suggested that variable glomerular recruitment may play a part in the regulation of urine flow. The fact that small increases in glomerular hydrostatic pressure result in substantial increases in urine flow would explain the ability of hagfishes to regulate their body weight when the external medium is diluted. On the other hand, the low rates of urine production in normal sea water, together with the apparent inability of the nephron to transport water from the glomerular filtrate, are consistent with the inability of the hagfish to regulate its body weight in concentrated sea water. It is uncertain whether these animals normally swallow sea water to replace the small water losses via the urine and there is no evidence of water absorption in the gut, whose contents appear to be isosmotic with the blood.

The significant differences in ionic composition of blood and sea water imply the existence of ionic regulatory mechanisms, still incompletely understood. In the blood, sodium and bicarbonate concentrations are higher than in the water, while the levels of magnesium, calcium and sulphate are higher in sea water. The role of the kidney in ionic regulation is shown by differences in the distribution of ions in plasma and urine. Thus, potassium, magnesium and sulphate are in higher concentrations in the urine, but, sodium levels are lower. In spite of this difference in sodium concentrations, it had generally been thought that the myxinoid kidney lacks the ability to recover this ion from the glomerular filtrate, although sodium concentrations in the ultrafiltrate are reported to decrease along the length of the kidney duct. At least some of the divalent ions are believed to be actively secreted from the nephron and potassium may be transported in the

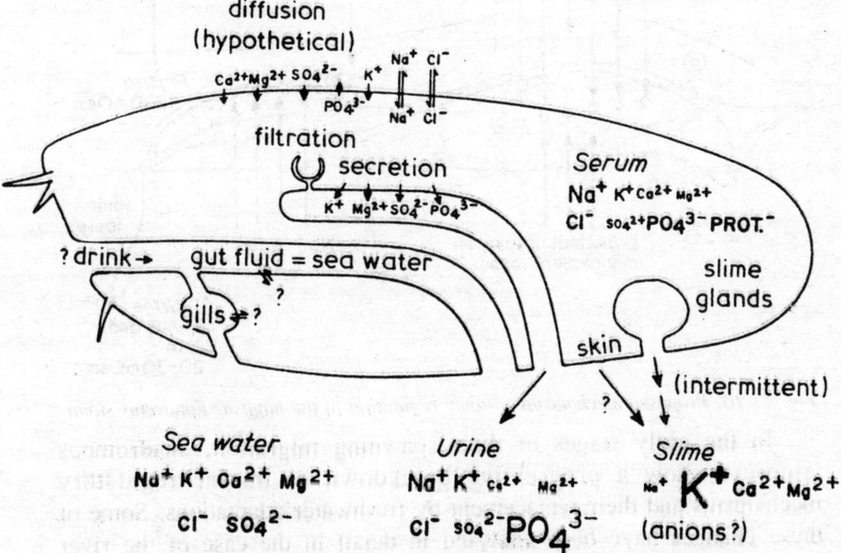

Fig. 11.9. Proposed mechanisms of ionic regulation in the hagfish, Eptatretus stouti.

narrow neck segment. Calcium and magnesium are in high concentration in the bile relative to the blood and unless they are subsequently reabsorbed in the gut, this could be a route for their excretion. Analyses of slime gland secretions have also raised the possibility that these organs may play a part in excreting calcium, magnesium and potassium.

Osmotic Regulation of Lampreys in Freshwater

The basic physiological problems confronting the lamprey in freshwater are the reverse of those that it encounters in the sea. The maintenance of an osmotic gradient of 200-300 mOsm between the body fluids and the environment must involve a very large influx of water through the relatively permeable body surfaces, whereas the ionic gradients in the same direction presuppose a continual loss of ions by diffusion; losses that can easily be replaced the animals are still feeding, but which must be counterbalanced by the active uptake of ions from the low concentrations in freshwater during the long periods of starvation that occur in the life cycle.

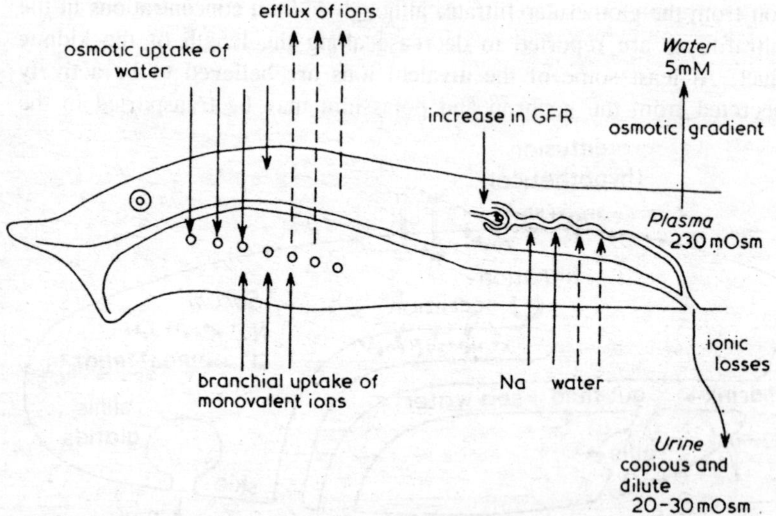

Fig. 11.10. Proposed mechanisms of ionic regulation in the hagfish, Eptatretus stouti.

In the early stages of their spawning migration, anadromous lampreys show a progressive breakdown of marine regulatory mechanisms and their replacement by freshwater adaptations. Some of these changes have been analysed in detail in the case of the river lamprey, *L. fluviatilis* which, when it first enters the rivers in the autumn may retain for a time some limited ability to regulate in saline media. This period of transition is marked by an increase in the permeability of the body surfaces to water and probably to ions. With the cessation of feeding there is a progressive atrophy of the intestine as the gonads mature, involving a dramatic reduction in its diameter and degenerative changes in its epithelial and muscular tissues. This is accompanied by a loss of the sea water swallowing habit and the ability actively to transport sodium ions across the intestinal wall, on which the absorption of the swallowed water depends. It is interesting to find that this degeneration of the gut and the loss of water swallowing and reabsorption can be inhibited by removal of the gonads. On the other hand, gonadodectomy does not prevent the increase in the permeability of the body surfaces during the period of the upstream migration. A further characteristic of this transitional stage in the development of freshwater adaptation is the replacement of the ion excretory cells of the early upstream migrants by a different cell type which appears to be identical with the presumptive ion uptake cells described by Morris and Pickering (1975) in the ammocoete gills. These cells differ from the ion transport cells of the marine phase in the absence of the extensive tubular, smooth endoplasmic reticulum and in the presence of characteristic cytoplasmic granules. These cells are believed to be responsible for the active uptake of monovalent ions in the gills that is such an important element in the physiological adaptation to freshwater life.

The osmotic flux of water through the body surfaces is balanced by the production of large volumes of dilute urine. According to the various techniques used to measure urine flow, estimates of the rate have varied widely from 48-423 ml kg^{-1} day^{-1}, but in the majority of cases the values represent volumes of 15-40% of the body weight day^{-1} and are considerably higher than those of freshwater teleosts. In freshwater, a mean single nephron filtration rate of 0.007 ml min^{-1} has been recorded for the kidney of *L. fluviatilis*. This is some 350 times higher that the comparable rates for the hagfish glomerulus. Regulation of urine flow in response to changes in the osmotic gradient are achieved through alterations in the glomerular filtration rate, brought about by changes in the hydrostatic pressure in the glomerular arterioles. Unlike the kidneys of teleosts or amphibians, urine flow is not adjusted by varying the rate of reabsorption in the tubule. This is said to remain constant for a given temperature and accounts for 36-60% of the volume of the glomerular filtrate. This difference in kidney function has been related to the absence in the cyclostomes of a renal portal system; an arrangement which makes the tubular circulation independent of the glomerulus and which can be regarded as an adaptation to water conservation. As in other vertebrates, tubular water absorption is a passive process, dependent on the active transport of sodium across the tubule. In the adult lamprey, the main site of water transport is thought to be the distal and collecting tubules, although it is possible that the kidney duct may also be implicated. In the ventral part of the kidney, a parallel arrangment of these ducts with respect to the vascular sinusoids recalls the arrangement in the medulla of the mammalian kidney and suggests that a countercurrent mechanism may be present. It may be significant that these structural arrangements are said to be absent in the kidney of the freshwater ammocoete stage.

An important aspect of kidney function in freshwater is its ability to conserve salts. This would be vitally important in the 6-8 months of fasting that occurs during the spawning migration of anadromous lampreys and in the long interval between metamorphosis and spawning in non-parasitic species. In the freshwater phase, the total concentration of the urine is from 7-15% that of the blood so that, in spite of the reabsorption of ions from the glomerular filtrate, the loss of ions, particularly sodium, potassium and chloride, represent a significant rate of ionic depletion for an animal that is no longer feeding. In addition, it is likely that extra-renal losses through skin and gills are even more serious than urine losses, emphasizing the crucial importance of the highly developed ion uptake mechanisms in the gill epithelium.

As regards freshwater adaptation, the non-parasitic lampreys are a rather special case. Their confinement to fresh water is a necessary consequence of the abandonment of an adult feeding stage. In these species, in which the growth of the gonads begins in the later stages of metamorphosis, the intestine is never functional and like that of the upstream migrant parasitic lamprey undergoes almost complete atrophy. Meanwhile, the adult foregut, which develops at metamorphosis, generally remains a solid cell cord without a lumen, so that both feeding and water swallowing are totally excluded, in all but perhaps exceptional or aberrant individuals; developments that would exclude any possibility of regulation in hyperosmotic media.

Osmotic Conditions during the Embryonic Development of the Lamprey

At the time of ovulation, the osmolar concentration within the egg of the lamprey is similar to that of the body fluids of the female. When deposited in freshwater therefore, the egg faces somewhat similar osmotic problems to those that confront the parent organism. Initially, the large osmotic gradient (about 210 mOsm) between egg and water leads to a osmotic influx of water, resulting in the accumulation of fluid in the perivitelline space between the egg surface and the outer egg membranes. Within the first few hours there is also a rapid uptake of water by the egg itself, but subsequently, the permeability of the vitelline membrane decreases sharply, remaining at about the same level throughout the first few days of embryonic development. The rapid initial influx of water into the perivitelline space and, to a lesser extent, the egg itself is reflected by a sharp drop in the osmolar and chloride concentrations. Since dilution of the egg contents does not entirely account for the observed decreases in concentration, some ions are presumably lost by diffusion.

The high permeability of the egg both to water and to salts is clearly only tolerable if a relatively short period elapses before the embryo is able to develop compensatory mechanisms, enabling it to take up ions from the water and eliminate water entering the egg osmotically. These characteristics would appear to be incompatible with marine development, where the osmotic and ionic gradients would be more severe. In fact in its osmotic relations and permeability, the egg of the lamprey bears a closer resemblance to that of the amphibians than marine teleosts and this would be consistent with the view that this group of cyclostomes has a freshwater ancestry.

Hormonal Factors

Vertebrate hormones principally involved in water and electrolyte balance are prolactin from the adenohypophysis, the neurohypophysial octapeptides and the steroids of the adrenal cortex. Among the target organs for these hormones are the body surfaces (including the gills), the bladder and the kidney tubules, which may be affected either by changes in permeability or in their ion transport mechanisms.

Such fragmentary indications as we have of the possible involvement in ionic and osmotic regulatory processes of cyclostome hormones have come mainly from observations on the effects of cyclostome pituitary extracts on the target organs of other vertebrate groups or the effects of exogenous (usually mammalian) hormones on certain aspects of cyclostome physiology. In either case, there is often a considerable element of doubt as to the significance that can be attached to what have often been essentially pharmacological procedures and in many instances the dosages employed have probably been quite unphysiological.

Prolactin, which plays an important role in the water and electrolyte metabolism of some teleost species has not been identified in cyclostome pituitary extracts, although injections of mammalian prolactin have been reported to alter the ionic composition of hagfish serum and tissues. The apparent absence of this hormone in the cyclostomes is all the more surprising in view of the very close parallels between the osmoregulatory mechanisms of teleosts and lampreys.

Among the range of vertebrate neurohypophysial peptides, only arginine vasotocin has so far been identified in lamprey pituitary extracts. Injections of this hormone in *L. fluviatilis* were found to have no effects on the rate of urine flow, neither did they produce an increase in body water as in the amphibians. On the other hand, injections of the mammalian preparation, pituitrin provided some evidence for increased sodium and potassium losses in the urine and similar effects have been reported following the administration of arginine vasotocin. Extracts of lamprey pituitaries has caused water retention in frogs, accelerated the transport of water across the toad bladder and exerted an antidiuretic effect on the rat. Arginine vasotocin has also been identified in the pituitary of the hagfish and pituitary extracts from *Myxine* had similar effects to those of the lamprey, when tested on the water and sodium balance of frogs. More direct methods, involving the administration of homologous pituitary extracts to the hagfish previously acclimated to dilute or concentrated sea water, have given some indication of changes in blood concentrations.

In view of uncertainty surrounding the steroidogenic capabilities of the lamprey interrenal and the, as yet, poorly understood adrenocortical homologue of the myxinoids, a possible role for corticosteroids in the salt and water balance of the cyclostomes must be very problematical. Of the range of vertebrate corticosteroids, the potent mineralocorticoid aldosterone has not been identified in cyclostomes. Nevertheless, injections of this steroid have produced slight reductions of urine flow and decreased losses of sodium in the lamprey, *L. fluviatilis*, perhaps through an effect on branchial sodium transport. In *Myxine*, only large doses of aldosterone have been found to be effective in reducing the sodium and potassium content of the muscles. Neither in lampreys nor in hagfishes have cortisone or cortisol produced anything other than very slight effects on salt concentrations or kidney function.

Cyclostomes and the Environment of the Early Vertebrates

Palaeontological and Geological Evidence

The oldest vertebrate fossils are the exoskeletal fragments of *Anatolepis* from marine Upper Cambrian deposits of North America.

The much more numerous remains from the Harding sandstones of North America represent later, middle Ordovician deposits, believed by White (1958) to be littoral deposits. This question has since been re-examined in detail by Spjeldnaes (1968) from the point of view of palaeoecology and salinity regimes. From the boron content of the clay minerals, he considers that these fossil heterostracans had been deposited under marine conditions, but in areas such as tidal flats, where shallow pools subjected to alternate evaporation or to dilution by rain, would have produced short-term changes in salinity; a type of habitat that might be expected to produce the necessary selection pressures favouring the development of regulatory mechanisms.

To what extent geological findings are relevant to the general problem of the environment of the vertebrate ancestors depends to a certain extent on the view that is taken of the relationships of the agnathans and the gnathostomes. If it is assumed that the agnathans were directly ancestral to the gnathostomes, or that the heterostracans in particular are close to the line from which the higher vertebrates originated, then the marine habit of the Cambrian or Ordovician heterostracans would be decisive. On the other hand, should we prefer to regard the agnathans and the gnathostomes as sister groups, both descended from a common ancestor, the geological evidence from these early heterostracans would scarcely be relevant and the absence of contemporary jawed vertebrates would be attributed to the imperfections of the fossil record and perhaps to their lack of calcification.

In the Silurian and Devonian, the anaspids and cephalaspids seem to have been predominantly freshwater forms and some of the heterostracans may even have developed euryhaline mechanisms. The earliest gnathostomes to appear in the fossil record — the acanthodians — were primarily a marine group in the Silurian, but had become an important freshwater group in the Devonian. These early penetrations of agnathans and gnathostomes into estuarine or freshwater habitats may have been connected with changes in their feeding or breeding habits. Just as today, estuarine regions are often a nursery ground for larval and juvenile stages of many marine fish species, so these habitats may have provided a safer environment for eggs and young and one where competition may have been less intense. If, as has been suggested, some at least of these agnathans were microphagous feeders, using the detritus deposited in slow-moving currents of rivers and estuaries, the adoption of these fresh or brackish habitats could hardly have occurred before the upper Silurian, when plant colonisation of these areas first occurred. At the same time we should not overlook the possible influence of global changes in palaeogeography on the environment of early vertebrate groups. For example, in a recent re-examination of the changes in the marine inundation of the Northern continents throughout the Phanerozoic, it has been claimed that the maximum extent of inundation was reached in the Ordovician and Silurian, when between 50-60% of North America and the USSR were covered by the sea. This was followed by a period of dramatic regression beginning in North America during the Silurian and continuing through the Devonian and Carboniferous, to fall to values of 20-30% during the Permian. Thus, the periods when the land areas were being elevated appear to coincide with the era when the early vertebrates, both agnathans and gnathostomes were beginning to leave marine environments and colonise freshwater habitats.

Physiological Considerations

The general picture that emerges from our information on the osmotic and ionic physiology of the myxinoids is that of a primarily marine group and as Robertson (1974) has pointed out, there are no precedents within the vertebrates for a reversion from osmotic independence to osmotic conformity. Many of the physiological characteristics of the hagfish are shared by marine invertebrates and may also have been common to the Cambrian or Ordovician agnathans. In this respect, the idea that the myxinoids stem from an early heterostracan stock would not be inconsistent with the physiological and palaeontological evidence. In their high concentrations of sodium and chloride ions and their isosmotic state, they have made the greatest possible economy in the energy costs of ionic and osmotic regulation, but at the price of evolutionary lability and the capacity for environmental adaptation. On the other hand, the plasticity of the petromyzonids manifested by the variety of their life cycles and migratory habits, may have been acquired at quite an early stage in their evolution, by adaptations to low or fluctuating salinities in brackish estuarine waters. This would not entirely conflict with the evidence from the Carboniferous *Mayomyzon*, whose fossil remains are said to be associated with a primarily marine fauna, but which might well have been transported from an estuarine or deltaic habitat. The very small body size of these fossil lampreys would inevitably have aggravated their osmotic stresses in water of high salinity and from what we know of the characteristics of present day lampreys, it would be difficult to imagine that they would have been capable of extensive migrations between fresh and salt water. Neither is it clear what selective advantages might have stemmed from such migrations. In present day forms, the seaward migration is able to take advantage of the greater availability and diversity of marine host fishes. For reasons discussed earlier, it is doubtful whether *Mayomyzon* could have been a parasitic feeder and the imperfectly developed sucker and tongue mechanism might have been related to a browsing feeding habit. The gradual development of predatory feeding may have accompanied a general increase in body size and the emergence of suitable host fishes, with an integument sufficiently delicate for the lamprey to penetrate. Parasitic feeding would in turn have been conducive to the development of hyposmotic regulation by facilitating the swallowing mechanisms. In addition, the ingestion of fish blood and tissues with similar concentrations to those of the lamprey itself, would make some contribution to its ionic and osmotic problems. This is all the more significant when we consider the fact that parasitic lampreys may spend most of their feeding life attached to host fishes.

The idea that the osmotic levels and ionic composition of vertebrates are a reflection of their early environment and marine history is mainly attributable to Macallum (1910). Impressed by the general similarities in the ionic composition of blood and sea water, this author interpreted this resemblance as an historic survival of an early period in vertebrate history when the body fluids would have been similar to the composition of the primeval seas. Such differences as now exist between blood and sea water were to be explained by the changes that were assumed to have occurred over geological time in the composition and salinity of the seas. Although there is now considerable doubt whether changes of this magnitude have occurred in sea water over the period covered by the evolution of the vertebrates, the idea that vertebrate body fluids have a 'historical component' has persisted and forms the basis of arguments advanced by Robertson.

Basing his case on the total ionic content of the plasma (in practice the sum of sodium and chloride ions), Lutz suggests that these reflect an original adaptation by the tissues to particular ionic concentrations and the development of a physiological dependence that would have limited the extent to which subsequent modifications could be tolerated when a particular vertebrate group moved from one type of environment to another. With the exception of the myxinoids, the range of vertebrate salt concentrations (200—500 mM l^{-1}) is very much smaller than that of the invertebrates 22—500 mM l^{-1}) suggesting that a commitment to a lower range of concentration may have been made at an early period in the history of the vertebrates, at the time of their initial radiation into brackish and freshwater environments. Groups that re-invaded the seas at a later time than the elasmobranchs and which therefore had a longer experience of freshwater life, might have become committed to lower salt concentrations, while the lowest values of all are to be

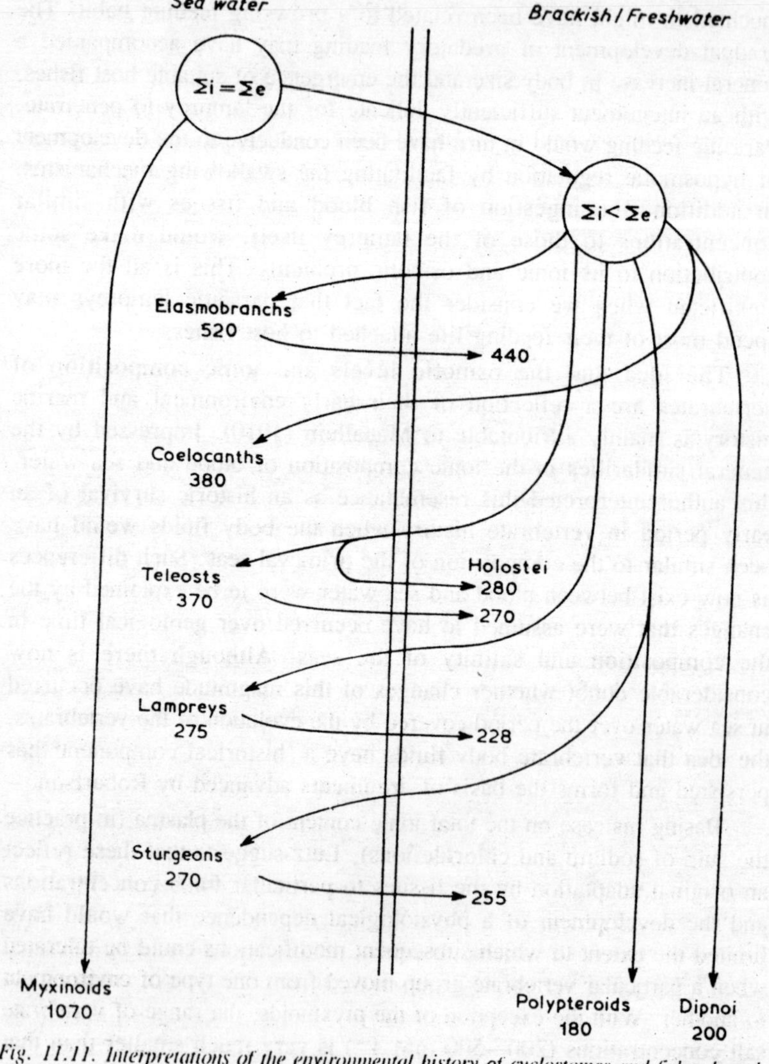

Fig. 11.11. Interpretations of the environmental history of cyclostomes and fishes based on ionic concentrations.

found in the lungfishes and polypteroids, which have had a long and continuous history of life in freshwater. Where vertebrate groups have subsequently developed a marine phase, the difference between their marine and freshwater blood concentrations are regarded as an indication of the antiquity of the marine habit; the most recent being the sturgeons where this difference is only (13 mM l^{-1}).

The lampreys do not fit easily into this hypothetical framework. On the limited evidence, their sea water values fall between those of the sturgeons and the teleosts, but the freshwater values are almost as low as those of the polypteroids and Dipnoi. Especially in view of other physiological evidence, the freshwater values of the lampreys are probably a safer guide to their evolutionary history than those of the marine phase. In very broad outlines, this picture, particularly as it affects the lampreys and freshwater teleosts, is in harmony with their respective freshwater and marine origins and with the assumption that the anadromous migrations of lampreys are, in geological terms, a comparatively recent development.

Reviewing the distribution within the vertebrates of osmoregulatory mechanisms, Sawyer (1973) has pointed out that the freshwater hyperosmotic mechanisms are much more uniform than those concerned with marine regulation. For example, the ability to take up sodium from the environment and to excrete an almost salt-free urine by re-absorption in the kidney tubule, is found in elasmobranchs, lampreys, actinopterygians, lungfishes and amphibians, although in elasmobranchs and lungfishes the recovery of sodium is barely adequate to maintain sodium balance in low concentrations of this ion. As regards the mechanisms of marine regulation, only lampreys and actinopterygians drink sea water, while the use of urea to maintain high blood concentrations has been achieved by elasmobranchs, holocephalans, coelocanths and anurans. This greater degree of uniformity in freshwater physiology would be consistent with an initial radiation of the vertebrates in freshwater and with their common ancestry, whereas the divergent modes of marine regulation reflect an independent return to the sea at varying geological periods.

Impressed by the adaptation of the glomerular kidney to the expulsion of large volumes of urine, it had been argued that this structure might have been first developed by vertebrates in freshwater to cope with the large osmotic influx. This argument appeared to be supported by comparative studies which showed that the glomeruli tended to be reduced or absent in some marine fish. However, the presence of a glomerular kidney in the myxinoids was a serious problem for this interpretation. What is more, in terms of body weight, the filtration surface of the hagfish kidney is apparently similar to that of freshwater teleosts and shows a high rate of filtration. As Robertson has emphasized, such a high rate of ultrafiltration is unlikely to have been developed solely as a means of eliminating nitrogenous compounds, since in fish, as in lampreys, this is mainly a function of the gill surfaces.

Further questions are posed by the peculiar features of the hagfish kidney—the very short rudimentary tubule and the apparent absence of water reabsorption—contrasting sharply with the highly developed nephron of lampreys. Is this simple structure a result of evolutionary regression or does it rather represent the primitive condition of the kidney in the marine vertebrate ancestors? Here it may be noted that in the kidneys of *Eptatretus* the ureters are described as serpentine, tending to form branched ducts leading to the renal corpuscle. This has been regarded as a more primitive condition than that of *Myxine* and suggestive of a phylogenetic regression from kidneys that originally consisted of branched ducts. However this may be, at least some of the functions normally associated with the more complex kidney tubule of the higher vertebrates are taken over by the archinephric duct of the hagfish, which is well able to cope with the small urine volumes that this animal normally produces.

The divergent kidney structure and osmotic physiology of lampreys and hagfishes presents a number of difficult evolutionary problems. In lampreys, the parallels between their freshwater physiology and those of higher vertebrates extend also to the cytology of the kidney tubule, its topography and vascular arrangements. Are all these structural and functional parallels to be attributed solely to evolutionary convergence or were they perhaps derived from ancestors common to both agnathans and gnathostomes? In view of the overwhelming evidence, there is no reason to doubt that the divergent patterns of marine osmoregulation have been acquired independently by different vertebtate groups and at different times. If, on the other hand, we accept that the freshwater regulatory mechanisms (and the differentiated nephron structure on which they partly depend) had already been developed in a common agnathan-gnathostome ancestor during the initial radiation of the vertebrates in freshwater, how are we to account for the rudimentary tubule of the myxinoids and for the persuasive arguments in favour of their marine origins? If the simplicity of the hagfish kidney has been due to regression this would presumably imply a change in the environment and in the physiological demands placed upon this organ which is quite inconsistent with the physiological evidence. A more economical hypothesis, would be that the myxinoids have always been marine and that they arose from a vertebrate line distinct from that which gave rise both to the lampreys and to the gnathostomes; a view which is in complete harmony with most of the morphological and non-morphological evidence discussed at greater length in the final chapters.

12

INTEGRATED SYSTEM

SPINAL CORD

Although some progress has been made towards a functional analysis of the cyclostome spinal cord, this task has been made more difficult by virtue of its peculiar characteristics which have impeded direct comparisons with the distribution of the cells and fibres in the spinal cord of higher vertebrates. The unique ribbon-like shape of the cord (said to be shared only by *Latimeria*) is almost certainly a secondary acquisition, since a more usual cylindrical form is present in earlier ontogenetic stages. Whiting (1972) considers that this flattened shape may have been due to the downward expansion of the dorsal fat column; a structure that he believes may already have been developed by heterostracans. In the absence of medullated nerve fibres there is no sharp division between white and grey matter, although the cells form a broad central mass surrounded by the fibres. Distinct dorsal and ventral horns are also missing and the dorsal and ventral spinal nerve roots of the lamprey are unique in failing to unite. In myxinoids, where the nerve roots do unite, these junctions lie far outside the spinal cord in the region of the parietal muscles.

Reticulospinal System

In both groups of cyclostomes, the principal descending fibre pathways consist of large unmyelinated axons in the ventral and lateral columns. These arise from cell bodies located in the medial regions of the brain stem and constitute the reticulospinal supra segmental motor control system. In keeping with their large diameters, the rate of conduction in these axons (6-10 m s^{-1}) is higher than in any other

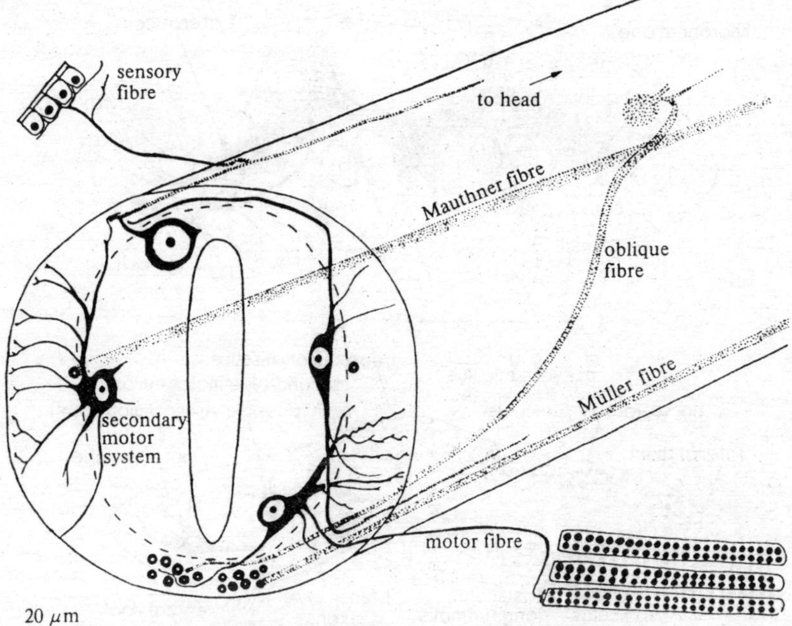

Fig. 12.1. The spinal cord of pro-ammocoete stage of lamprey.

nerve fibre. In the lamprey, about 50 of these reticular neurones have been identified in the brain stem, varying considerably in size. Some, larger than the usual somatic motor neurones, are more conspicuous and have been referred to as Muller cells, but there have been differences in the number of such cells that have been recognized by different authors. Muller cells are arranged bilaterally in pairs and are constant in their location. Leaving aside those cells on which there has been disagreement, at least 7 pairs have been identified within the brain stem. In *Myxine*, none of the axons in the medial longitudinal bundle (fasciculus longitudinalis medialis) can readily be distinguished from the others by size nor can they be homologized with the Muller cells of the lamprey. For this reason, Bone (1963) preferred to speak only of 'Muller-type' axons, considering true Muller cells as specialized elements of a reticular through conducting system to be absent in hagfishes. In addition, unlike the Muller fibres of the lamprey, the Muller type axons of *Myxine* appear to cross in the ventral commissure to descend in the medial longitudinal bundle of the opposite side.

Apart from the Muller neurones, Rovainen *et al.* (1973) have identified two pairs of Mauthner neurones in the lamprey brain stem,

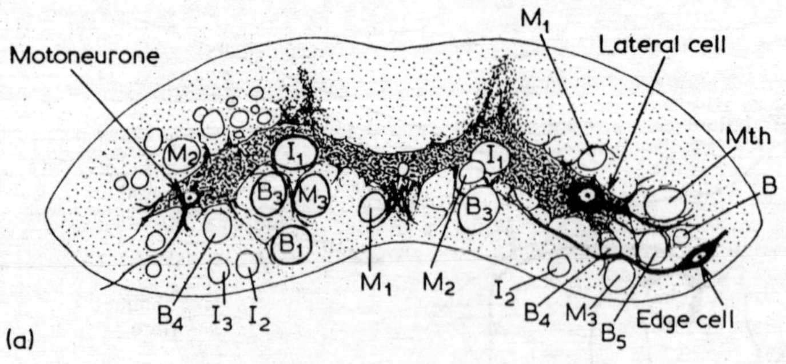

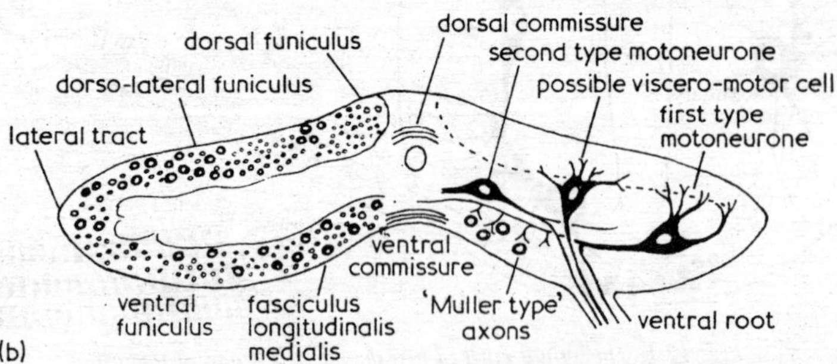

Fig. 12.2. The cyclostome spinal cord. (a) Spinal cord of a lamprey showing some of the identified neurones and fibres. (b) Fibre tracts and neurons of the spinal cord of Myxine.

one of which is regarded as the homologue of the Mauthner neurones of fish and amphibia. These cells, which appear to be absent in the hagfish, are distinguished from the Muller neurones by their proximity to the acoustic nerve (VIII) and the fact that their axons cross before descending on the opposite side of the cord. Throughout their passage down the cord, the Muller axons establish synaptic contacts with the dendrites of giant interneurones which show both chemical and electrical coupling. In addition they form monosynaptic connections with spinal motoneurones. Thus, like the giant command systems of invertebrates, they connect the brain directly to the motor system and provide a rapid pathway for the execution of predatory or escape reactions. Such frantic and indiscriminate escape responses are seen when resting lampreys are suddenly disturbed and have been attributed to prolonged 'afterdischarges' that have been observed in Muller cells after experimental stimulation of the cranial nerves. Unlike the invertebrate systems or the Mauthner neurones of teleosts which give all-or-none responses to single shocks, the Muller and Mauthner fibres of lampreys

Muller neurone
giant cell
giant cell
Axon of Muller neurone

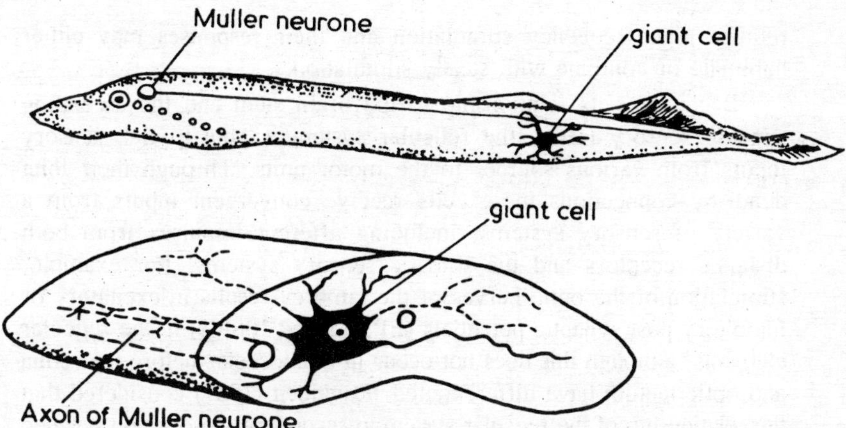

Fig. 12.3. Position of giant interneurones of the lamprey and their relationship to the Muller fibres.

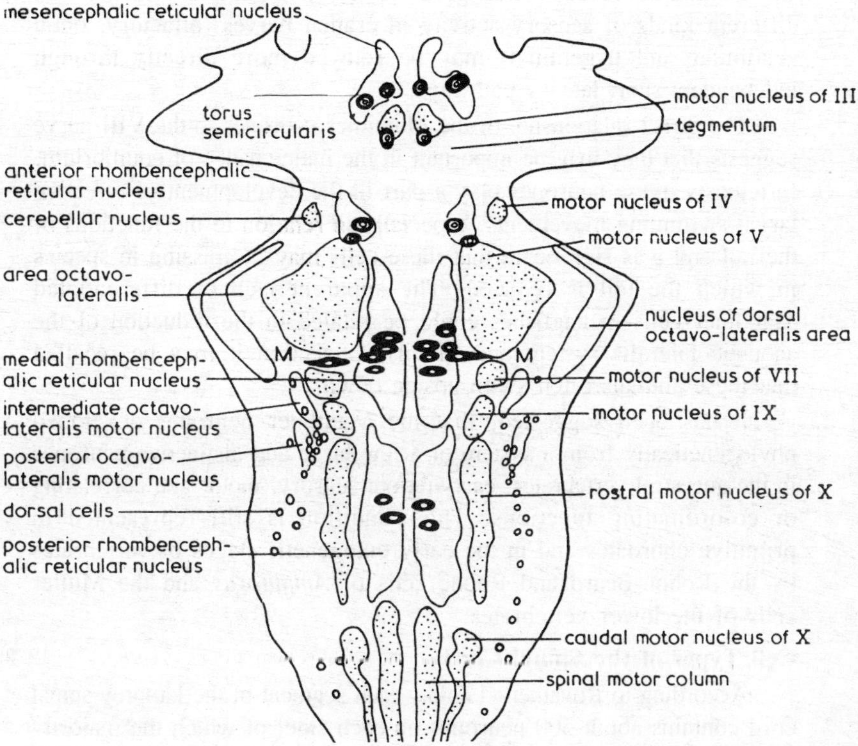

mesencephalic reticular nucleus
torus
semicircularis
motor nucleus of III
tegmentum
anterior rhombencephalic
reticular nucleus
cerebellar nucleus
motor nucleus of IV
motor nucleus of V
area octavo-
lateralis
nucleus of dorsal
octavo-lateralis area
M
M
medial rhombenceph-
alic reticular nucleus
motor nucleus of VII
intermediate octavo-
lateralis motor nucleus
motor nucleus of IX
posterior octavo-
lateralis motor nucleus
rostral motor nucleus of X
dorsal cells
posterior rhombenceph-
alic reticular nucleus
caudal motor nucleus of X
spinal motor column

Fig. 12.4. Topological chart of the brain stem of the lamprey, Lampetra fluviatilis showing the positions of the main cell masses.

require high frequency stimulation and their responses may either habituate or continue with steady stimulation.

Because of its relationship to the brain stem and the ascending somatic sensory tracts, the reticular system is able to link sensory inputs from various sources to the motor units. Through their long dendritic connections these cells receive convergent inputs from a variety of sensory systems, including afferent impulses from both distance receptors and the somatic sensory systems. For example, stimulation of the optic nerves of the lamprey results in excitatory or inhibitory post synaptic potentials (EPSP's and IPSP's) in the reticular elements, although this does not occur in ammocoetes before the retina and optic tectum have differentiated. Rovainen (1967) considered that the relationship of the reticular system to secondary sensory polysynaptic pathways was indicated by the long latencies of responses to cranial nerve stimulation, but Wickelgren (1977a,b) has found evidence that different kinds of sensory activity in cranial nerves (olfactory, optic vestibular and trigeminal) may be relayed more directly through independent short latency pathways.

The direct relationship of the Mauthner neurones to the VIII nerve suggests that they may be important in the maintenance of equilibrium. In teleosts, these neurones play a part in the development of effective larval swimming movements, especially in relation to the functions of the tail and it is significant that these cells may be missing in species in which the tail is reduced. The apparent lack of differentiated Mauthner cells in hagfishes could be related to the reduction of the acousticolateralis system and in this connection it may be recalled that these animals often swim upside down.

It has been suggested, that the Mauthner neurones developed phylogenetically from a system of a few large and distinctive neurones in the ancestral vertebrates, carrying out sensory, motor and correlating or co-ordinating functions. This condition is still represented in primitive chordates and in the early ontogenetic stages of vertebrates by the Rohon-Beard and Rhode cells of *Amphioxus* and the Muller cells of the lower vertebrates.

Cell Types of the Spinal Cord

According to Rovainen (1974a), each segment of the lamprey spinal cord contains about 500 neurones on each side, of which the majority are small cells less than 15 μm in diameter, whose functions have not yet been investigated. Of those that have been studied, the following types are recognized and most of them have their counterparts in the spinal cord of the hagfish.

1. Somatic motoneurones, innervating slow and fast muscle fibres, of which there are about 40 pairs in each segment.
2. Edge cells or 'Randzellen' numbering about 20 per segment.
3. Lateral cells — not more than one per segment.
4. Giant interneurones — not more than one per segment.
5. Dorsal cells — about two per segment.

Fast and slow motoneurones

In the spinal cord of *Myxine*, two types of primary motoneurone were described by Bone (1963). The first more numerous cell, often situated at the lateral borders of the grey matter, was believed to represent a somatic motoneurone innervating the fast muscle fibres of the myotomes. A second less abundant type, situated medially to the point of emergence of the ventral root of the spinal nerves, was thought to supply the slow myotomal fibres. From both cell types processes extend towards the ventral commissure and in the case of the 'fast motoneurones' these divide into ascending and descending branches on the opposite side of the cord. In addition, branches are also given off to the medial longitudinal fibre bundle and especially to the ventral area occupied by 'Muller type' fibres.

In the lamprey, the primary motoneurones are among a group of transversely orientated disc-like cells, immediately lateral to the Muller fibres and dorsal cells. Intracellular stimulation of these cells results in one-to-one contractions in the muscles of the same side, occasionally spreading over two or three segments and representing 5-12 myotomal sub-units. Contrary to Bone's suggestion in regard to *Myxine*, the slow motoneurones of the lamprey are believed to be the more abundant type, representing from two-thirds to four-fifths of the total number of motoneurones in each segments. Both types show similar membrane potentials (60-80 MV), but differences in input resistance and time constants may reflect variations in cell size and in the degree of dendritic arborization. In fact, the fast motoneurones are somewhat larger than the slow type, with diameters of 15-25 × 45-70 μm compared to 20-25 × 25 30 μm, and in addition they show more extensive dendritic branching. Unlike the motoneurones of *Myxine*, the processes of the lamprey motoneurones do not extend to the opposite side of the cord, although some dendrites approach the mid-line in the region occupied by the Muller fibres.

Edge cells or 'Randzellen'

This type of spinal interneurone is about 20-50 μm in diameter and located in the lateral tracts of the lamprey cord. Their axons extend forwards, in some cases towards the opposite side of the cord and appear to have their counterparts in the lateral arcuate cells of *Myxine*, which have decussating axons or a long axon on one side with dendrites ending on the limiting membrane. Because of their relationship to the peripheral dendritic network of the somatic motoneurones, Bone suggested that they may be involved in the reciprocal excitation and inhibition of motoneurones on either side of the cord during swimming movements. A somewhat similar conclusion was reached by Teravainen and Rovainen (1971b) who observed activity in the edge cells of the lamprey during myotomal reflexes when they inhibited the motoneurones of the opposite side. Some of the edge cells in the lamprey were excited by intracellular stimulation of one of the Muller axons, but no antidromic action potentials could be elicited in edge cells from more posterior regions of the cord.

Lateral cells

These are amongst the largest cells in the lamprey cord with transversely directed dendrites confined to the same side. Their long axons extend towards the tail and since lateral cells behind the brain have been stimulated in the tail region, these axons must in some cases extend to 8-10 cm in length. Intracellular stimulation of the lateral cells failed to elicit synaptic potentials in giant interneurones, edge cells or other lateral cells and indeed for the most part in motoneurones. However, they were found to inhibit certain unidentified medially placed cells some of which may have been motoneurones. Rovainen (1974a) suggested that these inhibited cells may be connected with local contractions in adjacent myotomes and that the function of the lateral cells may be to bring about the relaxation of myotomes behind them, in this way contributing to the production of undulatory waves of contraction during swimming. Lateral cells were consistently excited by Muller axons.

Giant interneurones

In *Myxine*, these cells are located in the lateral or medio-lateral areas of the grey matter; several pairs being alternately arranged in each segment. Their axons join the ventral commissure and run in the median longitudinal bundle or the ventral funiculus. Like the giant Rhode co-ordinating neurones of *Amphioxus*, with which they have been homologised, the axons of posteriorly placed cells ascend the cord, while the more anterior axons descend caudally.

The giant interneurones of lampreys (or large internuncials of Whiting), are located in the lateral grey of the posterior third of the cord making synaptic contact with a pair of Muller axons and sending dendrites into the dorsal or dorso-lateral fibre columns, where they would be in a position to intercept sensory axons. Their axons ascend towards the brain, but although it has been possible in only one case to trace them into the brain stem, the fact that stimulation produced excitatory post-synaptic potentials in a contra-lateral Muller cell shows that they must reach at least as far as the medulla. Regarded as higher order sensory neurones, these cells have not produced movements of the body or synaptic potentials in motoneurones when stimulated intracellularly. On the other hand, they are readily excited by mechanical stimulation of the tailor skin surface and the fact that they fire together suggests that they may play a part in swimming movements. Dorsal cells when stimulated, elicit both monosynaptic and polysynaptic EPSP's in the giant interneurones, whose role has been defined as part of a convergent multi specific sensory system extending towards the brain. The arrangement of the giant interneurones in the hagfish could allow them to play a part in coordinating the contractions of successive groups of motoneurones. This, together with the cranio-caudal directions taken by their axons, has led to a suggestion that they might be implicated in rapid responses in either direction of the cord following stimulation of trunk or tail, like as those observed in the knotting movements of the hagfish when it is escaping from slime.

Dorsal cells

On either side of the mid-line of the lamprey cord are two rows of dorsal cells or Hinterzellen, whose fibres emerge through dorsal roots to terminate in the skin. Most of these cells appear to have anterior processes extending at least as far forwards as the branchial region and after antidromic stimulation of their rostrally projecting axons, Martin and Bowsher (1977) found that in three-quarters of the neurones they investigated, these axons reached the isthmic region of the brain stem. Measurements of conduction velocities showed a decrease in the rostral direction, suggesting the existence of numerous collaterals in the spinal cord of brain stem. These cells, which have been homologised with the Rohon-Beard cells of earlier ontogenetic stages of lower vertebrates, have not been identified in myxinoids, although Kuhlenbeck (1975) has suggested that they may have their counterparts in certain comparatively large cells dorsal to the central canal. Dorsal cells are regarded as first order sensory cells in which action potentials are produced by stimulation of their sensory endings and not by synaptic activation of the cell body via other elements in the central nervous system. Individual dorsal cells are responsible for responses to touch, pressure or nociceptive stimuli and their receptive fields are located in the ventral or ventro-lateral body surfaces. Although differing in some electrophysiological properties, the trigeminal ganglion cells are also regarded as first order sensory neurones equivalent to dorsal cells and responding to the same range of sensory modalities, but with their receptive fields on the surface of the head. Dorsal cells have been shown to be involved in reflex movements of the dorsal fin after the skin of the same side of the trunk has been stroked and similar fin movement follows intracellular stimulation, in some cases on the opposite side of the body. Stimulation of single dorsal cells does not produce a response in single motoneurones and an appropriate response probably requires the simultaneous activity of a number of these cells.

Reflex Behaviour and Neuronal Interrelationships

The application of electrophysiological techniques has thrown some light on the interrelationships of some of the larger neurones in the spinal cord of the lamprey but we are still unable to form a clear picture of the neural mechanisms involved in normal swimming behaviour. Local or segmental reflexes which require the presence of intact dorsal roots, appear to involve only the slow motor units and in experimental investigations, this has generally been true also of the long reflexes that continue rostrally after destruction of the dorsal roots. Activity in fast muscle fibres has been noted only after intense mechanical stimulation as when the tail is pinched or pricked. These differences in the reflex behaviour of the two motor systems suggest that the slow units are more active and have a lower threshold. Stimulation of dorsal cells has produced myotomal movements and both inhibitory postsynaptic potentials (IPSP) and excitatory postsynaptic potentials (EPSP) in fast and slow motoneurones, probably via interneurones, although no response occurred in single motoneurones after intracellular stimulation of a single dorsal cell. In simple fin reflexes, only two groups of neurones—dorsal cells and fin motoneurones—have been implicated.

Intracellular stimulation ofneurones of the reticular system has shown that motoneurones are excited directly by certain Muller cells

(M^1, M^4) in the mesencephalon and further neurones in the isthmic region (I_1), as well as by the larger Mauthner cell. On the other hand, Muller neurones in the rhombencephalon excited lateral cells and only one activated motoneurones. During mechanical stimulation, this latter cell was excited by a giant interneurone. Swimming responses have been observed in an intact lamprey after stimulation of a Muller cell of the rhombencephalic group and in another case, stimulation at a frequency of 100 s^{-1}, resulted in short contractions on the same side of the body, although no undulatory waves were produced. Presumably, normal swimming requires the simultaneous activity of larger numbers of neurones, but as decapitated lampreys and hagfishes show swimming behaviour after intense mechanical stimulation, many of these cells must be intraspinal. Relevant to these wider effects is the fact that in addition to their monosynaptic connections with motoneurones, Muller axons may act through other relay pathways and that many other thinner descending fibres as well as interneurones and dorsal root afferents may provide synaptic inputs to motoneurones. That the integrity of the reticular neurones is not essential to the execution of normal swimming reflexes is also demonstrated by experiments involving complete severance of the spinal cord. Under these conditions, ammocoetes regain apparently normal co-ordination of their swimming movements after several weeks, although the large reticulospinal axons degenerate beyond the zone of transection. Functional recovery has been attributed to the short distance sprouting of axons of giant interneurones or dorsal cells, which may establish polysynaptic connections with unspecified interneurones across the severed cord.

The Mauthner cell of the lamprey produces composite EPSP's in a majority of the myotomal motor neurones and it seems likely that, as in teleosts, these cells may be involved in rapid escape reactions. In fish, a 'startle response' can be elicited by acoustic or vibrational stimuli, exciting the Mauthner cells through the acoustic nerve. This in turn activates the motoneurones over a larger area, producing a sudden, rapid flexing of the whole body to one side, followed by a less extensive and slower return flip in the opposite direction. Thus, in teleosts, the significance of the Mauthner system lies in its ability to activate directly a large part of the myotomal system and the same effect can be produced experimentally by antidromic stimulation of the Mauthner neurones or by electrical stimulation of the VIII nerve. In contrast to lampreys, where responses to vibrational stimuli are located within the labyrinth and could therefore activate the Mauthner neurones via the acoustic nerve, the labyrinth of hagfishes does not respond in this way to mechanical stimuli. This might well account for the absence in the hagfish of a differentiated Mauthner cell system. In *Myxine*, vibrational responses have been detected from a skin area below the degenerate eye and although it has been suggested that this might correspond to the rudimentary lateral line area of *Eptatretus*, it should be noted that it is innervated by the trigeminal (V).

Brain

In the linear arrangement of the fore, mid and hind brain, the lamprey (or more appropriately the ammocoete) has been claimed to show the closest approach to the archetypal vertebrate pattern and on Halstead's (1973 a, b) interpretation, a similar arrangement was also present in the heterostracans. Moreover, except for the presence of a bilobed cerebellum, the cephalaspid brain has been held to show marked parallels to that of the lamprey. At the same time, in the adult cyclostomes, these simpler brain patterns have been considerably distorted by the changes that occur during embryonic or larval development. For example, in the adult lamprey, the disposition of the

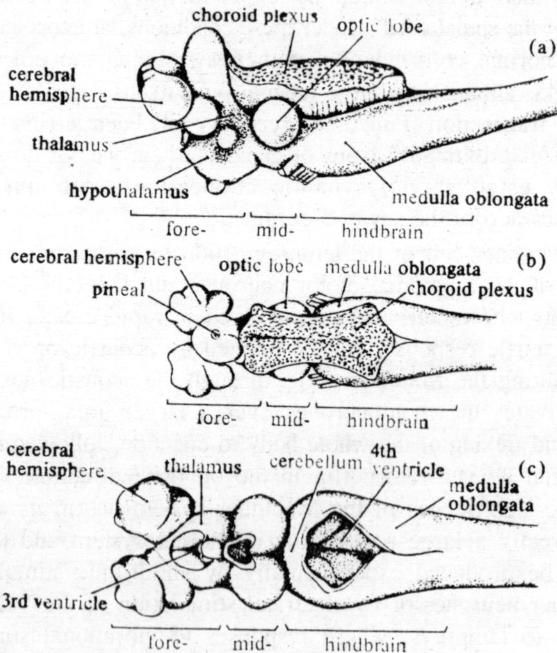

Fig. 12.5. Brain of the lamprey. (a) side view; (b) dorsal view with choroid plexus intact; (c) after removal of choroid.

telencephalon in relation to the diencephalon is affected by the development of the sucker and olfactory organ and in the hagfish, the dorso-ventral flattening of the brain has been related to its development within a rigid shelled egg. In addition, the relative development of various brain regions in the hagfish must reflect the impoverishment of certain sensory systems and its greater reliance on others. Some of its peculiar features, such as the reduction of the ventricular cavities and the absence of choroid plexuses, are secondary developments arising in the course of its embryonic life.

Medulla

According to classical interpretations of the topography of the central nervous system, the walls of the neural tube have been divided into longitudinal functional zones — a dorsal and primarily sensory alar plate area separated by a groove, the sulcus limitans, from a ventral, primarily motor, basal plate area. In further elaborations of this pattern, the alar plate has been further sub-divided into dorsal somatic sensory and ventral viscero-sensory zones and the basal plate into medial somatic motor and lateral viscero-motor areas. In its segmental arrangement of branchial nerves, dorsal roots and spinal nerves, the cyclostome medulla retains much of this primitive segmental pattern of the spinal cord, but superimposed upon it are intersegmental correlating mechanisms provided by the longitudinal functional systems. In his topographical analysis of the brain stem of the lamprey, Nieuwenhuys (1972) finds that the medial somatic motor and lateral

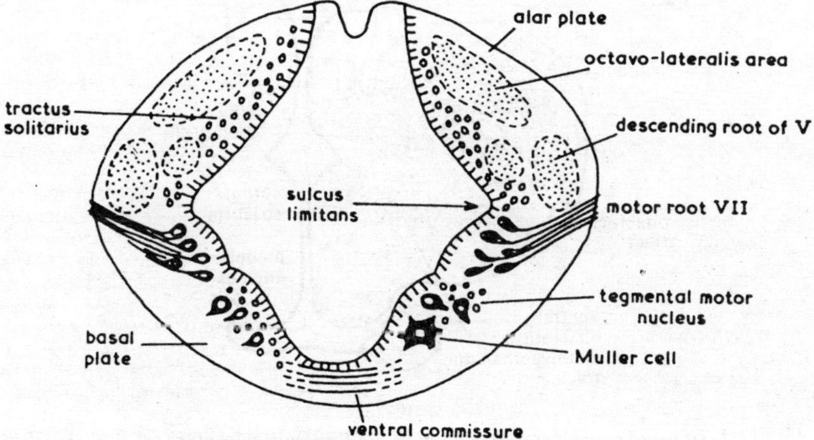

Fig. 12.6. Distribution of longitudinal functional systems in the alar and basal plate areas of an ammocoete medulla.

visceromotor columns are clearly recognizable in the rhombencephalon; the former consisting of caudal motor nuclei belonging to the most anterior spinal nerve roots, together with the three motor co-ordinating centres (tegmental motor nuclei) of the reticular formation. The lateral viscero-motor column consists of the five efferent nuclei of the so-called branchial nerves — cranial nerves V, VII and IX, together with the rostral and caudal nuclei of X. On the other hand, although afferent fibres representing the general and special somatic sensory and visceral

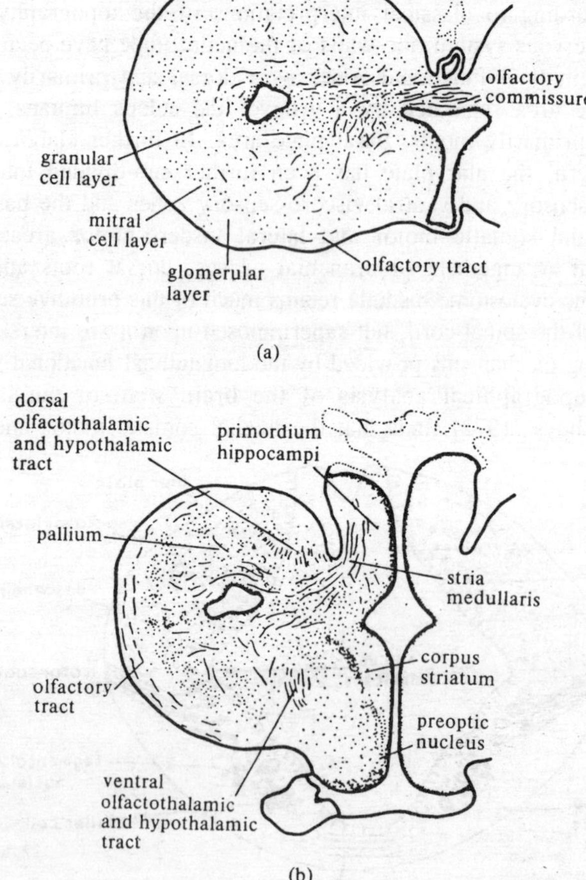

Fig. 12.7. Transverse sections through (a) the olfactory bulb and (b) the cerebral hemispheres of Lampetra fluviatilis.

sensory divisions reached the alar plate area, Nieuwenhuys concluded that the centres where these fibres terminate do not show the typical dorso-ventral pattern of the longitudinal functional zones. As is the case with the dorsal spinal nerve roots, the afferent fibres of the branchial nerves divide into ascending and descending bundles after they enter the medulla.

In the lamprey, the special somatic sensory division is represented by the so-called octavo-lateralis or acoustico-static area, occupying virtually the entire rostral alar plate region. This consists of three longitudinally arranged cell masses; the first two representing the end stations of lateralline nerves and the third those of the acoustic nerve. From these centres arise arcuate fibres which cross to form an ascending tract — the lateral lemniscus — passing to the tectum or cerebellum. In addition, other secondary fibres descending from the vestibular nucleus may join the longitudinal medial bundle to reach the spinal cord. The Mauthner neurones, considered by some to have arisen within the vestibular nucleus, are displaced in a medial direction towards the tegmental nucleus, although their axons descend in the lateral tracts, rather than in the main reticular bundle.

The lateral line system, represented in lampreys by a longitudinal row of neuromasts as well as tracts around the snout and eye regions,

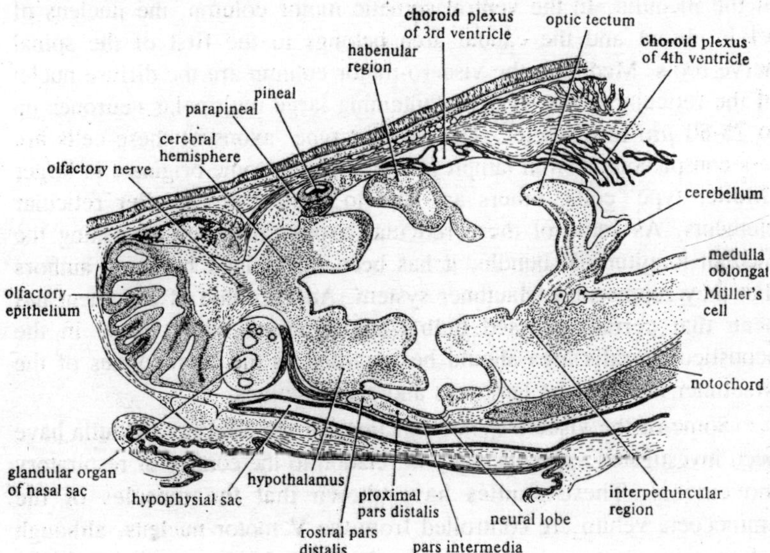

Fig. 12.8. Sagittal section through head of lamprey.

is virtually absent in the hagfish, which must rely to a greater extent on cutaneous touch or chemoreceptors, situated on the head region and especially on the tentacles. Correlated with this reduction of the lateral line system and vestibular organs and its greater dependence on cutaneous receptors, the myxinoid medulla shows some modifications from that of the lamprey. The main afferent pathways from the peripheral nerves enter the anterior projecting horns of the medulla on either side of the mesencephalon and here the dominant component is the general sensory division of V. The enormous size of the general cutaneous system, especially the sensory V, is one of the more conspicuous features of the myxinoid medulla and compared to this, the acoustico-lateralis and visceral divisions are insignificant. Although the primary sensory fibres are said to be arranged in functional systems, their boundaries are not well-defined. The reduced acoustico-lateralis area lies between the trigeminal and visceral afferent areas. The latter is made up of some fibres from VIII, but mainly those from the IX/X complex. Arcuate fibres from the various sensory areas constitute a conspicuous ventral commissure forming an ascending lemniscus, where they are joined by other fibres reaching as far forwards as the tectum of the mesencephalon.

The motor nuclei of the branchial nerves are situated at the sides of the medulla. In the ventral somatic motor column, the nucleus of VI is absent and the caudal area belongs to the first of the spinal nerve roots. Medial to the viscero-motor column are the diffuse nuclei of the reticular motor system containing large multipolar neurones up to 75-80 μm in diameter. The 'Muller type' axons of these cells are less conspicuous than in lampreys and although some originate in larger 'Muller type' cells, others are said to come from smaller reticular elements. As some of these reticular axons cross before joining the median longitudinal bundle, it has been maintained by some authors that they represent a Mauthner system. Another line of argument has been that certain crossed bulbar-spinal axons which arise in the acoustico-lateralis area should be regarded as the homologues of the Mauthner neurones of lampreys and gnathostomes.

Some of the viscero-motor functions of the lamprey medulla have been investigated experimentally in relation to the control of respiratory movements. These studies have shown that the muscles of the ammocoete velum are controlled from the V motor nucleus, although a few velar motoneurones may also be located in the nucleus of VII. In both adult and ammocoete, the muscles of the branchial basket, including those of the gill pouches and ectal valves, are controlled from the motor nuclei, of IX and X. Using isolated preparations of the brain, velum and branchial region of the ammocoete, Homma was able to show that during typical breathing activity, the potentials recorded from the surface of the velum preceded those of the branchial basket by about 117 ms, and were of longer duration — averaging 248 ms against 130 ms. During this respiratory activity, periodic bursts of spike discharges were recorded from the motor nucleus of V and these were synchronized with the discharges from the velum. Similar periodic discharges were also recorded from the motor nucleus of X, but these were of greater amplitude, shorter in duration and with longer latencies than the discharges from the velar musculature. Moreover, these bursts in the X nucleus were always preceded by activity in the velum. Transection of the brain stem behind the nucleus of V did not inhibit its regular discharges, nor the contractions of the velum, but activity did cease in the branchial muscles. Unilateral section of the brain stem stopped branchial contractions on the operated side, but velar movements continued. Homma suggested that pacemaker activity is generated near the V motor nucleus and that this is transmitted in part to the velar motoneurones of the opposite side and in part ipsilaterally to the motor nuclei of IX and X. However, the origin of this pacemaker activity has not been determined and the nucleus itself remains inactive when it is completely isolated from the rest of the brain. During metamorphosis, when the velum loses its respiratory functions, it seems that there must be a dramatic change in the site of the respiratory pacemakers. With the transition from the unidirectional respiratory current of the ammocoete to the tidal pumping system of the adult lamprey, respiratory functions are apparently transferred to the X motor nucleus. Thus, in isolated brain-gill preparations of adult lampreys, Rovainen observed powerful periodic bursts of EPSP's in the X motor nucleus, but similar discharges were not seen in the motor nucleus of V. These potentials occurred immediately before each branchial contraction, but their origin within the brain has not been established.

Velar motoneurones are the smallest motoneurones recorded in lampreys (5-10 μm) with the lowest conduction velocities (0.2 m s^{-1}), producing EPSP's with one-to-one contractions in the longitudinal or oblique velar muscles. Motoneurones to the branchial basket of the ammocoete are also small cells (less than 10 μm in diameter), but with somewhat higher conduction velocities than the velar motoneurones

(0.6 m s[1]). Most of the identified branchial motoneurones of the adult lamprey were located in the rostral part of the X motor nucleus. These were about the same size as the myotomal and fin motoneurones with similar input resistances and time constants. However, the much lower conduction velocities of these branchial motoneurones suggests that their axons must have much smaller diameters.

Cerebellum

The vertebrate cerebellum is a correlation centre developed from the dorsal sensory columns of the medulla and usually consisting of a median corpus cerebelli and laterally evaginated auricles. On one interpretation, it has evolved from a fusion or bridge between the acoustico-lateralis areas on either side, but an alternative view regards the corpus cerebelli as derived from the general somatic sensory area of the trigeminal.

The cerebellum of lampreys is a simple transverse plate, continuous laterally with the acoustico-lateralis area of the medulla, and at least one of the nuclei of the lateral line nerves extends into the grey matter of the cerebellum. The main inputs consist of primary fibres from the lateral line, and nerves V and VIII, as well as some secondary fibres from the last two systems. Other afferents come from the lateral columns of the spinal cord (spinocerebellar tracts) passing through the medulla, from the optic tectum of the midbrain (tecto-cerebellar) and from the hypothalamus (lobo-cerebellar). The output from the cerebellum arises from a larger cell type which has been regarded as a precursor of the Purkinje cells of higher vertebrates. Efferent fibres pass to the optic tectum (cerebello-tectal), to the tegmental motor nuclei of the

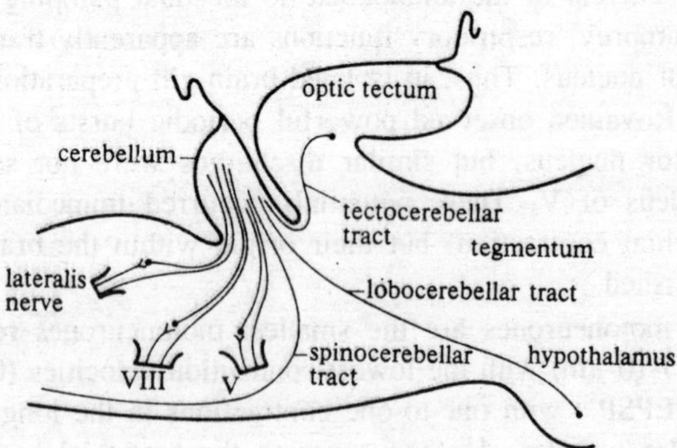

Fig. 12.9. Diagram of the afferent connections of the lamprey cerebellum, projected on a sagittal plane.

midbrain, to the motor nuclei of the medulla and the nucleus of the oculomotor (III). Others join cerebello-spinal tracts descending through the medulla or join the medial longitudinal bundle of the reticular system.

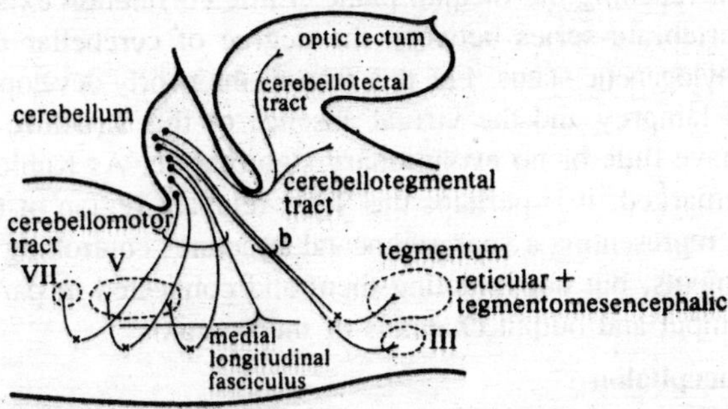

Fig. 12.10. Diagram of the efferent connections of the lamprey cerebellum, projected on a sagittal plane.

In vertebrates reflex movements co-ordinate the movements of the eyes in association with the labyrinth, allowing the animal to stabilize an object in its field of vision in spite of movements of the head. These vestibulo-ocular reflexes have already been developed by the lampreys, producing opposing eye movements and turning of the body itself, when the animals are rotated. Electrophysiological studies have shown that these reflex movements can be produced, both by direct mechanical stimulation of the labyrinth or by electrical stimulation of either the anterior or posterior branches of the vestibular nerve. For example, mechanical stimulation of the anterior part of the labyrinth caused a rotation of the ipsilateral eye in a dorso-caudal direction, accompanied by a ventro-rostral rotation of the opposite eyeball. Mechanical stimulation of the posterior labyrinth caused a similar dorsal and ventral rotation of the ipsilateral and contralateral eyes, but in this case, the former moved rostrally and the latter caudally. Unlike other vertebrates, fast recovery movements of nystagmus were not seen. Although the pathways for these movements are not known, Rovainen believes that they represent direct and specific connections between the individual ampullae and the various oculomotor muscles. The interneurones are thought to be certain large cells in the vestibular nucleus which are intimately related to the axons of the vestibular nerve and which have projections towards the oculomotor and other more posterior motor nuclei.

The existence of a cerebellar rudiment in myxinoids has been the subject of widely divergent views. It is possible that this region may be represented in *Myxine* by the posterior tectal commissure and, in *Eptatretus*, a small acoustico-lateralis commissure has been described, containing cell strands continuous with the acoustico-lateralis area, but not reaching the median plane. Little correlation exists throughout the vertebrate series between the degree of cerebellar differentiation and phylogenetic status. For this reason, the poorly developed cerebellum of the lamprey and the virtual absence of this structure in hagfishes, may have little or no evolutionary significance. As Kuhlenbeck (1975) has remarked, it is perhaps the 'least relevant region of the vertebrate brain, representing a suprasegmental structure, controlling or smoothing movements, but not initiating them and connected in parallel with the main input and output channels of the neuraxis'.

Mesencephalon

In the lower vertebrates this region has been regarded as the dominant directing centre of the brain; the supra-segmental tectal cortex being a centre where sensory inputs from wide areas including the optic system are processed, while the bulbo-spinal and reticulospinal tracts provide the efferent channels through which this processed information can be transmitted through the spinal cord. The relative importance of the visual system is shown by the fact that the tectum tends to be largest in those fishes in which visual cues are most important in courtship or hunting and smaller in nocturnal or abyssal forms or in genera that make more use of other senses in feeding or

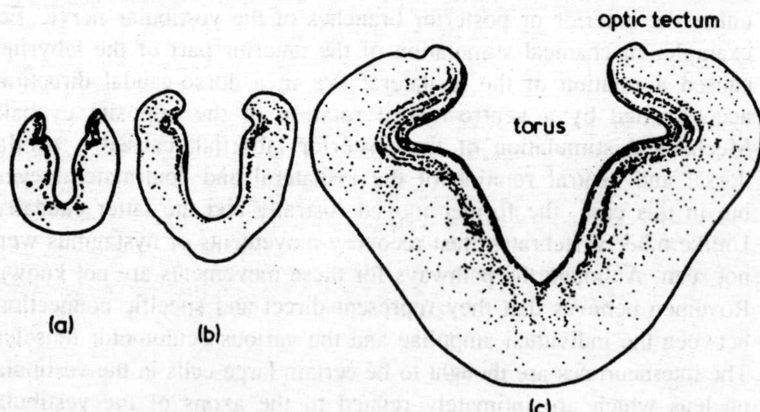

Fig. 12.11. Successive stages in the growth and differentiation of the optic tectum of the sea lamprey, Petromyzon marinus during metamorphosis. (a) 70 mm ammocoete. (b) 105 mm ammocoete. (c) Adult lamprey after the completion of transformation.

territorial behaviour. Its significance is also illustrated by the four-fold growth of the optic tectum during the metamorphosis of the lamprey accompanying the functional development of the functional eyes.

In its possession of a choroid plexus, the lamprey mesencephalon is unique amongst vertebrates and it is interesting to recall that a similar condition is thought to have existed in the Heterostraci. The posterior boundary of the mid-brain is marked by the posterior commissure and in the pretectal region there are paired differentiations of the ependymal roof — the sub-commissural organ. Laterally and caudally, the tectum takes the form of paired structures uniting behind the choroid plexus and projecting backwards over the cerebellum. In hagfishes, the mesencephalon is compressed into a wedge-shaped structure, partially divided into two halves by a dorsal groove and with its lateral sides covered by the lateral horns of the medulla.

In its general architecture, the mid-brain still retains the basic topography of the rhombencephalon, divisible into a ventral tegmental motor area, regarded as an extension of the rhombencephalic basal plate, and a dorsal sensory correlation area, representing a continuation of the alar plate. In the lamprey, the motor tegmentum of the mid-brain contains the nucleus of III and the cranial extension of the reticular system with its mesencephalic Muller cells. That part of the mid-brain dorsal to the motor tegmentum contains the two large somatic sensory correlation centres — the semicircular torus and the optic tectum. The tectal cortex consists of 8-10 layers of neurones with their dendrites directed towards the surface and overlying the fibrous layer. Its chief input are the optic fibres, but it also receives general somatic afferents, passing through the bulbar lemniscus and the torus from the spinal cord and other parts of the mid-brain. The main efferent pathways lead towards the medulla, the habenulae, thalamus and hypothalamus. The torus, situated between optic tectum and motor tegmentum, receives its main afferents from the acoustico-lateralis area.

Because of the rudimentary condition of the eyes and optic nerves, the tectum of myxinoids cannot function as an optic correlation centre. This region in *Myxine* shows ill-defined cell groups not clearly separated from the semicircular torus. The input to these cells is assumed to come from ascending fibres in the lemniscus and descending fibres from the fore-brain. Rostrally, the cells of the tectum cannot be separated from those of the posterior thalamus. Among the cell masses in the basal plate, the nuclei of the cranial nerves supplying the non-existent eye muscles are lacking and as in lampreys, the mesencephalic nucleus of the gnathostome trigeminal is also absent. This can be related to an absence of jaw muscles, to which in higher vertebrates these nerves supply proprioceptive fibres. The main efferent tract is the medial longitudinal bundle arising from the cells of the mesencephalic reticular nuclei. Further efferent fibres are carried in tecto-bulbar and tectospinal tracts.

In the lamprey, Schwab (1973) considers the caudal thalamus, pretectum and tectum to be a common functional system, with similar cytoarchitectural differentiation and relationships and above all with a common dominant efferent pathway — the optic tract. This view is supported by the simultaneous development of these areas at the time when the paired eyes are developing during metamorphosis. Thus, in young ammocoetes the two halves of the tectum are still entirely separate and even in quite large animals their fusion is still incomplete. Also, in young ammocoetes the caudal thalamus, pretectum and tectum

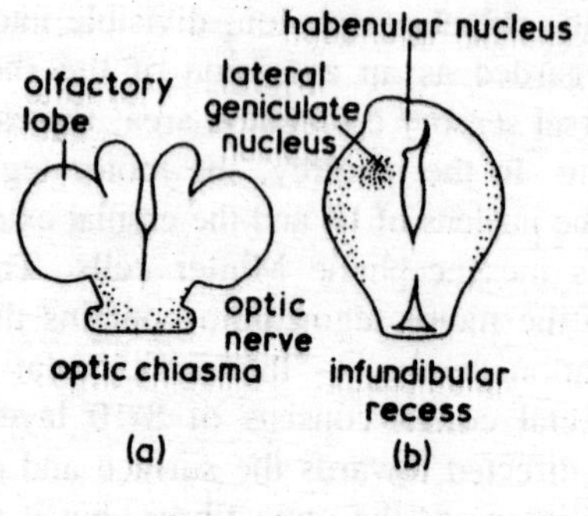

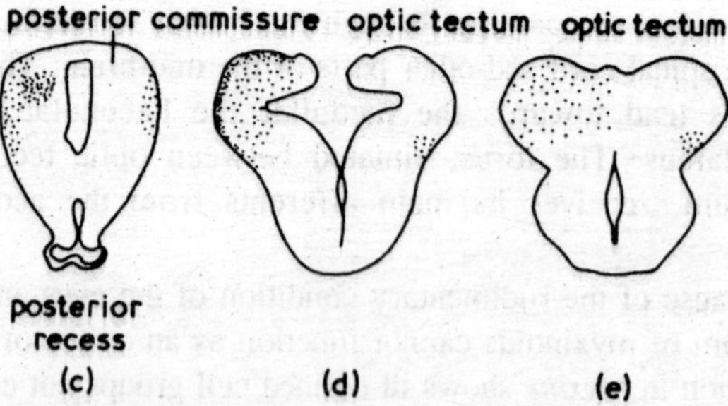

Fig. 12.12. Projections of retinal ganglion cells in the thalamus and optic tectum of the sea lamprey, P. marinus after section of the optic nerve and removal of the eye on one side, as seen in transverse sections of the brain at various levels. (a) Diencephalon at the level of the optic chiasma. (b) Diencephalon at the mid-thalamic level. (c) At the level of the posterior commissure. (d) At mid-tectal level. (e) Caudal tectal level in the region of the ventral commissure.

are continuous and without cellular differentiation. It is only when the optic tract is fully developed that these areas assume their definitive characteristics. Further support for these views has come from observations on the degeneration of the axons of retinal ganglion cells following section of the optic nerve and unilateral removal of the eye. The projections of these retinal axons have been identified contralaterally, in the posterior third of the dorsal thalamus, the pretectum and tectum. Especially in the adult, these axons in the dorsal thalamus have been observed within the area of the lateral geniculate nucleus. Similar results have been obtained by the use of the horseradish peroxidase technique, which has resulted in labelling of optic fibres in the tectum and geniculate nucleus as well as retrograde transport in cells of the dorsal and central tegmentum. In the optic tectum itself, the pattern of the optic nerve terminals is said to be typical of other non-mammalian vertebrates, in so far as they are distributed contralaterally and superficially, lying outside a central dense fibre zone. The tectum shows a distinct lamination of the medial and denser cell region, which becomes much more distinct at metamorphosis, when there is a massive enlargement of the whole area of the brain, due partly to cellular proliferation, and partly to the migrations of cells from the periventricular zone. Degenerating axons in the adult lamprey were located in the eighth and more especially the ninth and most superficial layer, containing the optic tract fibres. At metamorphosis, a small and additional ipsilateral projection of optic fibres appears in an area at the ventrolateral margins of the pretectum and tectum, which in its location is said to have no counterpart in other vertebrates.

The visual system of the lamprey has been investigated physiologically by stimulating the retina with light flashes of varying frequency or by direct electrical stimulation of the optic nerve. Light flashes have resulted in responses, not only in the optic tectum, but more widely in the medulla and spinal cord. The fast oscillations recorded in the medulla still occur after removal of the tectum, but are abolished by incisions made between mid-brain and medulla. For these reasons, responses in the medulla and spinal cord are thought to be mediated through the tegmental motor nucleus of the mid-brain and the system of Muller neurones, thus implying the existence of visual afferents that by-pass the tectum. This view is strengthened by recordings from electrodes deep in the mesencephalon, in the area of the tegmental nucleus, which showed potentials similar to those recorded from the tectum and with a similar latency. Medullary potentials have a latent period some 10-20 μs longer than those of the tectum and one or two of the later waves may be absent, suggesting the presence of an extra relay.

Responses in the tectum to light flashes show a positive wave of up to 200 μV with a duration of 60-80 μs and a latent period of 60-80 μs. This is followed immediately by a short negative wave and a series of variable oscillations. A number of observations point to a different origin of these components. With increased frequency of flash stimulation, the amplitude of the first wave in the tectal response decreases, whereas the amplitude of succeeding waves tended to increase but with decreased latency. More frequent stimulation thus selectively shortens the first wave. Anaesthetization with nembutal abolished all but the first wave and similarly, after interrupting the water supply irrigating the preparation, all components except the persistent first wave showed a decrease in amplitude and increased latency. Correlation with simultaneous recordings of activity in the optic nerve during light stimulation suggested that the primary tectal wave was due to the discharge of 'on units' in the retina and that the subsequent components represented the discharge of 'off units'. Direct electrical stimulation of the optic nerve results in a fast component interpreted as the presynaptic potentials of retinal ganglion cells, followed by a slow negative or positive-negative wave, thought to represent dendritic potentials within the radially arranged neurones of the tectal cortex.

These experimental observations on the retino-tectal visual system of the lampreys illustrate the primitively diffuse organization of its nervous system, in which responses are spread over wide areas of brain and spinal cord. Although earlier work had suggested that these

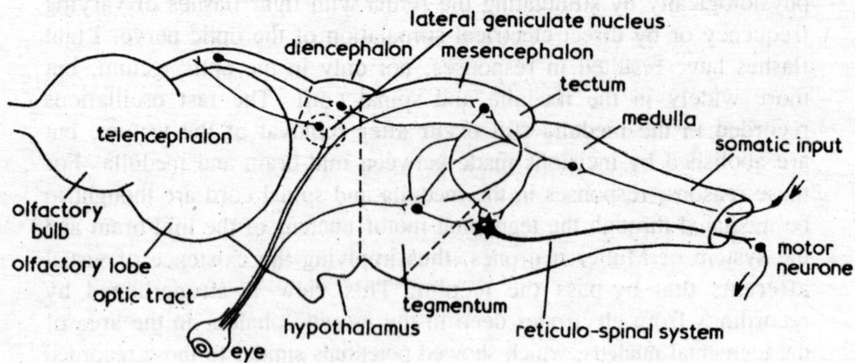

Fig. 12.13. The visual system of the lamprey and the integrative role of the mesencephalon.

responses did not extend to the telencephalon, more recent studies have indicated that a retino-thalamictelencephalic visual system may exist in lampreys and other lower vertebrates, in which the responses to photic stimulation of the retina have much longer latent periods than those of the tectum. No details are known of these pathways, but it has been suggested that, as in the elasmobranchs, this system may be independent of the tectum.

The relative importance of the cyclostome mesencephalon has been emphasized by Karamyan (1975) in a survey of the reflex mechanisms of the vertebrate brain. This author distinguishes five successive phylogenetic stages in the evolution of the integrative activity of the CNS, beginning with the purely spinal level of the cephalochordates and reaching a climax in the neocerebellar-neothalamic-neocortical stage attained by the mammals. In this scheme, the lamprey represents a second stage, in which the dominant role in the establishment of reflex linkages is provided by the mesencephalic-bulbar system, where distance receptor afferents are switched to motor neurones through a single system of links — the reticulo-spinal system. This may be contrasted with conditions in fishes, where the well-developed cerebellum shares with the mid-brain the main burden of integration.

Diencephalon

Comprising those parts of the brain surrounding the third ventricle, the diencephalon may be divided into a dorsal epithalamus with a thin ependymal roof, the dorsal and ventral thalami forming the walls, and a ventral hypothalamus. Apart from the epiphysis which is evaginated from the roof of the lamprey diencephalon, the most conspicuous features of the epithalamus are the habenulae, whose relatively large size reflect the dominance of the olfactory system. The main input to the habenulae are olfactory fibres of second and third order, reaching it in widely dispersed olfacto-habenular tracts. The main efferent channel is the conspicuous fasciculus retroflexus or Meynert's bundle, which appears to end around the interpeduncular nucleus, although some fibres may pass to the ventral thalamus and the posterior tubercular nucleus.

Developing from the epiphysis in the lamprey are the light-sensitive pineal and para pineal organs. An epiphysis, absent altogether in *Myxine*, is present in the embryo of *Eptatretus*, whereas in the adult it is represented only by a rudiment, often containing a small ventricular recess. In lampreys of the genera *Lampetra* and *Petromyzon*, the smaller parapineal lies below the pineal, whereas in the adult *Geotria* it is located somewhat forwards and slightly to the left.

However, this appears to be only a secondary condition and in the ammocoete of *Geotria*, the parapineal has a position very similar to that of the lampreys of the Northern Hemisphere. Above the pineal complex is an area free of myotomal muscles and covered by a translucent patch of unpigmented skin.

The pineal consists of a hollow vesicle connected to the pineal stalk. The vesicle has a dorsal pellucida, (which, in *Geotria*, has a lens-like shape) and ventrally a retina, consisting of an inner pigmented epithelial layer containing sensory cells and an outer layer of ganglion cells and nerve fibres. The comparatively rare photoreceptors are said to be of the cone type, with a few discs like those that support the photopigments of other retinal receptors. The ganglion cells form synaptic connections and have been implicated in the transmission of photic stimuli. Thus, although the structure of the pineal suggests that it retains some photoreceptive capabilities, it gives the impression of being an organ that is undergoing functional regression. In the smaller parapineal, pellucida and retina are much less distinctly differentiated and the lower wall may contain only a few poorly developed sensory elements. In *Lampetra*, the pineal stalk contains about 600 nerve fibres forming the pineal nerve and connected directly to the right habenular ganglion and posterior commissure and in *Geotria* at least, some branches are said to join the bundle of Meynert. The parapineal is connected to the left habenular ganglion through an adjoining ganglionic mass, some of whose neural elements may have been derived by rostral migration from the habenular ganglion itself. In the Southern Hemisphere genus *Mordacia* the parapineal is absent, but attached to the front of the pineal vesicle is a small structure — the pineal appendix. This contains a few ganglion cells and nerve fibres and has sometimes been regarded as a vestigial parapineal, although this possibility has been questioned by Eddy and Strahan, mainly on the grounds that the appendix has no independent connection with the left habenular ganglion. The 'paired' nature of the relationship between the two elements of the pineal complex and the right and left habenular ganglia has naturally given rise to the belief that the parapineal and pineal have been derived from an ancestral form in which these structures were paired and probably symmetrical. On the other hand, it has been pointed out that if such a condition ever existed in the vertebrate lineage, it must have predated the known fossil agnathans, none of which shows paired epiphyseal foramina.

The condition of these structures in the lampreys of the Northern and Southern Hemispheres raises interesting problems of phylogeny.

From the point of view of the histological differentiation of the pineal and parapineal and their photosensory functions, Eddy and Strahan suggest that *Geotria* represents the most developed form, *Mordacia* the least and that *Lampetra* (and presumably other Northern genera) would be intermediate in these respects. The complete (or almost complete loss) of the parapineal in *Mordacia* can hardly be other than a secondary regression, presumably indicating a long separation between this genus and the rest of the lamprey stock, but relative to *Lampetra* should we regard the more developed pineal of *Geotria* as a secondary improvement or as a more primitive state? However, the first of these alternatives would seem to imply a reversal of a previous evolutionary trend, involving the renewed elaboration of an organ that had already undergone considerable evolutionary regression.

Following illumination of the isolated pineal of *L. fluviatilis*, slow electrical discharges have been observed, as well as the impulse discharges of ganglion cells and nerve fibres. Weak light stimuli cause a slowly rising positive response and a smaller deflection after the light is switched off. Stronger stimulation produces a faster positive deflection preceding the slow wave. The dark adapted threshold intensity was about 10^{-3} lm m^{-2} and above this threshold spontaneously active units were inhibited by light pulses of all wavelengths. During exposure to constant light of greater intensity, these units respond with off responses. The pineal is therefore described as a photoreceptor adapted to functioning at low light intensities and under these conditions, light signals of all wavelengths result in the inhibition of nervous discharges. Following pineal illumination, a slow negative wave appears in the optic tectum with a long latent period of about 140-150 ms, but nothing is known of the precise pathways that may be involved.

In addition to its functions as a photoreceptor, it is now clear that the pineal is also a secretory organ, producing substances, which among other functions less well understood, are involved in the contraction of the melanophores and the diurnal changes in the skin colour of the ammocoete. These rhythms are abolished by pinealectomy, but the normal skin pallor in the dark can be elicited by administering pineal extracts or melatonin. A pineal enzyme, hydroxyindole-O-methyltransferase (HIOMT) converts serotonin to melatonin and the activity of this enzyme has been shown to reach a peak within 4 hours after the onset of the dark phase in *Geotria* ammocoetes. This short burst of melatonin produced melanophore contraction, but skin pallor did not persist over the whole of the dark phase and in fact skin

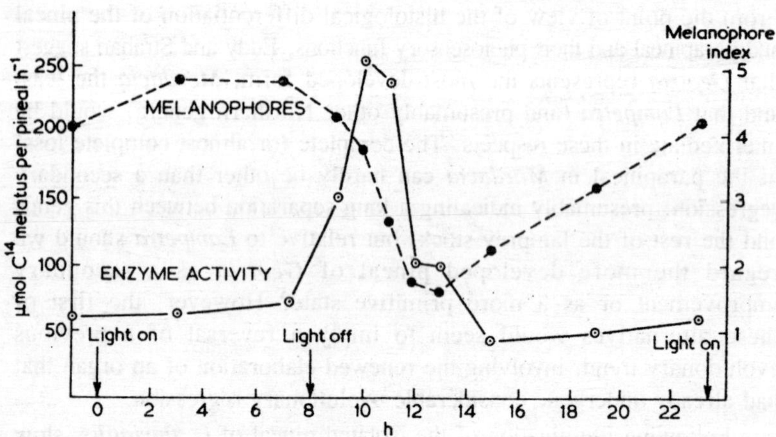

Fig. 12.14. Surge of hydroxyindole-O-methyl transferase activity in the pineal of the ammocoete of Geotria during a 24 hour cycle, correlated with changes in the contraction of the melanophores.

darkening began before the onset of the next light period. Since the melatonin surge was so short-lived, it is suggested that its wider function may be the timing of the circadian activity rhythms; the pineal acting as a biological clock or neuroendocrine transducer converting diurnal changes in light intensity (or even perhaps seasonal changes in day length) into hormonal signals with far-reaching effects on the animal's metabolism. This might help to explain reports of the inhibition of metamorphosis in pinealectomised ammocoetes, since this is an event that is highly seasonal in its incidence. Around the time of hatching and before the paired eyes are functional, the pineal of *Xenopus* tadpoles has been implicated in the reflex initiation of swimming activity; a response that can be produced by shading the pineal. Similar responses could be invaluable in the 'blind' ammocoete stage and might have been developed by the benthic ancestral agnathans as a protection against predators.

The asymmetry of the habenular ganglia, in which the larger right member lies in front of the left, has been related to the connection of the former to the more developed pineal and the two ganglia are said to be more equal in size in earlier ontogenetic stages. On the other hand, the same kind of asymmetry also occurs in the myxinoids, where it has been attributed to pressure exerted by the growth of the olfactory lobes. In this connection it may be recalled that in their interpretation of the heterostracan brain, Whiting and Tarlo (1965) have shown a symmetrical epithalamus, with equally sized habenulae.

Behind this region, the medial part of the dorsal diencephalon of the lamprey is assumed to have important optic functions, including the auxilliary optic tract to the hippocampal region. In this area, the posterior commissure connects the brain walls of the two sides, containing fibres ending in the pretectal region or running more ventrally towards the hypothalamus or the mesencephalic tegmentum. This dorsal thalamus also contains a dorso-medial cell group of bipolar neurones—the lateral geniculate nucleus, which receives the central projections of some axons of retinal ganglion cells.

With its reduced eyes, the myxinoid diencephalon is presumably dominated by olfactory afferents. Within the thalamus, the dorsal region is an important afferent area, receiving ascending fibres from the bulbar lemniscus and the tectum. In ventral areas there are important efferent centres with connections to the tegmentum via the post-optic commissure and with bulbar regions. In the hypothalamus, the post-optic nucleus receives important olfactory tracts and has connections with the tegmentum and medulla.

Telencephalon

The cyclostome telencephalon may be regarded as primarily, but not exclusively a rhinencephalon or nose brain, dominated by olfactory inputs through first order olfactory fibres entering the olfactory bulbs and by second or third order fibres to the olfactory lobes. As Andres (1975) has emphasized, there are striking parallels in the synaptic connections and neuronal circuitry of the olfactory bulb throughout the

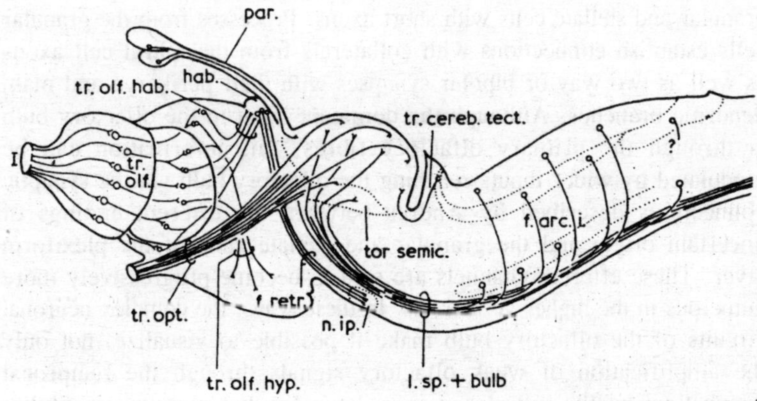

Fig. 12.15. Primary, secondary and some higher order olfactory connections and the principal afferents to the tectum mesencephali of the lamprey.

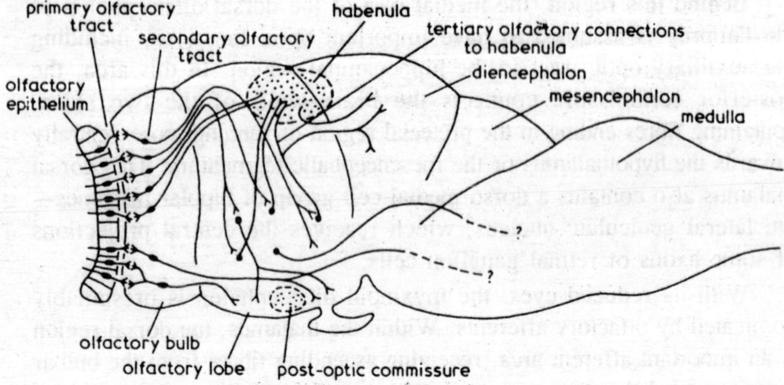

Fig. 12.16. Olfactory tracts of the hagfish.

vertebrate series from the cyclostomes to the mammals, although in the higher vertebrates there is an increasing trend towards greater complexity of its layered structure and the differentiation of its cellular elements.

Primary fibres from the olfactory epithelium enter the olfactory bulb, forming a superficial stratum nervosum. This olfactory information is conveyed to the dendrites of the mitral cells in what is termed the glomerular layer. Both the glomeruli and the mitral cells are said to be better differentiated in hagfishes than in lampreys. The mitral cell axons constitute the main, if not the sole, olfactory output from the olfactory bulb, forming the secondary olfactory tract to the olfactory lobes. Below the mitral cell layer is a plexiform zone containing granular and stellate cells with short axons. Processes from the granular cells establish connections with collaterals from the mitral cell axons as well as two way or bipolar synapses with their perikarya and main dendritic branches. Although the dominant input to the olfactory bulb is through the primary olfactory fibres, this information can be modulated by wider inputs reaching the olfactory bulb via the synaptic connections described by Andres between the efferent endings of uncertain origin and the granular and stellate cells of the plexiform layer. These efferent channels are said to become progressively more numerous in the higher vertebrates. In these ways, the complex neuronal circuits of the olfactory bulb make it possible to visualize, not only the amplification of weak olfactory signals through the reciprocal connections of the mitral and granular cells, but also the possibility for the damping down of strong stimuli through the activity of efferent impulses from the higher centres.

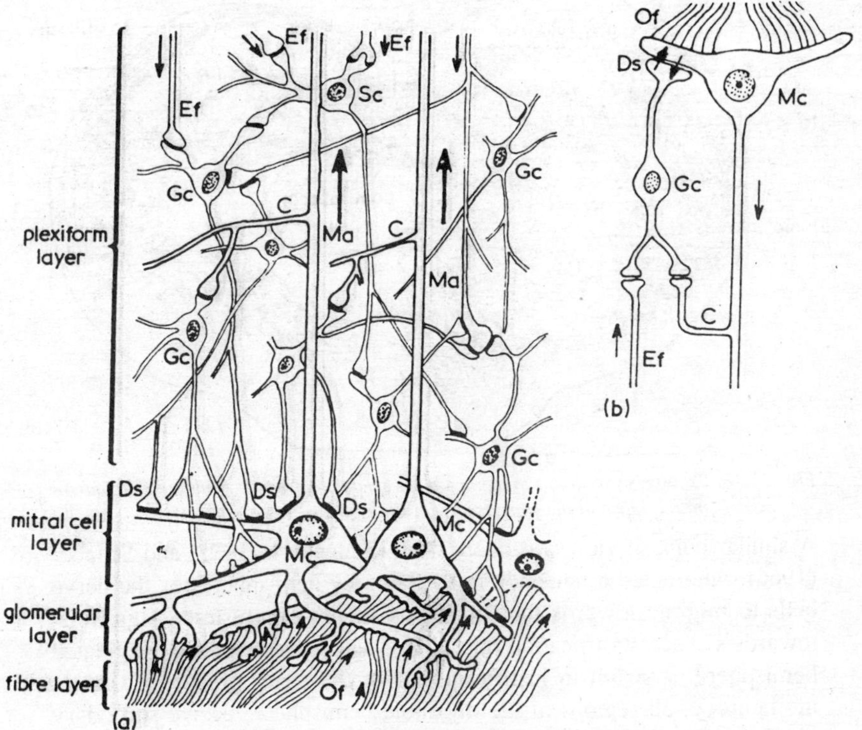

Fig. 12.17. Neuronal relationships in the olfactory bulb of a lamprey. (a) Section through the olfactory bulb. (b) Synaptic connections within a glomerulus.

Although there have been widely varying interpretations of the cellular areas of the lamprey hemispheres, Nieuwenhuys (1967b) identified only a primordium hippocampi, pallium, corpus striatum and pre-optic nucleus. Because it receives numerous secondary olfactory fibres, the evaginated pallial region is often referred to as the olfactory lobe. Some of these fibres decussate in the olfactory commissure and pass to the primordium hippocampi; others pass more deeply to the striatal and pre-optic regions. In addition to these secondary olfactory tracts ending in the telencephalon, others including mitral cell axons, are believed to reach the thalami and hypothalamus, or even as far as the tegmentum of the mid-brain.

In the solid hemispheres of the hagfish, Nieuwenhuys distinguished a basal area or striatal primordium, a hippocampal primordium and a pallial region; the latter showing a very distinct five-layered structure in which the second layer was considered to be a true pallial cortex.

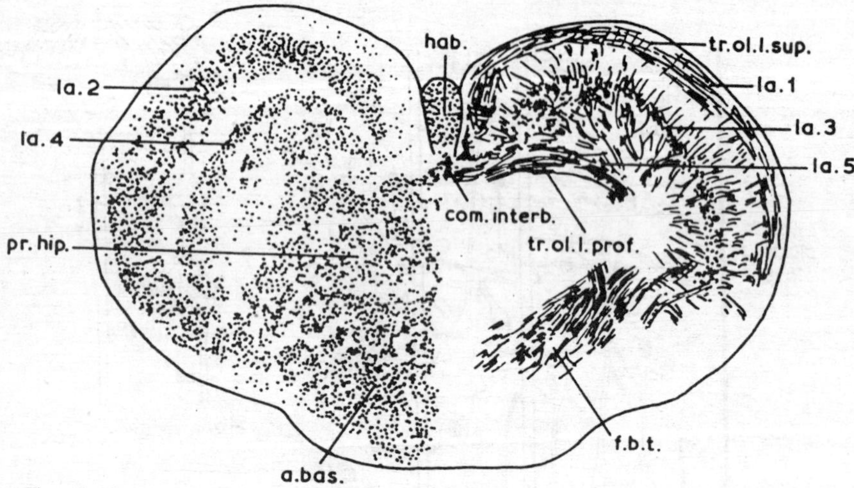

Fig. 12.18. Transverse sections through the telencephalon of Myxine, with the cell patterns on the left and the fibre tracts on the right.

A similar point of view was adopted by Kuhlenbeck (1975) and Schober (1966), who noted a tendency in the lamprey hemisphere for the nerve cells to migrate towards the surface as small cell clusters. This trend towards surface stratification outside the white matter of the evaginated hemisphere, was felt to justify the use of the term 'precortex'. As in the lamprey, all regions of the myxinoid hemisphere receive secondary olfactory fibres, carried in large lateral and smaller medial and ventral tracts. The lateral tracts have superficial and deep divisions corresponding to the first and fifth layers of the pallium and, from these, connections are established with cells in the second and fourth layers. Others pass to the opposite side in a commissure located immediately behind the habenulae. Fibres from the ventral tracts descend towards the pre-optic region, but a majority run in a basal forebrain bundle towards the hypothalamus. As in lampreys, there are also said to be olfactory connections with the tegmentum of the mid-brain. The absence of lateral ventricles in the hagfish forebrain is a secondary condition, arising during embryonic development by the invagination of the hippocampal primordia and their fusion with the rudiments of the olfactory bulbs.

The hippocampal primordium of the lamprey is the dorso-medial and unevaginated part of the telencephalon. This area is said to retain the more primitive architecture of the CNS with nerve cells confined to a periventricular layer, which in this region lacks a typical ependyma.

Lateral to this layer is a fibre zone. The main afferent channels of this area are in the stria medullaris, crossing from the dorso-lateral telencephalon. In addition, there are connections with the hypothalamus and possibly with the mesencephalic tegmentum, while at the caudal end there are further links with the caudal thalamus and the habenulae. Schwab (1973) believes that this region has primary olfactory functions and forms an integral part of an olfactory system, embracing also the olfactory bulbs and habenulae: A quite different view has been adopted by Bone (1963) in relation to the hippocampal primordium of *Myxine*. This he considered to be relatively free of the olfactory dominance that characterizes other regions of the telencephalon, although it was thought to receive a few third order olfactory afferents from the olfactory lobes. For this reason, Bone suggests that it has retained a relatively primitive and undifferentiated condition, which in the phylogeny of the vertebrate brain would have permitted its further development as an associative cortex.

In both groups of cyclostomes, as in fishes, the immediate electrical response of the olfactory epithelium to irrigation by water containing odoriferous stimulants is a monophasic action potential, consisting of a fast 'on response' followed by a slow exponential decline and a return to baseline levels when the stimulus is removed. In lampreys, these responses have been observed to amines or crushed fish tissues and in *Myxine* to various amino-acids including L-glutamine and L-alanine.

Electrical activity in various parts of the olfactory system of the lamprey has been monitored after stimulation of the olfactory nerve fibres. The potentials observed are clearly very similar to those recorded in other vertebrates, emphasizing the basic conservatism in the structural organization of the olfactory system. Recordings in the olfactory bulb showed four, predominantly negative components. The first of these, smaller than the others, was at its largest on the surface of the bulb and in the neighbourhood of the fibres of the olfactory nerve and has been interpreted as the presynaptic potential of the olfactory fibres. Estimates of the velocity of conduction in these fibres, based on their average length and on the observed latency of about 10 ms gave a figure of 0.1-0.15 m s^{-1} which is similar to values reported for other vertebrate olfactory nerve fibres. The second component is considered to be the potential of the secondary neurones in the olfactory pathway, i.e. the mitral cells, and at the depth of the bulb where these cell bodies occur, it is replaced by the earliest signs of spike activity. The third component is believed to represent the activation of periglomeruiar granular cells through two-way synapses connecting these cells with the mitral cells and also via recurrent mitral cell collaterals. The final fourth and slow potential, which is thought to originate in the nucleus of the olfactory bulb, has a latency of about 60 ms. It reaches its greatest amplitude in the centre of the olfactory bulb and shows indications of spike activity. A characteristic feature of the second and third components is the reversal in their polarity that occurs when the electrode reaches the level of the glomerular layer and it is significant that the sequence of these changes in polarity is reversed, according to whether the electrode is penetrating this layer from above or below.

Recordings from the olfactory lobe (hemisphere) show a negative wave at all depths with a latency of 60-80 ms and a maximum amplitude at the centre of the lobe, decreasing towards both dorsal and ventral surfaces. Recording from various areas gave no physiological evidence of a topographical differentiation of the olfactory lobe. After olfactory nerve stimulation, responses have also been observed in the hippocampal region. These consisted of a first rapid and second slow negative component with a latency of 70-90 ms and attributed to the activation of hippocampal elements by mitral cell axons.

Opposing the view that the telencephalon of cyclostomes is solely concerned with olfactory afferents and that the neocortex of higher vertebrates has evolved from the 'olfactory brain' of primitive forms, Karamyan *et al.* (1975) have adduced experimental evidence that both visual and somatic afferent systems are represented in the lamprey telencephalon and that stimulation of the spinal cord evokes excitatory potentials in this region of the brain as well as in the diencephalon and brain stem. Similarly, stimulation of the optic nerve or illumination of the retina also evokes telencephalic responses which these authors believe to be the result of the presence of visual afferents in the hippocampal primordium. A possible anatomical basis for this contention may be the small decussation of optic fibres described by Schwab (1973) immediately dorsal to the optic chiasma and which he was able to trace to their endings in the hippocampal region.

General Characteristics of the Cyclostome Central Nervous System

Anatomists and physiologists alike have been impressed by the apparently diffuse nature of the connections within the cyclostome CNS. In his morphological analysis of the myxinoid brain, Jansen (1930) remarked on the wealth of diffuse connections that exist in this structure, in addition to better organized and more distinct tracts. This condition,

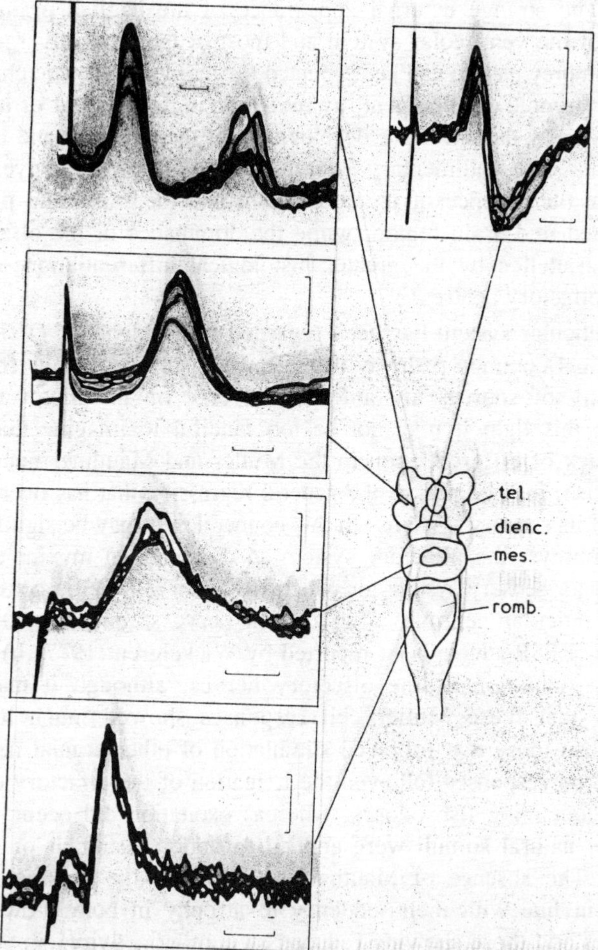

Fig. 12.19. Evoked potentials in the lamprey brain following stimulation of the spinal cord.

which he compared to the nerve nets of invertebrates, was attributed to the existence of long, branching dendritic processes and to the large numbers of collaterals that are given off along the course of the axons. A similar point was also made by Bruckmoser (1971), who stressed the comparatively undifferentiated and diffuse character of the lamprey olfactory system, as demonstrated by the uniform activation that occurs throughout the telencephalon after olfactory nerve stimulation.

The obvious similarities in the general organization of the CNS of myxinoids and lampreys, as well as the unique shape of the adult spinal cord, are sufficient reminders of a common, though remote ancestry. The strange external appearance of the hagfish brain, the reduction of the ventricular system and most of the other divergences from the lamprey pattern can be accounted for by morphogenetic changes in early embryonic development, by reduction in the eye, ear or lateral line organs or by the excessive development of the olfactory and tactile senses. Thus, the rudimentary lateral line and degenerate eyes are reflected in the absence of the cerebellum and the small and poorly differentiated mid-brain region, while the dominance of the olfactory sense is parallelled by the greater histological differentiation of the myxinoid olfactory centres.

The reticular system has been regarded by Kuhlenbeck (1975) as part of a final common pathway through which sensory inputs from a wide variety of sources are able to converge on the motor units. Viewed in this light it may not be too fanciful to imagine that the apparent lack of differentiation in the Muller and Mauthner neurones of the hagfish, may be part of the general regression that has overtaken so many of its sensory systems. In this connection it may be significant that in lampreys, the olfactory system so dominant in myxinoids, is the only major sensory system for which Rovainen (1967) did not detect effects in reticular neurones after cranial nerve stimulation. On the other hand, EPSP's have been reported by Wickelgren (1977a,b) after electrical stimulation of the olfactory nerves, although it may be significant that these Muller cell responses showed much longer latencies than those that followed stimulation of other cranial nerves. Moreover, no responses followed the irrigation of the olfactory organ by water containing fish odours, whereas excitation did occur when appropriate natural stimuli were applied to optic, vestibular or touch receptors. The absence of Mauthner neurones in the myxinoids is, however, in line with their tendency to atrophy in bottom-dwelling fish species and those showing reduced swimming activity.

13

MULTIPLICATION

Although the breeding habits of hagfishes remain almost a complete mystery and we know very little about their reproductive physiology, the histology of their unpaired gonad indicates that the sexual differentiation of these animals is in a curious transitional state between gonochorism and hermaphroditism. Whether the myxinoids have a determinate life span is unknown, but they certainly produce a succession of eggs; each ovulation being followed by the renewed growth of a reserve of smaller oocytes, which are presumably replaced by the division of oogonial stem cells persisting in the ovary throughout their reproductive life. At least within the genus *Eptatretus*, ovulation may be restricted to a definite breeding season, but in *Myxine*, females containing mature eggs have been found throughout the year, although the interval between successive ovulations is not known.

As is the case in the eel and some species of salmon, the single breeding season of the lamprey marks the end of its life span. Like birds and mammals, the proliferation of the germ cells is confined to an early ontogenetic stage and in the lamprey is restricted to a relatively short period in early larval life. As a result they produce a finite stock of oocytes, whose development is more or less synchronous and of which the vast majority are destined to undergo degeneration (atresia) at various stages of their development. Those eggs that survive are finally shed during the short breeding season in spring or early summer. Thus, the two groups of cyclostomes have adopted quite different reproductive strategies; the myxinoids repeatedly ovulating a small number of large, yolky shelled eggs throughout their life, the lamprey producing very large numbers of small eggs which mature simultaneously and which are all shed together at the close of the life cycle.

Sex Differentiation and Gametogenesis

Myxinoids

The inherent 'bisexuality' of the gonads of *Myxine* had already attracted the attention of 19th century zoologists, who remarked on the extreme rarity of 'true males' and the presence in the genital fold of both ovarian and testicular tissue; the former predominating in the cranial, and the latter in the more caudal areas of the gonad. Observations on the relative development of the two tissues in relation to body length suggested that these animals might be protandric hermaphrodites, functioning first as males and later as females. However, as a result of more detailed histological studies, Schreiner (1955) was unable to detect any constant relationship between length (or age) and sexual status and he concluded that *Myxine* was in fact, a dioecious animal. In those individuals where the cranial or caudal segments contained exclusively male or female germ cells, sex differentiation was believed to have occurred at body lengths of 170-180 mm. At this time, either the cranial region became a functional ovary with rapidly developing oocytes or the caudal testicular region became dominant. In the latter case, the cranial ovarian segment developed only a few oocytes and these rarely grew to diameters of more than 1 mm before becoming atretic. Such animals were destined to become functional males. A third category consisted of animals in which both eggs and testis follicles occurred together in the cranial

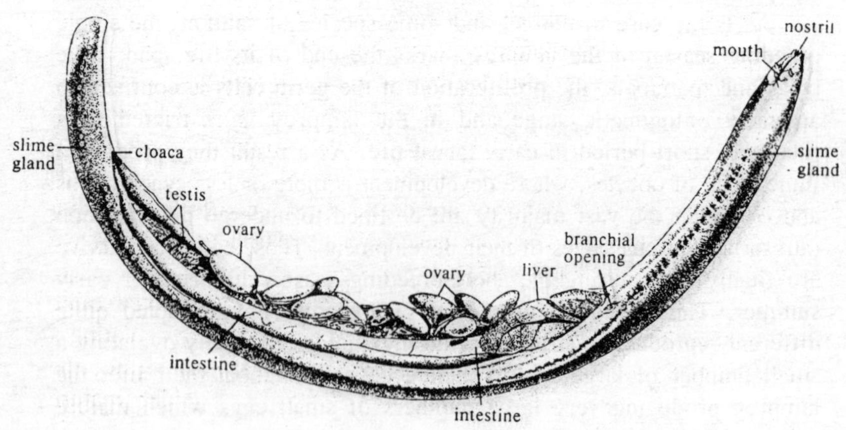

Fig. 13.1. Bdellostoma, partly dissected, showing ovary with eggs.

region and in this type, the direction of their future development would depend on the relative volumes of the testicular and ovarian tissues. Where the numbers of oocytes was small, the testicular tissue continued to develop, but the eggs did not progress far before undergoing atresia. Such animals would become functional males. On the other hand, if the volumes of male and female tissues were to be more equally balanced, both eventually degenerated resulting in the production of sterile individuals. These forms were surprisingly common and at body lengths of 25-29 cm accounted for about 13% of the 4000 hagfish examined by Schreiner. The picture that emerges is one of a delicately poised mechanism of sex differentiation in which the individuals in a population show a wide spectrum of sexuality and which probably vary widely in the length (or age) at which their sexual orientation is finally decided. From an initial common hermaphrodite stage the gonads of *Myxine* may develop in anyone of three directions:

1. The testicular region may continue to develop, while the ovarian region undergoes complete regression resulting in the development of pure males. Only 19 of this type were seen by Schreiner out of a total of 4000 animals.
2. The testis may remain more or less static, while the ovary continues to develop, producing a functional female.
3. The two areas develop together exercising mutual inhibition and resulting in a sterile gonad. Such animals were distinguished by Schreiner from others where sterility was attributable to a primary absence of germ cells.

In the *Eptatretidae* sexual status is more determinate than in *Myxine*. Early reports on *E. stouti*, indicated that sex ratios were approximately normal and that the two sexes did not differ in their range of body lengths. No female tissue occurred in the males and no testicular tissue in the females. Similarly, the male gonads of *E. burgeri* were said to show no traces of oocytes. Among 17 specimens of this species examined by Schreiner, there were five females showing some signs of rudimentary hermaphroditism, one animal was sterile and eleven were true males — a far higher proportion than he had observed in *Myxine*.

Information on the early development of the gonad of *Myxine* is limited to animals already over 100 mm in body length. In the smallest specimens described by Schreiner, germ cells were already present in the epithelium of the genital ridge (germinal epithelium). In the more caudal regions, germ cells (spermatogonia) were leaving the germinal epithelium to enter the underlying mesenchyme where they proliferated to produce testis follicles. Unlike the future female germ cells (oogonia) of the cranial region, which showed meiotic changes soon after they had left the germinal epithelium, the onset of meiotic prophase was delayed in the spermatogonia of the more caudal regions. Furthermore, the germ cells of the latter area invariably differentiated in a male direction, but in the cranial segments the further development of the germ cells was less constant. Here, they showed all transitional states between the two extremes of pure ovarian and pure testicular tissue.

The course of the ovarian cycle has been described by Patzner (1974) in comparatively young stages of *Myxine* with body lengths of 15 cm. Here, two regions of the genital fold could be distinguished. The proximal part continuous with the mesovarium consists of a mesenchymal core covered by a flattened peritoneal epithelium, but at the distal end this germinal epithelium is several cells thick and contains the oogonia, from which successive waves of oocytes are produced. With the onset of the meiotic prophase, the oocytes now move below the surface accompanied by follicle cells, and grow to diameters of 1-2 mm while remaining within the central region of the genital fold. Further oocyte growth is then arrested until the previous generation of mature eggs has been shed. The largest and oldest eggs are ultimately found in a row at the boundary between the gut and the mesovarium and at this stage the ripe eggs hang below the free border of the ovary. Throughout oogenesis, a high proportion of the oocytes are lost by atresia and, where this occurs in the later stages, the atretic follicles progressively move away from the distal border towards the proximal regions of the ovary. Thus, the ovarian cycle involves a movement of oocytes towards the proximal region adjacent to the mesovarium and away from the germinal epithelium, but as they become more mature, they increasingly sink downwards below the distal border. Finally, when the egg has reached its maximum dimensions (in *E. burgeri* about 20 × 8 mm) the shell and the adhesive filaments are secreted by the follicle cells. The filaments have characteristically expanded and lobed heads and are arranged in tufts at each pole of the egg, anchoring them together in a string and presumably attaching them to the substrate after ovulation. Below the animal pole the shell shows a thin suture — the opercular ring — marking the line along which the shell is believed to break open at hatching. At the animal pole and surrounded by the apical tuft of anchor filaments is the micropylar funnel, formed by a cord of granulosa cells. Electron micrographs reveal that at the bottom of this funnel, which is about 0.2 mm in

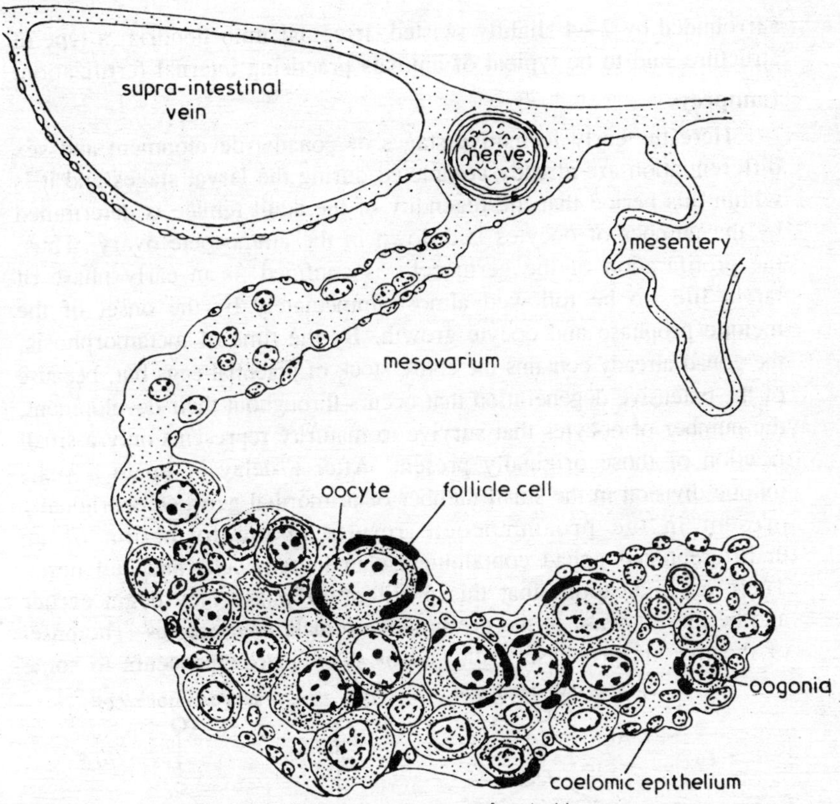

Fig. 13.2. The ovary of a young female Myxine.

diameter, there is a honeycomb-like structure consisting of some 3500 individual 'cells', only one of which was found to be open. This implies that the effective diameter of the aperture through which the sperm must pass is about 3.0-3.5 μm, or only twice the diameter of the sperm head.

Within individual testis follicles, the male germ cells develop synchronously but, even in the same animal, different follicles may represent widely varying stages of spermatogenesis from spermatogonia to spermatozoa. Amongst over 1000 specimens of *Myxine* examined by Jesperson (1975), 200 were adult males but, of these, only one contained ripe sperms. On the other hand, a much higher proportion of ripe males were found in the several species of *Eptatretus*. In his ultrastructural examination of the sperm of *E. burgeri*, Jesperson remarked on the long middle piece (20 μm) consisting of the flagellum surrounded by 2—4 slightly twisted, irregular mitochondria; a type of structure said to be typical of animals practising internal fertilization.

Lampreys

Here the early formative stages of gonadal development and sex differentiation are already completed during the larval stages and it is within this period that the fecundity of the adult female is determined by the number of oocytes laid down in the ammocoete ovary. Thus, the proliferation of the germ cells is confined to an early phase of larval life, to be followed almost immediately by the onset of the meiotic prophase and oocyte growth. By the time of metamorphosis, the gonad already contains the entire stock of potential eggs but, because of the extensive degeneration that occurs throughout their development, the number of oocytes that survive to maturity represents only a small fraction of those originally present. After a delay of up to a year, mitotic division in the small number of primordial germ cells originally present in the proammocoete results in the formation of an undifferentiated gonad containing isolated germ cells and cell nests. There are indications that this proliferative phase may begin earlier and be more intense in the presumptive female ammocoetes. The onset of the meiotic prophase, followed by oocyte growth occurs to some

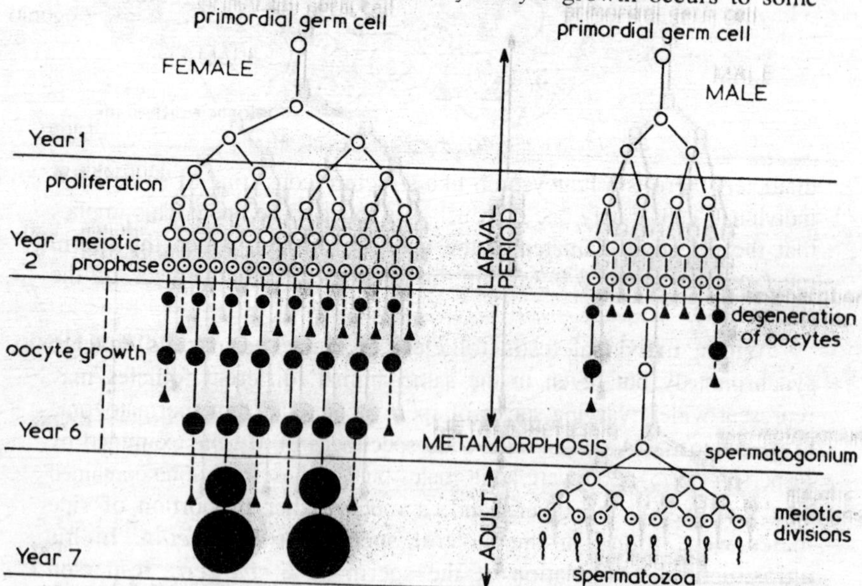

Fig. 13.3. Schematic representation of the course of gametogenesis in a non-parasitic lamprey, L. planeri.

extent in all gonads, irrespective of their future sex, although it is more syochronous and extensive in the future females. Before sex differentiation is complete the larval gonads will therefore contain variable numbers of oocytes, both normal and atretic, as well as undifferentiated cells, either occurring singly or in small groups. Since these cell nests were formerly regarded as male elements, this stage of gonadal development has been considered as a form of juvenile or rudimentary hermaphroditism whereas, in reality, these larval cell nests are not direct forerunners of the spermatogonial lobules of the adult testis, and cannot be said to be 'male-orientated'.

Although the presence in all ammocoete gonads of variable numbers of oocytes may prompt comparisons with the myxinoid gonad, the genital fold of the ammocoete shows no regional differences in their distribution. Even more importantly, unlike the male cells of *Myxine* which at an early stage proceed to form testis follicles, the vast majority of undifferentiated cells in the early ammocoete gonad show no such precocious male trends and either enter meiotic prophase and degenerate or temporarily differentiate as oocytes before undergoing atresia. In the lamprey, the differentiation of the male gonad is thus a 'negative process', consisting of the gradual elimination of growing oocytes and meiotic germ cells. These degenerative processes may be accompanied by some increase in the connective tissue stroma of the future testis but, because of the massive germinal destruction, the gonad may actually decrease in size at this stage of its development. The future male germ line can be traced to a very few stem cells, which for some obscure reason escape these regressive processes and their eventual proliferation to produce small testis lobules is delayed until late larval life and the approach of metamorphosis. In these lobules, the meiotic prophase and finally the production of secondary spermatocytes, spermatids and spermatozoa follow one another quite rapidly within a few months of the onset of spawning.

Reproductive Biology

Lampreys

In their patterns of sexual maturation there are important quantitative differences between the non-parasitic and the parasitic species of lampreys. In the latter, both the testis and the ovary undergo a slow and steady development throughout the phase of parasitic feeding and by the time and mature adults enter the rivers on their spawning migration, the testis lobules already contain masses of spermatogonia, while the oocytes in the female are in a comparatively advanced stage of vitellogenesis and growth. Lacking this interlude of adult feeding and growth, the non-parasitic lampreys enter the final period of sexual maturation from the lower baseline of gonadal development that they have attained at the end of their larval life and, within the six to nine months that elapses between metamorphosis and spawning, must be compressed the entire cycle of oocyte growth and spermatogenesis.

During these final periods of sexual maturation, all lampreys, whatever their type of life cycle, undergo metabolic and morphological changes, most of which can be related directly or indirectly to the cessation of feeding and to the metabolic demands of the ripening gonads. These include atrophy of the gut and loss of its osmoregulatory functions, changes in the dentition, shrinkage in the length of the body and loss of weight, degenerative changes in the liver, depletion of lipid stores and liver glycogen and regression of the haemopoietic tissues. Finally, as the breeding season approaches, secondary sex characters develop; the urinogenital papilla of the male is everted to form a 'penis-like' structure and the male sea lamprey develops a rope-like thickening in front of the first dorsal fin. In the female, the leading edge of the second dorsal fin becomes swollen and oedematous and in addition, a small anal fin develops immediately behind the cloaca. In ripe animals of both sexes the fins become higher and more vascular, while the cloacal labia also become oedematous. In the southern hemisphere species, *G. australis* and *M. lapidica*, ripe males develop a large gular sac hanging down behind the oral disc and in both genera there is a remarkable increase in the size of this disc, accompanied by an extension of the snout region.

Shortly before spawning, the oocytes are liberated from the ovarian follicles and lie free in the body cavity. The mature egg is enclosed in a double-layered chorion formed during the third and final vitellogenic phase. Although the mechanisms of ovulation are not completely understood, it has been suggested that the mucopolysaccharide granules in the disintegrating granulosa cells may cause an osmotic uptake of water, contributing to the rupture of the follicle. After ovulation, the remains of the granulosa cells still attached to the egg swell on contact with the water and assist in the adhesion of the eggs to the substrate of the nest. At the animal pole of the egg where the granulosa cells are absent is a tuft of mucoid fibres, which extends far into the water and is said to facilitate the movements of the sperms towards the egg, perhaps at the same time offering them some protection from this hostile osmotic environment.

As in the myxinoids, the sex products of the lamprey leave the body cavity through coelomic pores but, in this case, opening only just before spawning to connect the body cavity with the urinogenital sinus. This development is apparently controlled through the pituitary-gonadal axis, since injections of pituitary extracts have resulted in the opening of the pores in immature adults or even ammocoetes, while similar changes have occurred in the adult (but not in larvae) after the administration of steroid sex hormones.

Under laboratory conditions, the rate of sexual maturation depends on the temperatures at which the lampreys are maintained, but is independent of light. Animals kept in permanent darkness have reached breeding condition at the same time as those subject to normal illumination. However, under field conditions long-term temperature trends appear to influence both the onset and the duration of the spawning season and, once spawning has started, the behaviour of the spawning lampreys is markedly affected by relatively small transient changes in stream temperatures. In the case of the two *Lampetra* species, *planeri* and *fluviatilis*, spawning usually begins when spring water temperatues rise rapidly to about 11°C, but the sea lamprey *P. marinus* spawns later in early summer at temperatures of around 15°C.

At least in some species, the males are believed to reach the spawning grounds first to begin the preliminary nest building activity. Here, when conditions are favourble the animals may gather in large numbers within relatively restricted areas which remain in use from one year to another. The possibility that these local assemblages are influenced by olfactory signals is raised by a technique used by French lamprey fishermen. This involves placing a completely ripe male in a conical wickerwork basket to attract the migrating females. In addition to the presumptive ion uptake cells which are found in the gills of both sexes of the river lamprey, *L. fluviatilis*, the interplatelet areas of the spawning male show a distinctive type of mitochondria-rich glandular cell, containing large lipid bodies with translucent centres. Could these cells be responsible for the secretion of a pheromone-like sexual attractant?

The early stages of nesting activity involve the removal of larger stones from the nest site. Attaching themselves to the stones, the body is raised vertically while the tail makes vigorous vibratory movements. This has the effect of loosening the stones, which tend to be carried downstream with the current. However, even in sites where the current is slack, small stones lifted by the lampreys may be carried to the

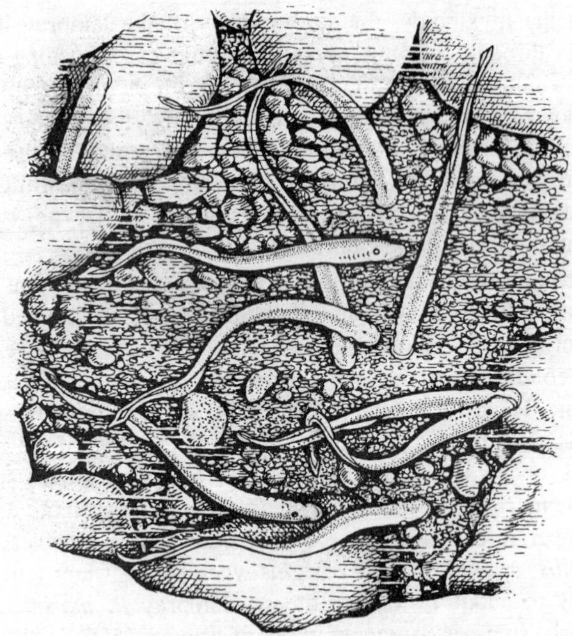

Fig. 13.4. Spawning lamprey seen in their nest.

sides or even upstream from the nest. As a result of this activity, a depression is produced in the stream bed cleared of larger stones and varying in its diameter with the body size of the spawning lampreys and their numbers. Because of the vortex created by the structure of the nest, the animals are usually able to move freely within it without being swept away by the current. The smaller non-parasitic lampreys tend to be more gregarious than the larger parasitic species and it is by no means uncommon to find 30 or more animals associated with a single nest. The nests of the large sea lamprey, on the other hand are generally occupied by only a single pair and under these conditions, a form of territorial behaviour has been described, in which a resident male is said to drive off others by attaching with its sucker and making violent vibratory movements.

Lampreys practice a form of sexual congress in which the male attaches to the female usually on the head or branchial region, rapidly coiling his tail into a loop round the female so that the 'penis' is directed towards the cloaca. This is immediately followed by a short burst of intense vibratory activity, during which a few eggs are expelled. This pairing act has two functions. It assists in ovulation through the pressure exerted by the male on the pre-cloacal region of the female and it also increases the chances of successful fertilization by directing the jet of milt directly on to the eggs. This is important since the sperms are said to remain active for less than a minute once they have made contact with freshwater. Only a small number of eggs are extruded at each spawning act so that this must be repeated over and over again, perhaps for several days before the major part of the egg stock has been exhausted.

Death of Lampreys after Spawning

The failure of lampreys to survive more than a few weeks at the most, after their first and only spawning puts them in the same category as the eel and the Pacific salmon, whose death after the completion of a single reproductive effort has sometimes been cited as an example of a genetically programmed event — a description that does little to answer any of the significant questions that confront the gerontologist. Obviously, for those species like the lamprey in which no reserve of undifferentiated germ cells remains after their single breeding season, further survival would be biologically meaningless, but this does not explain the physiological mechanisms that result in such relatively synchronous natural death.

As described elsewhere, spawning is preceded by a variable, but prolonged period of starvation, during which the animal can be regarded as a closed system in which energy reserves built up over the parasitic feeding phase are in part diverted to the developing gonads or used to furnish the energy expended in the rigours of the upstream migration and the activities associated with nest-building and spawning. One of the manifestations of these processes is the slow and continuous reduction in body length and weight, followed in the terminal stages of sexual maturation by a phase of more rapid shrinkage. At this period, the lipid reserves of the spawning or spent lamprey have dropped to about 4% of the wet weight of the body but, as Beamish *et al.,* (1978) have pointed out, it seems unlikely that this depletion is the sole or even the main factor in death since these terminal lipid levels are still higher than they are in the macrophthalmia stage, at a time when it is just beginning to feed after the metamorphic period. As in the spawning salmon, there is some evidence that when lipids have been almost exhausted, the lamprey may begin to catabolize its tissue proteins and perhaps it is at this stage that degenerative changes reach a critical and irreversible point in some vital tissues. Among such degenerative changes are the massive destruction of red cells, the atrophy of the haemopoietic tissues, a reduction in haemoglobin concentrations and degeneration of the liver. Nevertheless, not all tissues are equally affected by such regressive changes. This has been demonstrated in experiments involving the transplantation of eyes from spawning lampreys into ammocoetes, where the grafts remained viable for at least 6 months beyond the normal time of death of the donor animals and even then their further survival was only limited only by the termination of the experiments. In addition, the possibility of halting or even reversing degenerative changes has been illustrated by the effects of castration in inhibiting the atrophy of the intestine.

The effects of endocrine factors on the time of death of the lamprey, *L. fluviatilis*, have been studied by Larsen (1973). The possibility that hypersecretion of the adrenocortical tissues might be involved in the post-spawning mortality of Salmonid fishes has been widely discussed but, at least in lampreys, this can be discounted in view of the absence of evidence for the usual range of vertebrate corticosteroids, much less for their presence in high concentrations. In an extensive series of experiments, Larsen has shown that prolonged survival can be induced by maintaining the lampreys at constant, low temperatures, by castration and more especially by hypophysectomy. Thus, it would appear that any factor that interferes with the metabolic changes involved in sexual maturation or which retards their progress, is capable of prolonging the life span. Hypophysectomy is known to reduce the metabolic rate of the lamprey and it is probably this effect rather than a more specific influence exerted through the removal of gonadotrophins, that is responsible for delaying the time of natural death. Although in Larsen's experiments, some of the hypophysectomised lampreys survived for periods up to 10 months beyond the normal time of spawning, they eventually suffered a shrinkage similar to that of normal animals, although this had obviously occurred at a reduced rate. It would seem therefore, that death is ultimately due to degenerative changes induced by starvation and that, as Larsen has established, any procedures that slow down tissue mobilization are effective in prolonging their natural life span.

Myxinoids

In spite of the prize offered in 1854 by the Copenhagen Academy of Sciences for information on the reproductive habits of *Myxine* (which remains unclaimed), there has been little advance towards a resolution of the many problems posed by these enigmatic animals. How are we to explain the wide variations in the sex ratios of European populations and how can fertilization be achieved in areas where functional males

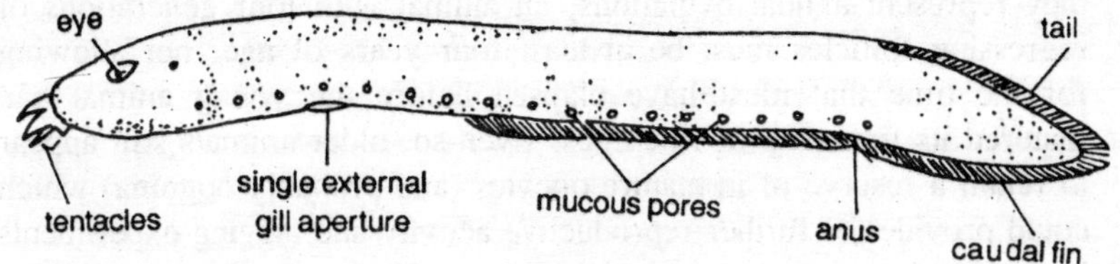

Fig. 13.5. Myxine.

are rare? Why have fertilized eggs so rarely been found, how are they fertilized and where are they deposited? Is breeding confined to some inaccessible spawning grounds?

Data collected towards the end of the last century suggested that *E. stouti* was markedly localized in its distribution off the Californian coasts and that males and females tended to occur in separate areas. This raises the possibility that the disparities in the numbers of male *Myxine* in different localities might reflect sex differences in habitat, feeding behaviour or patterns of movement. Furthermore, it is possible that information on numbers or distribution based on captures on baited lines or traps may be misleading. For example, appetite or feeding responses may well vary at different periods in the breeding cycle or as Holmgren (1946) suggested, the ripe hagfish may, like many other breeding fish, burrow deeply during spawning and so be out of the reach of trawls.

Fernholm's (1974) accounts of the seasonal movements of *E. burgeri* agree with earlier records for the same species. These indicated that these animals were not caught off the Japanese coast between the end of July and the end of October. Towards the start of this apparent breeding season, the proportion of males increased, but when the animals reappeared in October all the mature males were found to be spent. Similarly, in late autumn a high proportion of the females had empty follicles or corpora lutea, indicating recent ovulation. Although a number of earlier workers had postulated a seasonal breeding cycle for *Myxine*, this cannot be substantiated either by the state of the gonads or the times of the year when infertile eggs have been recovered in trawls, some from depths of 100-200 m. Living as it does in such deep water, the cyclic processes of oogenesis in *Myxine* probably owe more to endogenous rhythms than to seasonal factors.

In the ovary of *Myxine*, it is possible to recognize as many as four generations of spent follicles or corpora lutea in various degrees of involution. These must represent successive ovulations, but at what time intervals is not known. If, as has been suggested in *E. burgeri*, they represent annual ovulations, an animal with four generations of regressing follicles must be at least four years of age, not allowing for the time that must have elapsed before the young animal had matured its first crop of ripe eggs. Even so, older animals still appear to retain a reserve of immature oocytes (and probably oogonia) which could provide for further reproductive activity and tagging experiments have given some indications that egg production may continue even when growth has slowed down or ceased altogether.

The absence in myxinoids of sex dimorphism or secondary sex characters is easily understood in the context of their relatively undifferentiated sexual status and the poverty of their sensory equipment. Although the ovarian tissue has the ability to produce a range of steroids, it seems more likely that these may be involved in the regulation of gonadal activity, rather than in the development of breeding signals. Sex recognition, whatever the form it takes, is presumably dependent on olfactory stimuli.

Metamorphosis and the Significance of the Larval Stage

At normal stream temperatures, the hatching of the lamprey egg occurs at 10-14 days after metamorphosis and the pro-ammocoete as it is called at this stage still retains considerable quantities of yolk, although the stomadaeal depression has broken through to the mouth cavity and the gills have become functional. The first movements to develop are lateral undulations involving almost the whole body, but with the increasing absorption of the yolk (now restricted to the gut lumen), typical swimming movements appear, although at this stage the pro-ammocoete does not yet exhibit light avoiding responses. When disturbed, they swim vigorously towards the surface, then subside towards the bottom where they lie motionless on their sides. Quite abruptly, this behaviour changes and by about the third week they begin to develop burrowing responses, ceasing their surface swimming and diving rapidly into a soft substrate. These successive phases of larval behaviour obviously ensure, first their emergence from the sand of the nest and their distribution downstream by the current and secondly, their establishment in suitable ammocoete habitats.

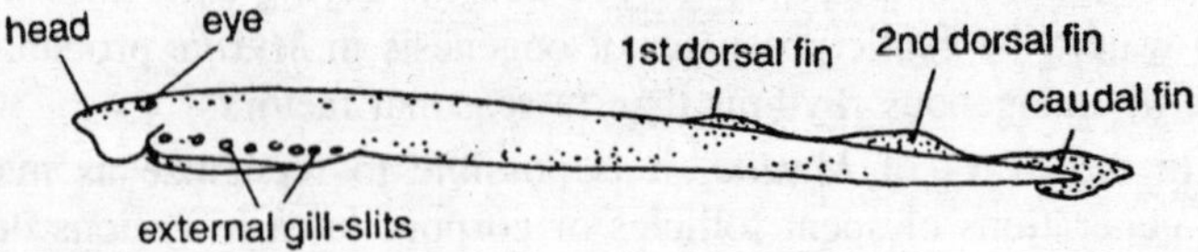

Fig. 13.6. Ammocoete larva.

In relation to the embryonic development of the myxinoids, opinion has been sharply divided, between those who like Løvtrup (1977) maintain that the large yolky egg represents a primitive feature and others who believe that the absence of a larval stage represents a secondary consequence of an evolutionary increase in the volume of the hagfish egg and its yolk content. On the other hand, a majority of authors have tended to accept without question, the primitive nature and evolutionary significance of the larval stage of the lamprey, assuming a similar ammocoete stage to have already been present in some at least of the fossil agnathans. Watson (1954) for example, argued that the development of the upper lip and the resulting dorsal position of the nasohypophysial opening must have preceded the formation of the oral hood as part of the feeding mechanism of the ammocoete. Since the nasohypophysial complex occupied the same position in both cephalaspids and anaspids and was assumed to develop in the same way through the hypertrophy of a post-hypophysial fold, Watson inferred that these groups would have passed through an ammocoete stage. Because the myxinoids also show a post-hypophysial fold, although with no obvious functions, he considered that they too must once have possessed a larval stage, which was subsequently suppressed by their secondary evolution of large yolky eggs. On account of the paired nostrils opening into the mouth, Watson excluded only the heterostracans from this common possession of an ammocoete-like larval form. More recently however, Whiting (1972) in his comparisons of the morphology of the proammocoete and the fossil agnathans has concluded that both the heterostracans and the cephalaspids may have passed through a proammocoete or ammocoete stage.

Somewhat different conclusions were reached by Strahan (1958) in his analysis of probable growth patterns in the agnathan head and, if these are valid, Watson's argument would lose much of its force. Thus, Strahan suggests that the most direct transformation from embryo to adult lamprey must involve a stage in which an ammocoete-like upper lip is present. This structure was therefore regarded as an inevitable stage in the growth processes involved in the development of the oral funnel, rather than as a unique adaptation to larval feeding. For this reason he excluded the ammocoete stage from the life cycle of the cephalaspids, although he believed that it may have been present in the anaspids.

Certain features of the fossil *Mayomyzon* raise considerable doubts whether a larval stage was present in these Carboniferous lampreys. In a number of respects this fossil lamprey appears to have retained

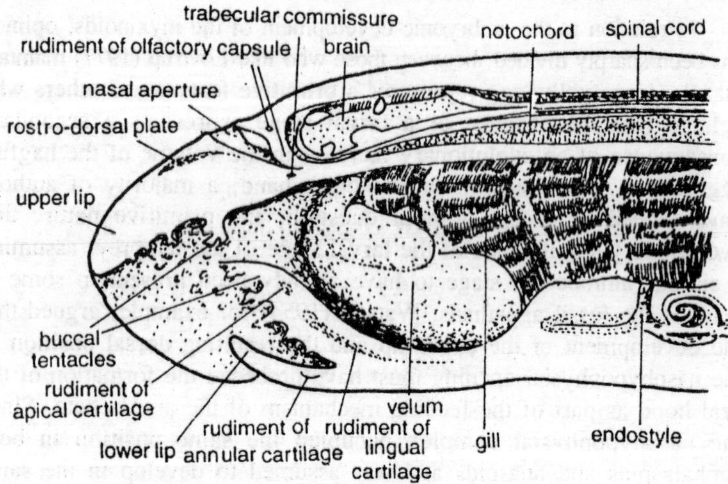

Fig. 13.7. Ammocoete. Section of anterior region.

primitive characteristics which in present day lampreys are to be found only in embryonic or early larval stages. These include the anterior position of the branchial region, the comparatively short pre-oral region, the tubular shape of the snout, the probable absence of disc teeth, the less extensive piston mechanism and the lack of a direct connection between the sense capsules and the neurocranium. Moreover, the somewhat expanded annular cartilage suggests that the preoral region may have resembled the oral hood of the ammocoete. An even more convincing argument is the small size of *Mayomyzon* (30-60 mm). If these very small creatures were, in fact, mature adults, this would seem to imply that they had not yet evolved the parasitic method of feeding, that alone has made possible the much greater body size of present-day lampreys. It may of course be argued that these fossils represent recently metamorphosed or marcrophthalmia stages of a much larger animal. However, this would imply that even if a larval stage were present, it must have been of very short duration, since the smallest of these fossils is no larger than present day ammocoetes after little more than six months of larval life. Moreover, if we assume that both the size of the eggs and the patterns of oogenesis in *Mayomyzon* were at all like those of present day lampreys, their small body size would imply very low levels of fecundity, quite incompatible with an extended life cycle like those of living forms.

Damas (1944) has suggested that in their biology and anatomy, the ancestors of the lampreys would have been intermediate between the present-day ammocoete and the adult and that these subsequently became more divergent and specialized. As de Beer (1952) has emphasized, it is a corollary of larval adaptation that the better suited it becomes to its mode of life, the greater will be the divergence between the form of larva and adult and the more violent the metamorphosis. This would be all the more so in the lamprey, where the larval stage has been prolonged to an exceptional extent and occupies such a large part of the life cycle. As the specialized larval adaptations are further developed, the future adult tissues and structures remain as undifferentiated embryonic primordia, whose further development is suspended until the time of metamorphosis. Thus, what was originally a continuous sequence of development from embryo to adult is now interrupted by the interpolation of a larval stage and morphogenetic processes previously confined to embryonic life, are now postponed for periods of several years. Examples of this kind of delayed differentiation are to be found in the persistence of mucocartilage, the development of the eye, kidney, thyroid and adenohypophysis. At metamorphosis, the larval kidney regresses and is replaced by an adult mesonephros arising from undifferentiated nephrogenic tissue in the nephric fold of the ammocoete. The curious history of the ammocoete eye involves the early appearance of visual cells forming the so-called retina A. This is surrounded by other visual elements that do not differentiate until metamorphosis (retina B). As we have seen earlier, it is also possible that the adult thyroid follicles may be derived from relatively undifferentiated elements of the endostylar epithelium, perhaps playing little or no part in the thyroidal biosynthesis of the ammocoete. The significance of these conditions had already been appreciated by Leach (1944) who noted that, whereas in a frog tadpole the thyroid and pituitary are well-developed embryonic structures, in the ammocoete both thyroid follicles and hypophysial development are concurrent with a metamorphosis taking place some seven years after the embryonic period. As further examples of delayed differentiation Leach cited the undifferentiated tubular muscles of the larval pharynx and the retention of a rudimentary eye.

No doubt it is true of the ammocoete, as it is of larval amphibians, that the timing of their metamorphosis is dependent on both age and size. Thus, within a particular larval population there may be a considerable range of ages amongst metamorphosing ammocoetes depending on differences in their growth rates. While therefore we can only speak of an average age at metamorphosis, this shows characteristic species differences that must be assumed to have a genetic basis. Although there are some conspicuous exceptions, parasitic species tend to have a shorter larval life than the nonparasitic forms and these differences are particularly evident among the members of paired species. This ability to vary the duration of the larval phase must have been an important factor in the adaptability of the group.

We have as yet no firm evidence of hormonal involvement in the initiation or control of metamorphosis, although the switch that has been reported in larval metabolism towards lipid accumulation in the period before transformation could be related to hormonal changes. That environmental factors are in some way implicated is shown by the fact that throughout the Northern Hemisphere, metamorphosis is generally only observed within a restricted period from June to September. This degree of synchrony strongly suggests the involvement of temperature or daylength, perhaps mediated through the pineal complex. However, whatever may the precise factors that are concerned in triggering transformation, these probably operate long before external morphological changes appear, perhaps early in the spring. There is now some evidence that certain internal changes may begin some time before the external signs of metamorphosis appear and this has led to attempts to define a pro-metamorphic period, analogous with that of the anuran tadpole. Among these precocious internal changes are the cavitation of the nasohypophysial stalk, the development of the accessory olfactory organ and the replacement of the larval mesonephros by that of the adult. Once metamorphosis has begun, it proceeds very rapidly and most of the external changes are completed within 3-5 weeks, although some of the physiological transformations like those of the haemoglobins may take much longer, especially in the non-parasitic forms.

Reference has been made earlier to the discovery in one locality of Northern Italy of neotenous female ammocoetes of the brook lamprey, *L. zanandreai*. At first sight, the occurrence of neoteny in a nonparasitic lamprey might hardly seem surprising. These species, although not dispensing with metamorphosis have nevertheless sacrificed most, if not all of the biological advantages of adult life. Since they never feed, their body size cannot be increased beyond the larval stage and as a result they are unable to mature more than a very small proportion of the oocytes in the ammocoete gonad (in *L. planeri* only about 20%). The brook lampreys therefore, suffer the additional and relatively heavy mortality associated with the hazards of adult life, but without compensatory gains in the shape of increased fecundity. Why then have these dwarf forms not more often become neotenous and abandoned altogether this apparently functionless adult stage? Here, the answer is to be sought in the physiological, morphological and behavioural characteristics of larval lampreys which would preclude their ability successfully to complete the reproductive cycle. Apart from its photophobic burrowing responses and low metabolic rates, ill-adapted to the vigorous free swimming activities associated with spawning, the cardiovascular system and low heart ratio of the ammocoete would be quite unable to cope with the sustained locomotory activity involved at the breeding season. Lacking the better developed fins of the adult or its paired eyes and rheotactic responses, it would be incapable of making its way upstream towards suitable spawning sites and, once there, possession of a suctorial oral disc would be essential for spawning activity, copulatory pairing and the ability to maintain station in the face of rapid currents.

EVOLUTIONARY ASPECTS

Nature of Myxinoid Sexuality

In one of the earliest investigations carried out on the European hagfish, *Myxine glutinosa*, Nansen (1888) commented that this animal 'still seems to be seeking, without yet reaching that mode of reproduction most profitable for it in the struggle for existence.' This picturesque phrase aptly describes the major evolutionary problem that is raised by the sexual status of these animals. Are they evolving towards a hermaphrodite condition or moving towards dioecism from an originally hermaphrodite state? An answer to these questions might be easier if we had more complete information on other myxinoid species. There are some indications that a more stable dioecism may exist within the genus *Eptatretus*, where the proportion of pure males appears to be much higher than in *Myxine*. From the morphology and habits of the former genus, it has been argued that *Myxine* may represent a more aberrant (or apomorphic) genus and that the species of *Eptatretus* may be more primitive and closer to the ancestral myxinoids. On the other hand, Schreiner considered that the ancestors of the hagfishes were probably self-fertilizing hermaphrodites, subsequently evolving towards pro tan dry and the postponement of ovarian function as the eggs increased in size and developed a higher yolk content. Throughout these trends, a mutual antagonism developed between male and female gonadal territories, finding its extreme expression in the production of sterile gonads. These sterile individuals might be regarded as a means of eliminating unfavourable genotypes that otherwise might upset the delicately balanced mechanisms of sex differentiation. This interpretation would imply that, in their reproductive physiology, the *Myxinidae* are the more primitive or plesiotypic forms and the *Eptatretidae* the more advanced or apotypic.

Hermaphroditism is most likely to develop in species with low mobility, for example in parasitic or sessile forms, where there will be little sex dimorphism and the male will not evolve mechanisms for seeking out and holding the female. Low mobility limits male reproductive success (measured by the numbers of eggs that can be fertilized competitively) and this will rise at less than a linear rate with the input of resources into male function. Schreiner's view was based on the rather dubious assumption that the ancestors of the myxinoids were parasitic and wide-ranging forms with limited opportunities for cross-fertilization. Later, they adopted what he called saprophytic modes of feeding, leading them to collect together in large numbers, favouring cross-fertilization and the development of protandric hermaphroditism. This change was, in turn, associated with a reduction in fecundity and the development of large and more yolky eggs. As evidence for these changes, Schreiner cited the presence in many specimens of *Myxine* of a rudimentary left genital fold and also pointed out the relationship between increased oocyte numbers and more intense atresia. However speculative these views may be, some relationship between the habits and ecology of hagfishes and their peculiar sexual condition seems highly probable. Although they are said to be present in considerable numbers within restricted areas, it is possible that these local populations may be isolated from one another and that individual hagfishes do not move very far from their 'home' territory. The absence of sexual dimorphism would also be consistent with a relatively recent escape from a condition of functional hermaphroditism.

Sex Differentiation

The various inductive theories of sex differentiation stress the primary role of the somatic gonadal tissues in controlling the sexual orientation of the germ cells. Schreiner believed that there was evidence for this kind of inductive effect in the onset of meiosis in the oogonia of *Myxine* immediately they have acquired an envelope of follicle cells and migrated from their original site in the germinal epithelium. The somatic tissues were also held responsible for the antagonistic effects of male and female gonadal territories in the potentially hermaphrodite genital fold. A similar incompatibility also appears to be present in the early ammocoete gonad, where the oocytes that develop in the presumptive male gonads invariably degenerate before reaching the final stages of oocyte growth. It appears that with the progressive development of the somatic substrate, the cellular environment of the future testis may become increasingly hostile to the further development of germ cells that have differentiated in a female sense. This may be compared with the situation in hermaphrodite males of *Myxine*, where variable numbers of oocytes are present which fail to reach maturity. In both hagfish and lamprey, oocytes occur almost universally whatever the future sex of the gonad and, in *Myxine*, they are only absent in the rare so-called true males. In hermaphrodite males they appear in variable numbers and their further development and the size that they attain is related to the relative volumes of testicular and ovarian tissue.

In *Myxine*, it is appropriate to speak of a range of sex phenotypes ranging from pure males to pure females through a graded series of hermaphrodite gonads containing both ovarian and testicular tissue of which only one will become dominant and functional. This does not appear to differ fundamentally from the situation in ammocoete gonads before and during sex differentiation, except that, in this case, the potentially male germ cells remain as undifferentiated elements and never form recognizable testicular structures. Nevertheless, in regard to the numbers of growing oocytes, the ammocoete gonads show a continuous spectrum in which the extremes of higher and lower numbers are believed to correspond to the future male and female gonads. The conditions of the gonads in both groups of cyclostomes can probably best be explained by postulating a sex-determining mechanism involving a delicately poised and multifactorial system, capable of producing a wide range of sexual phenotypes. Such a system might be very susceptible to the kind of chromosomal variability that has been reported in a Swedish population of *Myxine* where enormous variability was observed in chromosome counts and the existence of haploid individuals and even parthenogenetic reproduction has been suggested.

In lampreys, the appearance of a prolonged stage of sexual indeterminacy could be a consequence of the introduction of a larval stage. This could have the effect of spreading out over a longer period developmental processes that were originally completed during the embryonic phase. By delaying the differentiation of ovarian and testicular tissues, a slowing down of gonadal development might then give the appearance of a prolonged intersexual stage. It is interesting to find that a similar explanation was put forwards in the late 19th century to account for the hermaphrodite gonads of myxinoids. Unlike Schreiner,

Dean supposed that the hagfishes were derived from dioecious ancestors with right and left genital folds, but that with the loss of the left gonad, development of the right gonad became precocious thus prolonging the undifferentiated stage common to all vertebrates, allowing its development to proceed to a point where both male and female tissues are already present when final sex differentiation took place.

Fecundity and Survival

As relict groups, the cyclostomes have survived from the Palaeozoic by adopting completely contrasting survival strategies. The route followed by the myxinoids has involved the production of small numbers of large yolky and shelled eggs from which the young hatch in a relatively advanced stage of development. Lampreys, on the other hand, have maximized fecundity to a point where egg numbers are limited only by the body size that the female is able to attain; the small eggs hatching to produce an immature larval stage with habits and mode of feeding completely different from those of the adult lamprey.

Fecundity in lampreys varies widely in different species and is broadly correlated with body size. The smallest non-parasitic forms (in which the larval stage tends to be extended and the more vulnerable adult stage curtailed) have egg numbers varying from 400-9000, whereas in the largest parasitic species like the sea lamprey, egg numbers may reach several hundred thousand. Although the frequency of ovulation in hagfishes has not been determined, at least in *E. burgeri*, there are indications that only a single crop of eggs may be shed annually. Eight females of this species kept in cages on the sea bottom throughout the period when they would normally spawn i.e. from September to November, deposited a total of 178 eggs. Thus, the average number of 22 ovulated by each female corresponds quite closely with the number of mature eggs that are normally found in the ovaries of other species. In both hagfishes and lampreys a considerable proportion of the oocytes are lost by atresia. In *Myxine*, this has been put at about half all the oocytes between 2-23 mm in diameter and, in the non-parasitic lampreys, atresia may account for 60-90 % of all the oocytes in the ovaries of metamorphosing ammocoetes.

The enormous disparity in the fecundity of hagfishes and lampreys must of course imply the existence of corresponding differences in mortality rates. By virtue of their concealed and relatively inactive burrowing habits, the myxinoids are presumably subject to only low rates of mortality from predation and this is probably true also of the concealed larval stages of lampreys. In this respect, hagfishes conform to the pattern that characterizes other members of the deep sea community, where predation is thought not to be the main mechanism of population control. Thus, compared to shallower water species, these forms are said to be distinguished by small brood sizes and with slow growth rates and high longevity, show a higher proportion of adult to younger stages. Based on the annual cycle of oocyte growth, Patzner (1978) has calculated that the body length of *E. burgeri* only increases by about 4-5 cm a year. Assuming slow growth rates of this order, the largest hagfishes measuring up to a metre in length must presumably have attained a very considerable age. Low mortality and reproductive rates have been regarded as expedients adopted in populations where long term or evolutionary diversity is a product of past biological interaction within a physically stable environment such as that of the myxinoids.

Such conditions are in sharp contrast to the life cycles of lampreys, where mortality rates are certain to be high during the more adventurous and exposed episodes of migration and spawning. Statistics from the albatross population of South Georgia have provided some idea of the severity of the losses that may be sustained by lampreys during their oceanic migration. On Bird Island, the lamprey *Geotria australis* (presumably originating from spawning populations in South American rivers) forms an important part of the diet that the Grey-headed Albatross supplies to its young. Analysis of the diet of the 5000 albatross chicks on this island has indicated that the lamprey accounts for about 10% of their solid food, representing an annual consumption of 20 tons, or about 350000 lampreys. If we extrapolate from these figures to the entire albatross population of South Georgia, the total loss through bird predation would work out at 120 tons or over two million lampreys.

Although we are ignorant of the breeding habits of hagfishes, it is unlikely that these would rival the complex breeding patterns of lampreys or involve the same endocrine mechanisms of control and integration. At the same time, in spite of these more complex adaptations, lampreys do not appear to be conspicuously successful in regard to their breeding efficiency. Experiments conducted on female sea lampreys under natural conditions have shown that the number of ammocoetes successfully hatched may not exceed more than 5.3-7.8% of all the eggs that the female produces. Whatever may be the mode of fertilization or egg protection adopted by hagfishes (and in spite of the apparent rarity in some areas of functional males), their very low reproductive rates must necessarily imply a vastly greater degree of reproductive efficiency.

14

Ammocoete Larva

In this chapter we shall study in detail two forms, the ammocoete larva of the lamprey and the dogfish. The purpose of this is to gain a foundation for the discussion of the anatomy of the chordates. Both the ammocoete larva and the dogfish are generalized forms and the specializations of higher chordates can be understood as modifications of their more basic structure.

There is no better way of introducing the chordates than by a study of the ammocoete larva. This form has long been regarded as one of the closest approaches to the archetypal chordates that we possess. We have previously noted these two facts: the Agnatha are the most primitive living chordates, and embryos frequently exhibit characteristics believed to be of a primitive type. Since the ammocoete larva is the embryonic stage of the lamprey, which is a member of the most primitive vertebrate class, it is small wonder that we consider it of such importance in understanding the basic structural plan of the phylum to which we humans belong.

The larva of the lamprey is an eel-like organism that spends most of its time buried in the muddy bottoms of streams. When it was first discovered; the true nature of this little animal was not appreciated. It was given the generic name Ammocoetes. Somewhat later, when it

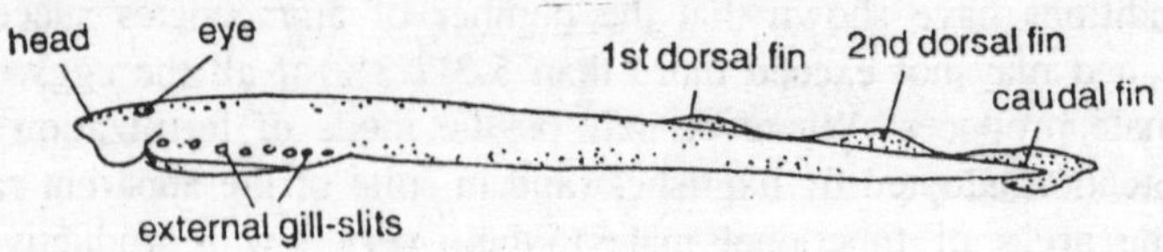

Fig. 14.1. Ammocoete larva.

was realized that this was merely the larval stage of the lamprey, the name Ammocoetes was replaced by the proper generic name of the adult. Ammocoetes is now used as a common name for the larval stages of all genera of lampreys.

Petromyzon is a common genus of lampreys. The adults, which may be several feet in length, spend most of their life in the ocean. Shortly before breeding, they enter rivers and swim upstream in their search for spawning grounds. They generally select small streams with sand or gravel bottoms, and fashion a crude nest by removing the stones. The male encircles the body of the female and sheds sperm when the eggs pass out of her body. The developing embryos usually become buried in the sand. Death comes to the adults shortly after spawning, and their bodies are washed down toward the sea. The embryonic stages require several years and it is not until this time has passed that the ammocoete larva metamorphoses into an adult. A return migration to the sea then occurs.

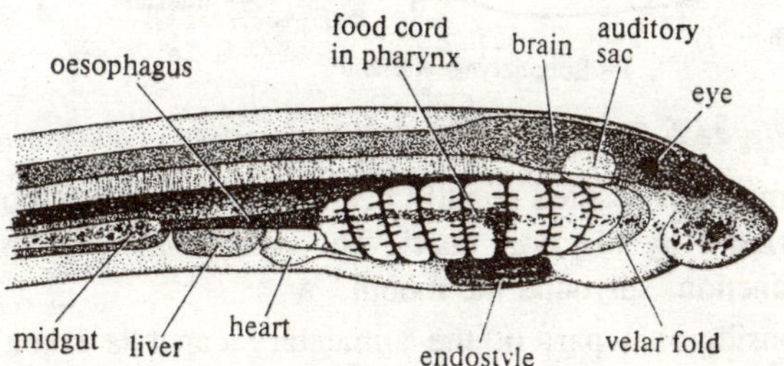

Fig. 14.2. Young ammocoete larva of lamprey fixed while feeding on green flagellates and detritus and then stained and cleared.

The brook lamprey, Entosphenus, spends its entire life in fresh water. It is much smaller than Petromyzon, but its breeding habits are essentially the same.

The three diagnostic characters of the chordates are prominent. The *dorsal nerve tube* occupies its usual position, and its anterior portion is enlarged to form the brain. The *notochord* is very large. The *pharyngeal gill slits* are seven pairs of openings between the pharynx and the outside. Indications of segmentation can be seen in the muscular system. In ammocoetes, the muscles in the dorsal part of the body are in the form of blocks extending from the head to the tail. This muscular segmentation can be best observed at the level of the notochord. (The notochord is beneath the muscles, though in the nearly transparent specimen the two appear to be at the same level.)

Digestive and Respiratory Systems

The alimentary canal is a straight tube, nearly as long as the body, that extends from the mouth to the opening of the cloaca. It develops from the archenteron. Its mucosa lining and the glands derived from it are endodermal. The connective tissue, parts of the circulatory system, and smooth muscle that make up much of the wall of the alimentary canal are mesodermal. Its nerve supply is ectodermal.

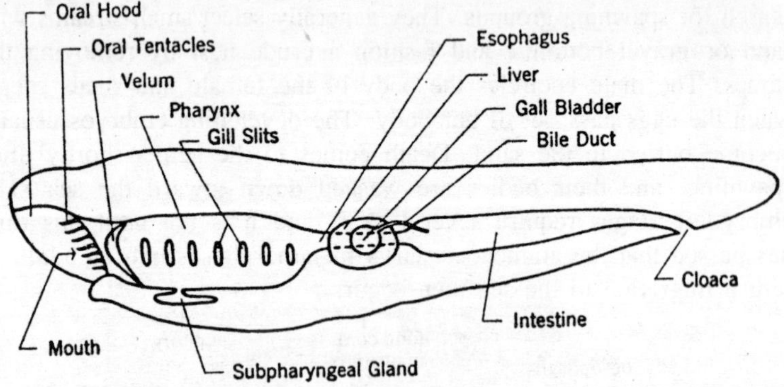

Fig. 14.3. The digestive and respiratory systems of ammocoetes.

The mouth is located slightly behind the anterior end of the body. It is overhung by the *oral hood*. The *oral tentacles*, which have a sensory function, surround the mouth.

A considerable part of the alimentary canal is taken up by the *pharynx*. This division has important respiratory as well as digestive functions, which is why the two are treated together in this section. The *velum* is located in the anterior part of the pharynx. Its movements help to create a current of water that enters the mouth, passes through the pharynx and then leaves the body through the gill slits. There are seven pairs of *gill slits*, which are openings extending from the pharynx to the outside. The wall of the pharynx between adjacent gill slits is strengthened by rods of cartilage, the *gill bars*. Each gill slit has a *gill* on both its anterior and its posterior surface. The gills, which are constantly bathed by a current of water, are richly supplied by capillaries. It is here that the blood loses carbon dioxide and gains oxygen.

The *subpharyngeal gland*, which passes into the pharynx. It opens into the lumen on the pharynx by a small duct. One of its functions is to secrete mucus, which passes into the pharynx. Because mucus is sticky, tiny particles of food adhere to it. Both food and mucus are forced posteriorly into the intestine by cilia lining the pharynx. The subpharyngeal gland has some resemblance to the endostyle found on the floor of the pharynx in amphioxus. When metamorphosis occurs, a portion of the subpharyngeal gland is converted into the *thyroid* which, as in all vertebrates, is able to accumulate iodine.

At its posterior end, the pharynx narrows to form the *esophagus*. This portion, which is relatively short, terminates in the *intestine*.

In the embryo, the *liver* arises as an outgrowth from the gut wall in the region that will form the anterior part of the intestine. It maintains a connection by means of the *bile duct*. Note that the *gall bladder* in ammocoetes is unusually large.

The *pancreas* is not present as a separate organ, but cells similar to those found in the pancreas of higher forms are present in the anterior part of the intestine. In higher forms the pancreas has two main functions are secreted by zymogen cells; and the hormone insulin, which helps to regulate the amount of glucose in blood, is secreted by the islets of Langerhans. In ammocoetes, the zymogen cells are found in the mucosa of the anterior portion of the intestine. The islet cells bud off from the intestine and form clusters in the submucosa. It is interesting that these cells have the same physiological function as they do in higher vertebrates, namely, the regulation of sugar concentration in the blood.

It is probable that nearly all digestion and absorption occur in the intestine. Can you think of any reason why the extent of digestion might be limited in the pharynx? The intestine is horseshoe shaped when seen in cross section. At the posterior end the kidney ducts join the intestine, thus forming a *cloaca*.

The alimentary canal of ammocoetes is of interest not only for its simplicity but also for the structures it lacks. There are no teeth of the typical vertebrate type. The adult possesses horny tooth-like structures that are formed entirely by the ectoderm, while the true teeth of higher vertebrates are entirely different structures, which are derived from both ectoderm and mesoderm. There are no jaws in the larval or adult lamprey. The alimentary canal shows relatively little regional differentiation in the post-pharyngeal portion. There is no stomach, no separate pancreas, and no division of the intestine into a small intestine and large intestine. The alimentary canal of the ammocoete larva is probably of as simple a type as we could expect in the early chordates.

CIRCULATORY SYSTEM

The blood system of ammocoetes appears to be very primitive, and once again we lean heavily on this form for our concept of what the circulatory system may have been like in the early chordates.

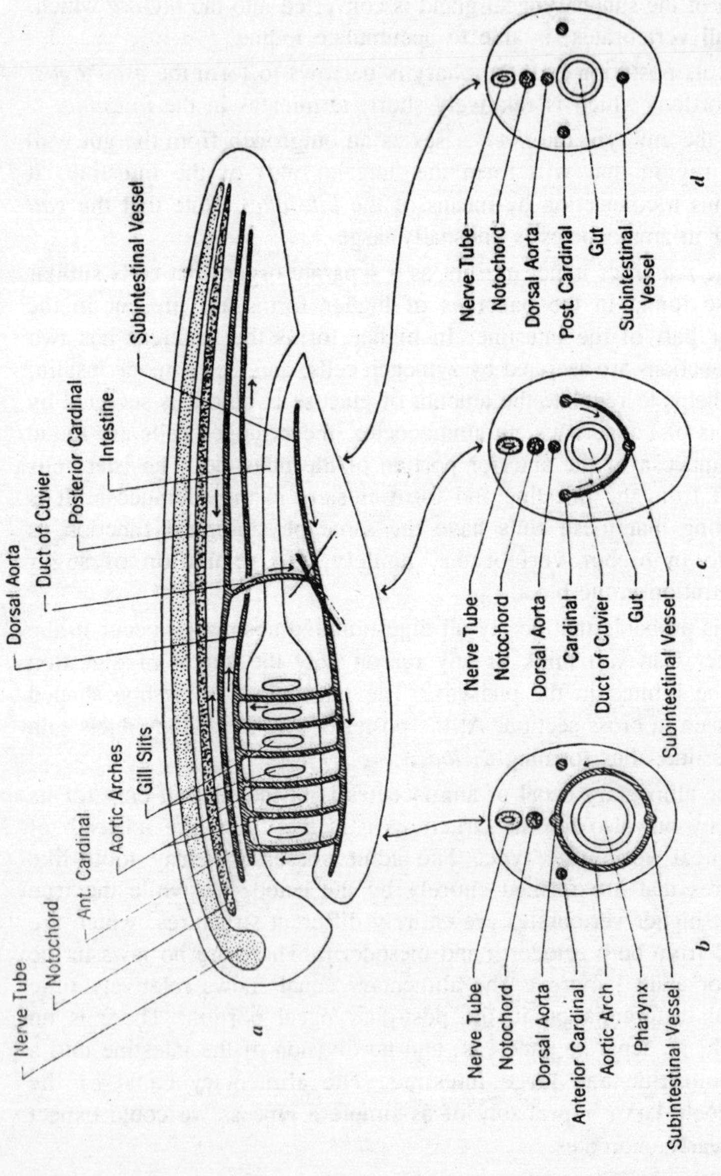

Fig. 144. The hypothetical basic plan of the chordate circulatory system: (a) is a lateral view; (b), (c), and (d) are cross sections of the body at the levels indicated by the arrows.

Basic Plan of the Circulatory System

The basic plan of the circulatory system of the hypothetical early chordates might be imagined to consist of four longitudinal vessels with two types of interconnection. The four longitudinal vessels would be:

1. The *dorsal aorta*, which extends along the dorsal side of the alimentary canal immediately below the nerve tube.
2. The *subintestinal vessel*, which runs along the ventral side of the alimentary canal.
3. The *right cardinal vein*, which is in the body wall on the right side.
4. The *left cardinal vein*, which occupies a similar position on the left side.

The two types of connections, excluding capillaries, are:

1. *The right and left ducts of Cuvier*. These carry blood from the cardinal veins to the subintestinal vessel. That portion of the cardinal vein anterior to the duct of Cuvier is the *anterior cardinal*. The portion that is posterior is the *posterior cardinal*.
2. *The aortic arches*. These vessels are found in the region of the pharynx. They carry blood from the subintestinal vessel to the dorsal aorta. In the lower chordates the aortic arches form a capillary network in the gills, and it is here that oxygen is obtained and carbon dioxide lost.

Circulatory System of Ammocoetes

The general plan just given applies more-or-less to the embryo of all chordates, including ammocoetes. In ammocoetes, the *dorsal aorta* is the chief distributing vessel. It carries oxygen and digested foods to all the tissues. The direction of blood flow is toward the tail, except in the region anterior to the gills where it flows toward the head. It gives off numerous small branches. These are evenly spaced and correspond to the myotomes, thus exhibiting segmentation. There is only one large vessel branching from the aorta. It supplies the intestine, and is known as the *intestinal artery*.

The primitive condition, in which there was a continuous subintestinal vessel, is modified in ammocoetes. The posterior portion, in the region of the intestine, begins in capillaries in the gut wall and runs to the liver, where it forms capillaries. This is the *hepatic portal vein*. After passing through the liver capillaries, the blood is carried by the *hepatic vein* to the sinus venosus. The next more anterior portion

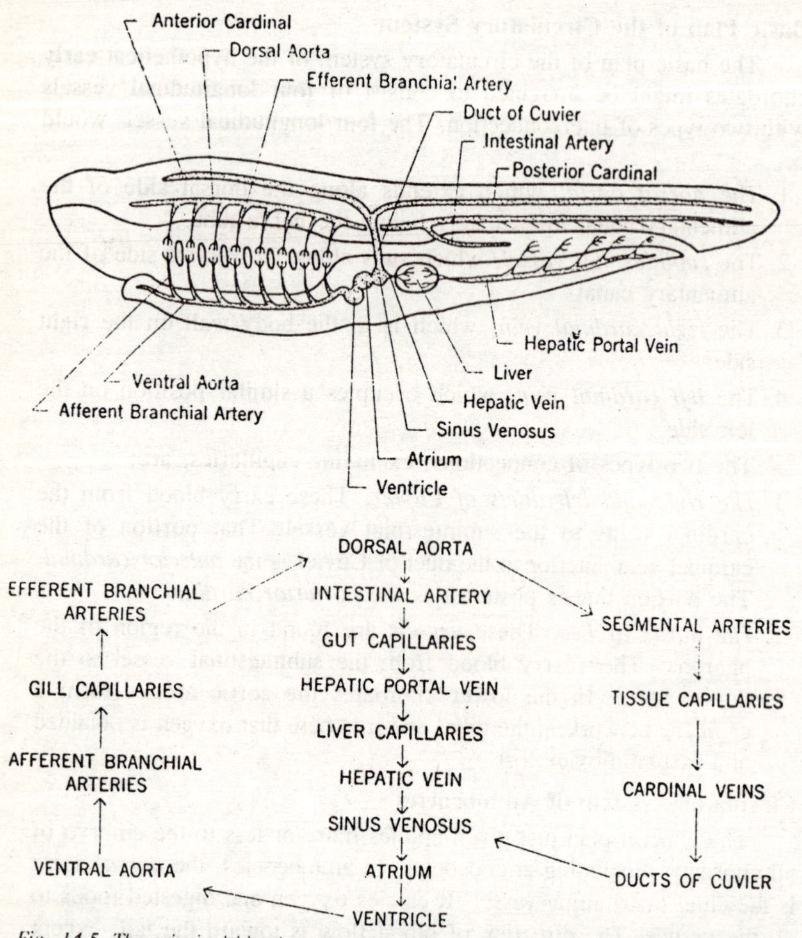

Fig. 14.5. The principal blood vessels and the route of blood movement in ammocoetes.

of the subintestinal vessel is the heart. The first portion is a thin-walled chamber, the *sinus venosus*, into which the hepatic vein and the ducts of Cuvier open. Blood passes from this into the *atrium* and then to the *ventricle*. The heart of the ammocoete has but one atrium and one ventricle, the latter having thick muscular walls which, by contracting, force the blood out through the *ventral aorta*. The ventral aorta is located immediately below the pharynx. It gives off eight pairs of *aortic arches*—eight vessels for the right side and eight for the left side. There is a single vessel between each gill slit, one anterior to the first gill slit, and another posterior to the seventh gill slit. The portion of the aortic arch that carries blood to the gill slits is the *afferent branchial artery*. The portion of the aortic arch that carries blood from the gill capillaries to the dorsal aorta is the *efferent branchial artery*.

The cardinal veins collect blood that was carried to the tissues by branches of the dorsal aorta. The *anterior cardinals* are lateral to the notochord and the *posterior cardinals* are lateral to the dorsal aorta. Blood from the cardinals flows into the right and left *ducts of Cuvier*, which carry it to the sinus venosus.

Finally, in the late larva the left duct of Cuvier disappears. The anterior and posterior cardinal on this side then open into the right duct of Cuvier. This change is a specialized feature of the lamprey and has no phylogenetic significance.

Excretory System

The excretory organs of the vertebrates are derived from the mesoderm. For this reason a study of this embryonic layer will aid in understanding the morphology of the kidneys and their ducts.

Derivatives of the Mesoderm

Trunk approximates the structure of the ammocoete and other primitive vertebrate embryos. The dorsal portion of the mesoderm is segmented, each block of tissue being known as a *somite*. The greater portion of the somite is the *myotome* region, which will give rise to most of the striated muscles. The outer portion of the somite is the *dermatome*. The dermis layer of the skin comes from it. There is much evidence suggesting that the somite had a cavity, the *myocoel*, at an early stage in its evolution and it is found in the embryos of some chordates. The inner, ventral portion of the somite is the *sclerotome*. The vertebrae develop from its cells.

The *nephrotome* area consists of a series of connections between the somites and the ventral portion of the mesoderm. The kidneys and their ducts originate from the nephrotome.

The mesoderm of the ventral region forms the *coelomic lining* and the *mesenteries* that cover the organs suspended in this body cavity. There is some evidence to suggest that the coelom may have been segmented at a very early evolutionary stage, but the primitive vertebrates living today have an unsegmented coelom except for the myocoel.

Excretory Organs of Ammocoetes

The excretory organs of ammocoetes first develop from the nephrotomes in the heart region. The most anterior nephrotome involved

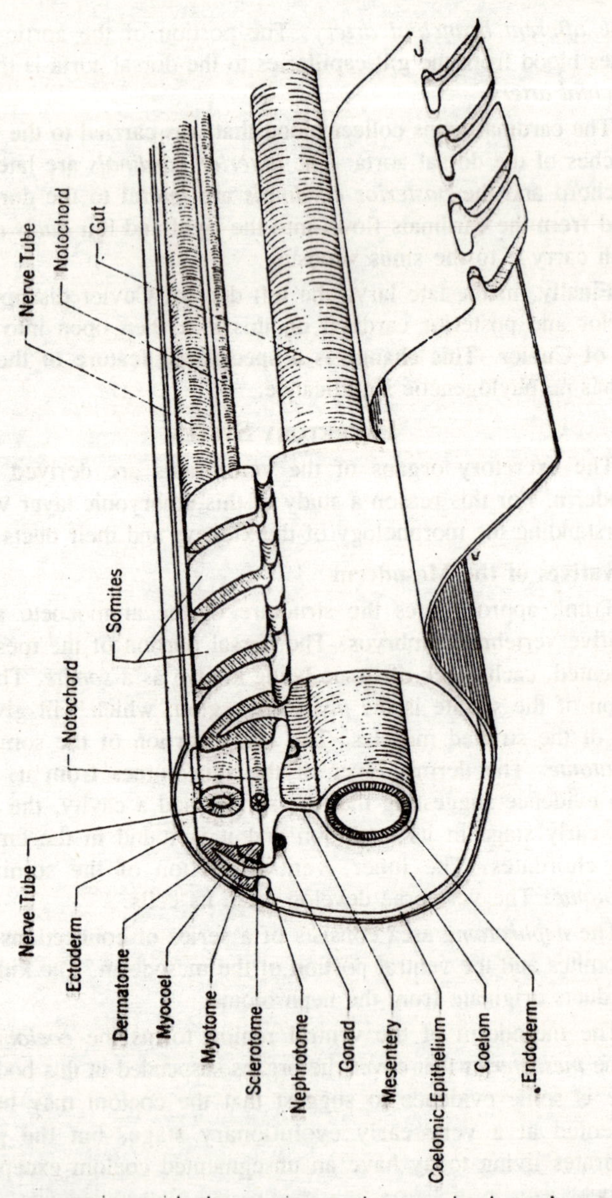

Fig. 14.6. A schematic representation of the trunk region of a chordate embryo.

breaks its conneetion with the somites and grows backward as a duct. The next posterior nephrotome connects with this duct. This process occurs on both sides of the body; a total of four pairs of nephrotomes are involved. They form the *pronephros* ('first kidney'), which is the

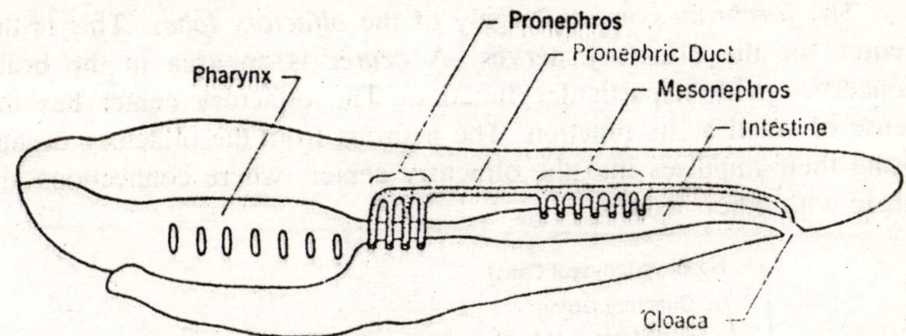

Fig. 14.7. The excretory system of ammocoetes.

functional kidney of the early embryo. The ducts, which are known as the *pronephric ducts*, grow posteriorly and enter the cloaca. Later in development additional tubules are formed in a more posterior region, opposite the intestine. These are the *mesonephric tubules*. They open into the already formed pronephric duct, which then becomes known as the *mesonephric duct*. Eventually, the pronephros degenerates leaving the mesonephros as the functional kidney of the adult.

Reproductive System

The gonads arise as paired ridges on the dorsal side of the coelom. They fuse subsequently to form a median structure. It is thought that this fusion is a specialization of the lamprey, since in all other vertebrates the gonads are paired. The lamprey lacks genital ducts, the gametes being shed directly into the coelom. In sexually mature individuals, there is an opening from the coelom into the posterior part of the mesonephric duct. It is through this opening that the gametes leave the coelom. The mesonephric ducts break their connection with the alimentary canal at the time of metamorphosis, and the right and left ducts unite, forming a urinogenital papilla. It is through this that urine and gametes leave the body. The urinogenital papilla and the intestine open separately into a shallow depression on the ventral side of the body. The adult, therefore, does not have a cloaca. The ammocoete is, after all, an embryonic stage, and the organs of excretion and reproduction do not reach their full development.

Nervous System

The central nervous system of ammocoetes consists of the brain and spinal cord. The brain exhibits the basic chordate plan. It is composed of three main divisions: forebrain, midbrain, and hindbrain. These three divisions are functionally associated with the three main sense organs, namely, olfactory organs, eyes, and auditory organs.

The *forebrain* consists largely of the *olfactory lobes*. This is the center for the olfactory nerves. A *center* is an area in the brain concerned with a particular function. The olfactory center has the sense of small as its function. The neurons from the olfactory organs bring their impulses into the olfactory center, where connections are made with other neurons.

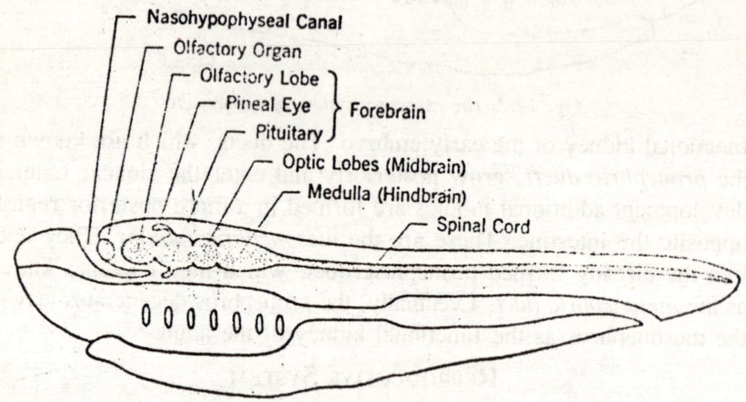

Fig. 14.8. A schematic representation of the central nervous system of the ammocoete larva.

Outgrowths on the dorsal side of the forebrain gives rise to the *pineal eye* and *parapineal eye*, both of which appear to be concerned with colour changes in the ammocoete. The paired *eyes* develop from the forebrain of the embryo. The *infundibulum* is formed from the ventral portion of the forebrain. The *hypophysis* is formed of cells and it is an ingrowth of the superficial ectoderm. The infundibulum, together with the hypophysis, forms the *pituitary gland*.

The *midbrain* is composed largely of the *optic lobes*. These are the center for the nerves originating in the retina of the eye.

The *hindbrain* consists largely of the *medulla*. A tiny *cerebellum* is formed from the hindbrain roof. The hindbrain is a center for nerves coming from the *ears* which, in the lamprey, are organs of balance.

The *spinal cord* begins at the hindbrain and extends to the tip of the tail. In every segment a pair of *dorsal roots* and a pair of *ventral roots* connect with the spinal cord. The dorsal roots consist mainly of sensory neurons, while the ventral roots consist of motor neurons. The nucleus of each sensory neuron is located in a ganglion on the dorsal root. In contrast to the condition in higher vertebrates, the dorsal and ventral roots of the spinal nerves do not join. The nerves of the lamprey have no myelin covering.

Skeletal System

The skeleton of the adult lamprey is very poorly developed, as compared with higher chordates. This is in part the result of degeneration. The skeleton consists of cartilage cases that partially surround the brain and sense organs. The gill region is supported by cartilaginous bars that form the branchial basket, and the vertebrate are represented by tiny pieces of cartilage adjacent to the notochord. The notochord is the main portion of the trunk skeleton of both the ammocoete and the adult. Jaws and paired appendages are lacking, as are those portions of the skeleton associated with these structures in higher forms.

15

ENDOCRINE GLANDS

EMBRYONIC DEVELOPMENT

Throughout the vertebrate series the pituitary shows a consistent pattern of development, involving two distinct components; a neurohypophysis, derived from the neural tissue of the floor of the diencephalon, and the adenohypophysis, an epithelial component originating from the epithelium of the embryonic mouth cavity. The cyclostomes diverge from the typical vertebrate pattern in that the adenohypophysial tissue develops from a naso-hypophysial anlagen, which in early ontogeny, lies immediately in front of the stomodeal depression. In all probability, this mode of development was also shared by the fossil agnathans and was already established at the time of the agnathan-gnathostome dichotomy. Nevertheless, although by slightly different routes, both vertebrate groups have attained the same association of nervous and glandular tissue that has made possible the integration of endocrine functions and the adaptation of hormonally controlled physiological processes to changes in the external environment.

In the gnathostome embryo, the stomodaeum develops a dorsal evagination — Rathke's pouch, directed upwards towards the floor of the brain — from which the adenohypophysial tissue is differentiated. Meanwhile, the pouch has made contact with a ventral extension from the floor of the hypothalamus — the infundibulum — from which the neurohypophysis subsequently develops. Although in the course of ontogeny, the oral connection of Rathke's pouch is normally lost, in some fish species an open connection with the mouth cavity may persist in the adult as a hypophysial duct. Neither in myxinoids nor in lampreys is a Rathke's pouch of the gnathostome type developed. Here, the adenohypophysial tissue originates from the epithelium of the nasohypophysial canal or in lampreys, from a cellular cord, within which this canal is subsequently formed. A strict homology between the nasohypophysial tract of the cyclostomes and the Rathke's pouch of the higher vertebrates remains somewhat controversial, but it would not be difficult to visualize the derivation of both cyclostome and gnathostome modes of pituitary development from a common ancestral pattern.

In lampreys, that part of the nasohypophysial tract that extends posteriorly from the olfactory organ remains, throughout the larval period, a solid epithelial cell cord. The differentiation of glandular tissue from this structure seems to be a continuous process, which continues or is renewed during metamorphosis at the time when this solid cord of cells is being converted into a hollow tube by the formation and fusion of intracellular vacuoles. No information is available on the earliest development of the pituitary in myxinoid embryos, but in the comparatively late stage described by Fernhorm (1969) the naso-hypophysial rudiment was an open canal, from which adenohypophysial tissue appeared to be differentiating. Even in the adult, diverticula of the canal may penetrate into the glandular tissue.

ADENOHYPOPHYSIS

General Morphology

Because of the absence of any division into distinct regions and the undifferentiated state of its cellular elements, the myxinoid adenohypophysis is unique amongst the vertebrates. Such simplicity is unlikely to be a primitive feature and is consistent with the regressive trends that have affected so many aspects of myxinoid organization.

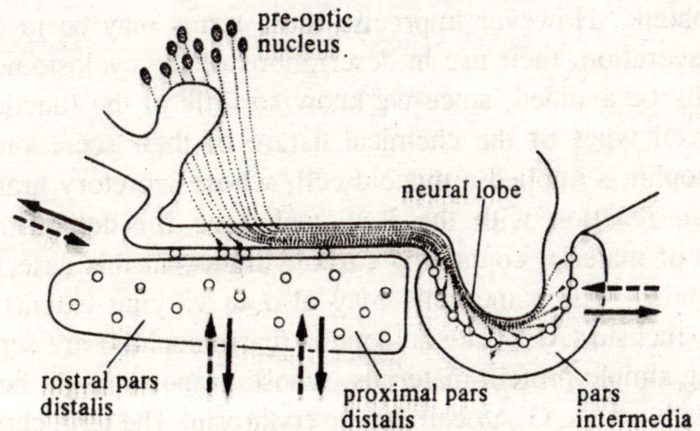

Fig. 15.1. Pituitary of lamprey.

The glandular tissue is in the form of a flattened plate lying in front of, and below the infundibular process, from which it is separated by a dense zone of connective tissue.

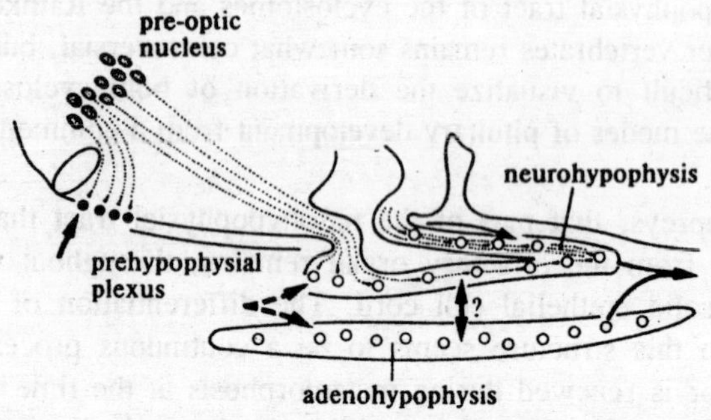

Fig. 15.2. Myxinoid pituitary.

In contrast to this simple organization, the adenohypophysis of the lamprey is (except in early larval stages) divided into three distinct regions, separated from one another by connective tissue septa. The two more anterior lobes, the pro- and meso-adenohypophysis are separated from the floor of the brain by a layer of connective tissue, but this thins out over the posterior region — the meta-adenohypophysis — and in this region there is a capillary plexus (mantle plexus) at the interface between the adenohypophysis and the neural tissue of the infundibulum.

Cytological Differentiation

In the description that follows, the cells of the adenohypophysis are referred to the three broad groups, whose staining reactions are described by optical microscopists as basophilic, acidophilic and chromophobic. However imprecise these terms may be in relation to cellular secretion, their use in descriptions of the cyclostome pituitary can hardly be avoided, since we know so little of the functions of the various cell types or the chemical nature of their secretions. By the term basophil is implied a mucoid cell, whose secretory granules give a positive reaction with the PAS technique for demonstrating the presence of material containing carbohydrates, in this case, muco- or glycoproteins. The same cells may also to varying extents, react to aldehyde-fuchsin (AF). The acidophils (carminophils) are serous cells, producing simple protein materials, whose granules might be expected to stain with orange G, azocarmine or erythrosin. The term chromophobe is ambiguous in the sense that cells which fail to respond to dyes,

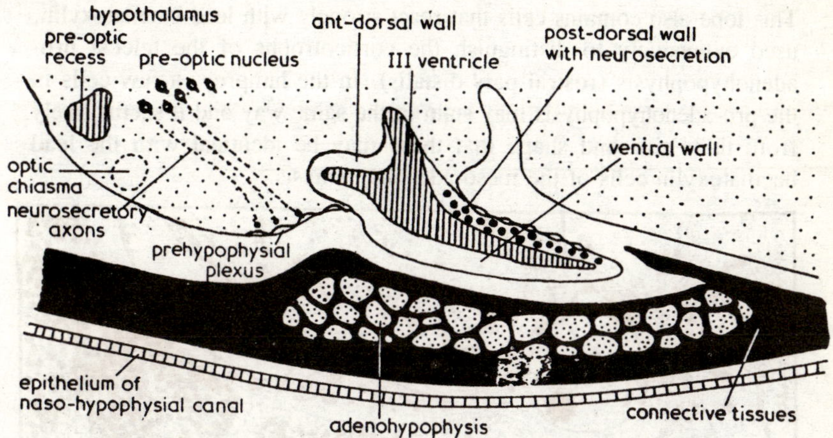

Fig. 15.3. Diagrammatic sagittal section through the pituitary of a hagfish.

may nevertheless show secretory granules in electron micrographs. These may be described as 'active' chromophobes or granular chromophobes to distinguish them from cells that show no indications of secretory activity.

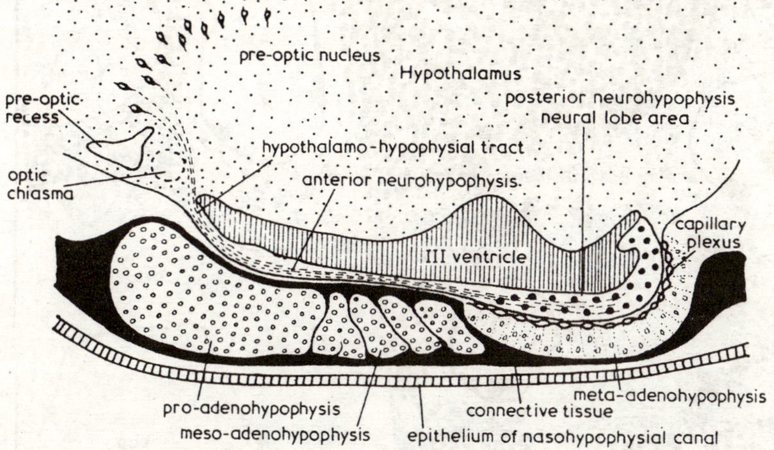

Fig. 15.4. Diagrammatic parasagittal section through the pituitary of a lamprey.

Lampreys

Both pro- and meso-adenohypophysis contain basophils, reacting to PAS, AF or alcian blue staining. Apart from the presence of chromophobes, many authors have also observed acidophils or carminophils, generally but not exclusively in the meso-adenohypophysis.

This lobe also contains cells that react strongly with lead haematoxylin, used extensively to distinguish the corticotrophs of the teleost pro-adenohypophysis (rostral pars distalis). In the lamprey, a few cells in the pro-adenohypophysis may stain in the same way and it seems likely from their size and shape that these may be identical with the lead haematoxylin cells of the meso-adenohypophysis.

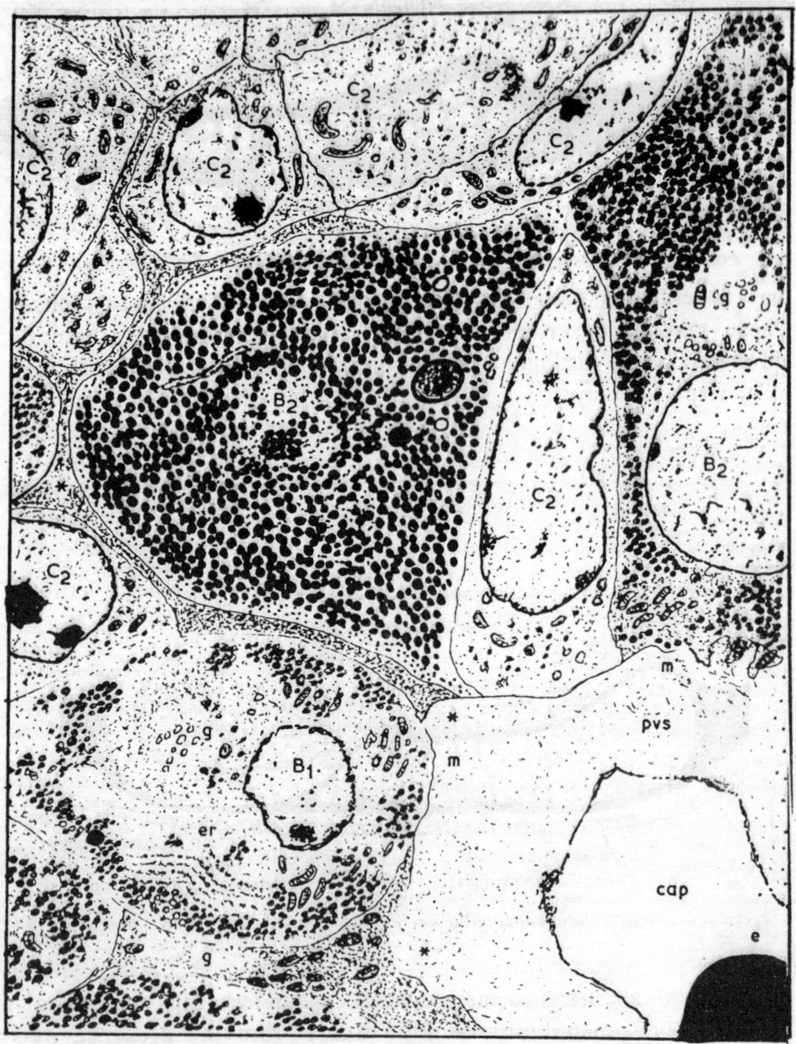

Fig. 15.5. Cell types in the pro-adenohypophysis of the lamprey, Lampetra fluviatilis, during the period of the upstream migration × 5700.

Compared to this limited range of cells types that can be distinguished by optical microscopy, the application of the electron microscope has produced a great variety of so-called cell types, in striking contrast to the apparently limited hormonal repertoire of the cyclostomes. These have been distinguished on the basis of their general morphology, position within the gland, the development of their synthetic organelles and above all the size and character of their secretory granules. Unfortunately, these methods of analysis often make it difficult to decide whether particular cells are merely stages in the secretory cycle of the same cell, rather than distinct functional elements. With this in mind, it is safe to assume that the number of functional cell types will always be less than the number that have been distinguished on the basis of these criteria.

The majority of cytological studies on the lamprey pituitary have been confined to the upstream migrant stage of the parasitic species, when the animals are already approaching sexual maturity. For this reason the discussion that follows has been based mainly on the condition of the gland at this stage of the life cycle.

Pro-adenohypophysis This lobe consists of vertically arranged cell cords, separated by connective tissue. In the sexually maturing stage, the dominant cell type is the basophil, whose granulation tends to become more intense as the animals approach the time of spawning. Although some authors have described several basophilic cell types, it now seems likely that these are only different stages in the development of a single cell. For example, in upstream migrants of *L. fluviatilis*, Båge and Fernholm (1975) described three basophils B^1-B^3 with different maximum and mean granule diameters, but with almost identical minima. The synthetic organelles — Golgi and endoplasmic reticulum — were best developed in the B^1 cell with the smallest granules, but were reduced in the B^3 cells with the largest granules. Moreover, the B^1 cell predominated in the autumn and the B^3 cell in the winter and spring. Clearly these three cell types can be interpreted as successive stages in the development of a single cell, which accumulates its maximum store of secretory materials in the final stages of sexual maturation. This example should also serve as a warning against placing too much reliance on the size of secretory granules in the identification of cell types. In the same species, Larsen and Rothwell (1972) described two types of pro-adenohypophysial basophil (III, IV), which can almost certainly be equated with the basophils of Båge and Fernholm.

Two types of chromophobe have been identified in the proadenohypophysis. The first of these shows no signs of secretory activity and is a small cell with a thin rim of cytoplasm surrounding the nucleus. This can be confidently described as an undifferentiated stem cell. A second type of chromophobe described by Båge and Fernholm appears from its fine structure to be an active secretory cell, becoming prominent in late winter and spring and containing granules, whose diameters form a double peak at 80 and 105 nm. An additional, but rare, chromophobe was also seen. This cell (D) which has small granules, appears to reach its maximum development early in the migratory period and may be identical with the scarce lead haematoxylin cells of the pro-adenohypophysis and their more numerous counterparts in the meso-adenohypophysis.

In its general morphology and the size of its granules it can probably be equated with the second type of chromophobe (type II)

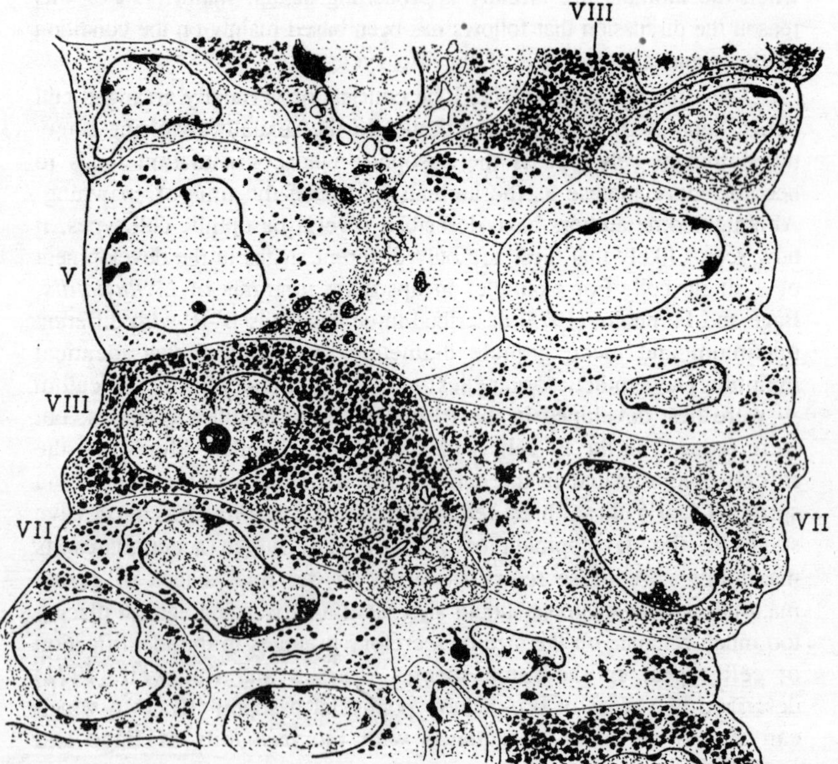

Fig. 15.6. Cell types in the meso-adenohypophysis of the lamprey, Lampetra fluviatilis.

described in the same species by Larsen and Rothwell (1972). Stellate cells, sometimes described as chromophobes, have been identified in the pro-adenohypophysis. These are relatively small, nongranular elements which are apparently capable of amoeboid and phagocytic activity. This is shown by their elongated cytoplasmic extensions (containing bundles of microfilaments) that are to be seen amongst the secretory cells and along the outer surfaces of cell cords. According to Båge and Fernholm these cells are found only in the pro-adenohypophysis of *L. fluviatilis*, although in *P. marinus*, Leatherland (1975) has also found them closely associated with the mesoadenohypophysial basophils, and with fibrovascular septa, to which they are attached by desmosomes. Apart from their phagocytic functions it has been suggested that they may regulate the secretion of the basophils by opening and closing the gaps between the cell surface and the perivascular space.

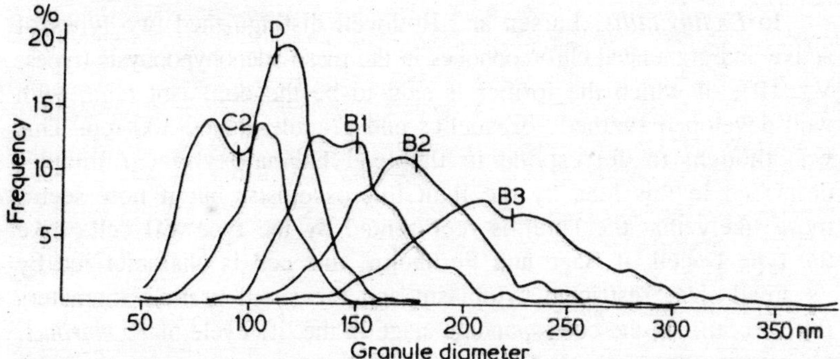

Fig. 15.7. Distribution of granule diameters in the basophils of pro-adenohypophysis of Lampetra fluviatilis.

Although ultrastructural studies have also been made on *L. (Entosphenus) tridentata* and *P. marinus*, it is difficult to compare these with the pituitary of *L. fluviatilis* in the absence of adequate light microscopic correlations. In the pro-adenohypophysis of *L. tridentata*, two types of granular cell were described; the first, (type I) thought to be in the storage phase and the second (type III) an active synthetic stage. In the first the granule diameters were 230—500 nm and 280-380 nm in the second. Similarly, two highly granular cells are described in the pro-adenohypophysis of the sexually mature *P. marinus* with granules 130—250/260 nm in diameter. In both species it seems likely that we are dealing with a single functional cell type.

Meso-adenohypophysis The cells of this region are more difficult to interpret than those of the pro-adenohypophysis and this is especially true of the several kinds of active chromophobes. In *L. fluviatilis*, the mesoadenohypophysial basophil appears to have distinctive staining properties and its granules are somewhat smaller than those of the basophil of the proadenohypophysis (80-200 nm). These cells, unlike those of the proadenohypophysis, only develop their granulation after metamorphosis and in the spawning migrant are said to become depleted after breeding. In *L. tridentata*, the ultrastructural counterpart of the meso-adenohypophysial basophil of *L. fluviatilis* may be the type IV cell described by Tsuneki and Gorbman, which has a similar vacuolated cytoplasm and is packed with granules with diameters of 240-300 nm. A type V cell has been described by these authors with vacuolated cytoplasm, but without prominent granulation and may represent the same cell in a depleted state.

In *L. fluviatilis*, Larsen and Rothwell distinguished two types of active and granulated chromophobes in the meso-adenohypophysis (types, V, VII), of which the former is said to be the dominant type, with well-developed synthetic organelles and granules of 60-300 nm. This was thought to correspond to the lead haematoxylin carminophil described in this lobe by the light microscopists, but it now seems more likely that the latter is repr sented by the type VII cell. Like the type D cell of Båge and Fernholm, this cell is characterized by its small size, restricted cytoplasm and the small granule diameters (80-150 nm). In the corresponding stage in the life cycle of *P. marinus*, two types of active chromophobe are present in the mesoadenohypophysis; both small cells with small granules, either of which could correspond to the type VII cell of *L. fluviatilis*. In *L. tridentata*, no fewer than four types of granular chromophobes have been listed, of which the closest equivalent to the *fluviatilis* type VII appears to be type VI of Tsuneki and Gorbman, which has comparatively small granules of 110-180 nm.

Meta-adenohypophysis This is separated from the overlying pars nervosa by only a thin connective tissue layer, containing capillaries, although the meta-adenohypophysis itself is avascular and is not penetrated by nerve fibres. In the ammocoete, this lobe of the pituitary is composed of chromophobic cells and the secretory granules (diameters 160-240 nm) do not appear until metamorphosis. In the adult lamprey, a variety of cell types can be observed, but there is general agreement that these represent stages in the secretory activity of a single cell

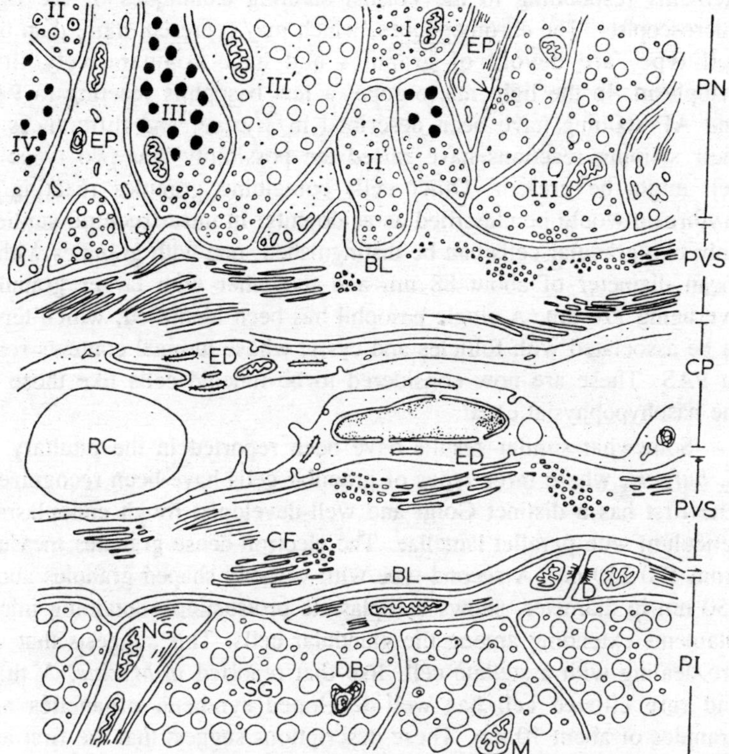

Fig. 15.8. Structure of the boundary region between the meta-adenohypophysis and the neural lobe region of the lamprey neurohypophysis.

type. Both the granular and agranular cells are elongated and arranged in a palisade-like fashion, with their cytoplasmic processes frequently extending vertically upwards to reach the fibrovascular space separating the meta-adenohypophysis from the pars nervosa above. In what is interpreted as their active secretory state, the secretory granules are concentrated in the apices of the cells, while the extensive lamellated rough endoplasmic reticulum is confined towards the base of the cell containing the nucleus. At spawning time the cells are said to be depleted of their secretory granules, which are presumably discharged into the capillary space between the neuro- and meta-adenohypophysis.

Myxinoids

The myxinoid adenohypophysis consists of cell nests and follicles, some with a central lumen. Its outstanding features are the predominance of chromophobic cells and the scarcity of granular elements responding to the normal staining techniques of the light microscopists. The chromophobes, which may represent more than one cell type, are devoid of granules and have comparatively little cytoplasm. In the light microscope, a few basophils reacting to PAS and AF staining have been described in *Myxine*, and differences in their staining reactions have raised the possibility that two types of cell might be present. Other cells containing granules reacting to erythrosin would be classified as acidophils. In ultrastructural studies, only two granular cells can be distinguished; one with granules with a mean diameter of about 88 nm and the other with larger granules averaging 176 nm. A single basophil has been identified, which tends to be associated with follicles and cysts, whose luminal contents react to PAS. These are now considered to be mucoid cells like those of the nasohypophysial canal.

Somewhat similar results have been reported in the pituitary of *E. burgeri*, where three types of granular cells have been recognized. The first has a distinct Golgi and well-developed rough endoplasmic reticulum with parallel lamellae. The electron-dense granules measure from 170-250 nm. A second type with variably shaped granules about 150 nm in diameter, shows cytoplasmic protrusions containing micro filaments extending among the glandular cells. This suggests that we are dealing with a stellate cell, like that reported in *Myxine*. A third and rare ciliated cell has well-developed synthetic organelles and granules of about 70 nm. These descriptions suggest that the first and third of these cells in *Eptatretus* may correspond to the two types of granular cell in *Myxine*. Based on average granule sizes, Tsuneki (1976) has also distinguished three granulated cell types in *E. burgeri*, although the extent of the overlap in granule diameters made it difficult to characterize them with precision. These types had granule diameters of 220-310, 170-220 and 100-170 nm respectively, presumably corresponding to the cell types described by Fernholm and Oota. In none was the rough endoplasmic reticulum well-developed and the degree of cytoplasmic activity, as judged by the state of the Golgi, mitochondria or the lysosomes, was very variable.

It now seems doubtful whether the myxinoid pituitary has any cellular elements comparable with the meta-adenohypophysis of other vertebrates. In larger (and older) specimens of *Myxine* some histological differentiation was noted at the caudal extremity of the adenohypophysis, where the connective tissue separating it from the neurohypophysis was thinner than elsewhere. In the gnathostomes, the aboral part of

Rathke's pouch, first coming into contact with the infundibulum, differentiates as the pars intermedia and by analogy it was suggested that the region of the *Myxine* adenohypophysis in closest contact with the neural lobe might have a similar connotation. However, some doubt has been thrown on this view by the fact that similar regions of contact have not been seen in *Eptatretus*.

Functional Interpretations

Lampreys

In attempts to identify functional cell types in the adenohypophysis, a variety of experimental techniques have been attempted, including the use of hormonal inhibitors, changes in environmental conditions and the surgical removal of all or part of the gland (hypophysectomy). In addition, cytological comparisons at different points of the life cycle might be expected to yield some clues to the functional activities of particular cell types in relation to such events as metamorphosis or spawning.

Perhaps the most obvious example of a correlation between cellular activity and a particular phase of the life cycle, is the apparent increase in the secretory activity of the pro-adenohypophysial basophils during sexual maturation. This leaves little room for doubt that these cells are involved in some aspects of reproductive physiology. The interpretation of the meso-adenohypophysial basophil is more difficult. These cells are said to become depleted of their secretory granules shortly before, or during the spawning season. Earlier suggestions that they might be thyrotrophs are difficult to sustain in view of the absence of any direct evidence for thyrotrophic activity in lampreys. On the other hand, the effects of hypophysectomy on sexual maturation are indicative of some kind of gonadotrophic activity in the meso-adenohypophysis. Assuming that this gonadotrophic activity is due to a glycoprotein, the meso-adenohypophysial basophil must be the most likely source. Furthermore, if as seems probable, the basophils of the pro- and meso-adenohypophysis are distinct functional cell types, we are forced to the conclusion that the lamprey pituitary produces two gonadotrophic hormones with different target organs and different physiological roles

Of the various active chromophobes that have been described ultrastructurally, one of these almost certainly corresponds to the lead haematoxylin carminophil of the light microscopists. These have been found to be degranulated after injections of aldactone (an inhibitor of corticosteroid biosynthesis), or when lampreys have been kept in chlorinated tap water. These responses to stress or corticosteroid inhibition suggest that we are dealing with a corticotroph and it may be significant that these cells are reported to increase in number at the time of spawning when some heightening of interrenal activity might be expected. On the other hand, to postulate the existence of a corticotroph is not without its difficulties in view of the absence of any evidence that lampreys are able to synthesize the usual range of vertebrate corticosteroids. Although the type V cell of the meso-adenohypophysis has been suggested as a possible corticotroph, the location and cytological characteristics of the type VII cell make this a more likely candidate. It appears that only a single functional cell type is present in the meta-adenohypophysis and that this is involved in the secretion of an MSH type hormone, but it is doubtful whether pigmentary control is its sole function. The suspicion that the meta-adenohypophysis may be involved in wider metabolic functions is strengthened by its comparatively large size, its close vascular relationships to the overlying neurohypophysis and the fact that its secretory activity becomes intensified at the time of metamorphosis.

In the sea lamprey *P. marinus*, Percy *et al.* (1975) have studied the fine structure of the pituitary from early larval stages through metamorphosis. In younger ammocoetes within two years of hatching, the adenohypophysis is a thin plate of chromophobic cells, without internal fibrovascular septa or regional divisions into distinct lobes. In electron micrographs, granular cells can be detected in the anterior third of the adenohypophysis, corresponding in position to the site of the future pro-adenohypophysis and underlying that part of the neurohypophysis where a median eminence-like structure has been described. Differentiation of the gland into its three component lobes is delayed until the ammocoete reaches lengths of about 90 mm and at this stage, the active cells of the pro-adenohypophysis contain granules of a similar size to those of the younger ammocoetes (140-230 nm) or to the basophils of the upstream migrants. On the other hand, it is not until the ammocoete is approaching metamorphosis at lengths of 130-150 mm that the basophils can be recognized in the light microscope by their affinity for PAS and aniline blue staining.

Up to the onset of metamorphosis, the meso-adenohypophysis appears inactive and is almost completely devoid of granular cells. At this time there is a rapid growth in the adenohypophysis, which in the non-parasitic *L. planeri* is due mainly to a rapid enlargement of the meso-adenohypophysis. On the other hand, in the closely related and parasitic form *L. fluviatilis*, little or no pituitary growth occurs in the earlier stages of metamorphosis and it is not until the end of metamorphosis that significant increases have been noted in the volume of the adenohypophysis. The earlier and more marked increase in the volume of the meso-adenohypophysis in the non-parasitic form, raises the possibility that this may be related to the almost immediate onset of sexual maturation that occurs at this period in the brook lamprey species.

From the outset, transforming ammocoetes of *P. marinus* show an increase in both the numbers of pro-adenohypophysial basophils and in the amount of stainable secretory material that they contain. In the meso-adenohypophysis however, basophils do not appear until metamorphosis is well under way. In keeping with its earlier differentiation in the very young ammocoete, this development of synthetic activity in the pro-adenohypophysis might be due to inductive factors from the hypothalamus, reaching it through the so-called 'median eminence' region of the neurohypophysis.

Myxinoids

In *Myxine*, the response of the erythrosinophil with the smaller type of granules to inhibitors of corticosteroid biosynthesis has suggested that this cell may be a corticotroph, or alternatively, that it produces a protein hormone related to the MSH, ACTH and LPH family, all of which share common aminoacid sequences and may have evolved from a single parent molecule. This possibility is not entirely at odds with some experimental evidence for ACTH-like activity in the hagfish pituitary. On the other hand, the suggestion that the second type of erythrosinophil might be involved in the secretion of a hormone related to growth hormone (STH) or prolactin (LTH) has little or no experimental support and prolactin-like activity has not been detected in myxinoid pituitary extracts, either by bioassay or immunological techniques.

Evidence for MSH-like activity in the hagfish pituitary is similarly unconvincing. Unlike the lamprey, the body colour of *Myxine* is not affected by hypophysectomy. Extracts of hagfish pituitaries when tested on the melanophore of frogs produced only variable and slight effects which could have been due to the presence of a melanophore dispersing substance from some source other than the adenohypophysis. Furthermore, no cytological differences have been noted in the adenohypophyses of dark or light-adapted animals and the glandular tissue apparently contains no cells that give a positive response to lead haematoxylin staining. On the other hand, an earlier report recorded strong melanophore dispersing effects of hagfish pituitary extracts on the melanophores of crabs from which the eye stalks had been removed and this activity seemed to be concentrated in the posterior part of the adenohypophysis. However, it is difficult to visualize any function for a melanophore dispersing agent at least in *Myxine*, which in the most recent experiments has failed to show any signs of pigmentary changes when subjected to illumination on black or white backgrounds.

Neurohypophysis

The vertebrate neurohypophysis develops as an extension of the hypothalamic floor (saccus infundibuli) in the region immediately behind the optic chiasma, which establishes contact with the aboral part of Rathke's pouch. In amniotes, the pars nervosa or neural lobe differentiates from the dorsal side of the infundibular sac. This is a region characterized by its content of neurosecretory material, passing into it through the axons of hypothalamic neurosecretory cells. Part of the hypothalamic floor lying immediately behind the optic chiasma is differentiated in the higher vertebrates as the median eminence; a neurohaemal area, where axons from other hypothalamic neurosecretory axons end in contact with capillaries. These drain towards the adenohypophysis, forming a hypophysial portal system through which the activity of the glandular cells can be controlled by humoral factors.

Lampreys

The cyclostome neurohypophysis has rightly been regarded as the most primitive vertebrate type, representing only a slight modification of the hypothalamic floor. This is especially true of the condition in the ammocoete, where the infundibulum shows no marked evagination and is recognizable only by the concentration in this area of stainable neurosecretory material. This is particularly marked in the region of the infundibular floor lying immediately above the meta-adenohypophysis.

On the infundibular floor of the adult lamprey, two distinct areas may be recognized. Above the meta-adenohypophysis and separated from it only by thin strands of connective tissue and a capillary plexus, is a thickened region, rich in neurosecretory materials. This is regarded as the parallel of the neural lobe or pars nervosa of higher vertebrates. In front of this region and separated from it by the infundibular commissure, is an area where the infundibular floor shows comparatively little modification and which contains relatively small amounts of neurosecretory material. However, on the basis of its fine structure, this whole area of the anterior neurohypophysis has been divided into anterior and posterior regions, differing in structure and in the relative content of neurosecretory material. Thus, the more anterior section lying over the pro-adenohypophysis and separated from it by a thinner zone of connective tissue, contains greater amounts of stainable neurosecretion than the more posterior areas.

Most of the axons carrying the neurosecretory material to the posterior neurohypophysis or neural lobe originate from neurones in the pre-optic nucleus. These extend in a broad arc from the pre-optic recess to the lateral walls of the posterior hypothalamus at the level of the meta-adenohypophysis. A majority of these neurosecretory axons enter the neurohypophysis at its anterior end, passing in two bundles from the ventral and lateral regions of the hypothalamus. Others run to the anterior end of the neurohypophysis in the areas where the connective tissue zone is attenuated.

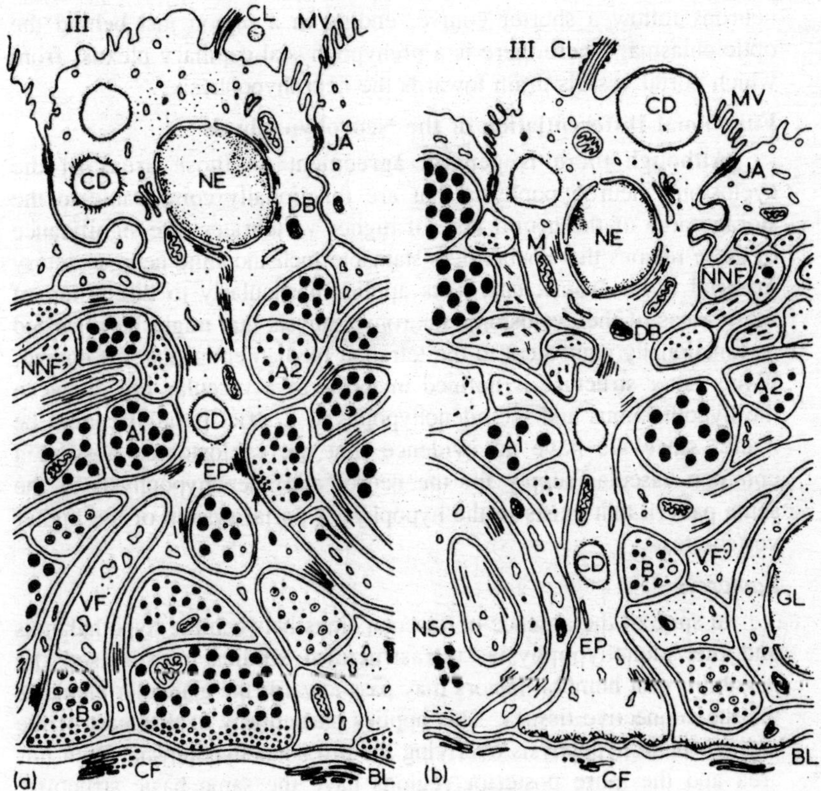

Fig. 15.9. Structure of the anterior neurohypophysis of the lamprey.

Myxinoids

The infundibular process of the hagfish is much more developed than that of the lamprey and has been aptly described as like a shoe, with the toe pointing backwards. The inner layer has a lining of ependymal cells whose elongated processes extend towards the external surface. Most of the neurosecretory fibres terminate in the outer layer, mainly in the dorsal wall of the infundibular process, where the neurosecretory substances are chiefly concentrated. These axons contain a variety of secretory granules ranging in diameter from about 65-400 nm. As in the lamprey, the main centre for the hypothalamic neurosecretory cells is the pre-optic nucleus, from which the axons pass through the neck of the infundibular process to reach the dorsal wall of the neurohypophysis. In addition, the hagfish has a unique vascular route for the passage of neurosecretions to the storage areas of the neural lobe. In this system, axons from some of the pre-optic neurons follow a shorter course, ending in a region just behind the optic chiasma, where there is a prehypophysial capillary plexus, from which portal vessels drain towards the neurohypophysis.

Functional Differentiation of the Neurohypophysis

Although there is general agreement on those areas of the cyclostome neurohypophysis that are functionally comparable to the storage area of the neural lobe of higher vertebrates, the significance of those regions that contain less stainable fuchsinophilic neurosecretory material is far from clear. This applies particularly to discussion of those areas of the cyclostome neurohypophysis that might be regarded as functionally equivalent to the tetrapod median eminence. Of course, if this latter structure is defined in terms of a vascular link between the hypothalamus and the adenohypophysis a strict parallel cannot be made, since we have no evidence that the cyclostomes possess a common vascular supply for the neuro- and adenohypophysis of the same pattern that exists in the hypophysial portal system of the higher vertebrates.

Lampreys

In spite of the absence of vascular links between the hypothalamus and the adenohypophysis, ultrastructural studies have raised the possibility that humoral factors may pass towards the glandular elements via the connective tissues. This applies particularly to that part of the anterior neurohypophysis overlying the pro-adenohypophysis. Both this area and the more posterior regions have the same basic structure, consisting of an inner layer of ependymal cells lining the third ventricle and below this a fibre layer containing the processes of the ependymal cells, intermingled with axons and axonal endings. These axons contain secretory granules of three different size classes, 140-220, 95-140 and 65-100 nm. Especially in the anterior zone, some of the axons that contain the larger granules make synaptic-like contacts with the branching processes of the ependymal cells. Functionally, the most significant difference between the anterior and posterior areas is that in the former, the axonal endings and the end feet of the ependymal cells actually make contact with the basal lamina of the connective tissue separating the neurohypophysis from the proadenohypophysis, whereas in the posterior region, the axonal endings are separated from it by a layer of end feet and the occasional glial cell processes. In this posterior region therefore, direct passage of neurosecretions from neurohypophysis to meso-adenohypophysis would be less likely, although it could conceivably be transmitted via the cytoplasm of the ependymal cells. These structural and functional differences between the anterior and posterior regions have been regarded as comparable with the median eminence and neural lobe of higher vertebrates even though the cyclostomes lack the same vascular organization as the tetrapods.

The median eminence of the tetrapods contains high concentrations of monoamines, believed to be implicated in the regulation of the secretory activity of the adenohypophysis. In the anterior neurohypophysis of the lamprey, *L. japonica*, strong monoamine oxidase activity has been detected in that part of the neurohypophysis overlying the pro- and meso-adenohypophysis, whereas the activity of this enzyme was much weaker in the posterior neural lobe area over the meta-adenohypophysis. In addition, in the anterior neurohypophysis of *L. tridentata*, axons are present that contain granules differing in size from those containing neurohypophysial peptide hormones and which might represent releasing hormones or monoamines. Unfortunately, this clear picture of the parallels between the anterior neurohypophysis and the median eminence has not been sustained by fluorescence studies on the actual distribution of monoamines themselves, which appear to be present in high concentrations in the rostral region of the posterior neurohypophysis, but in only low concentration in the anterior region. Nevertheless, it is possible that the weaker monoamine fluorescence in the anterior neurohypophysis may reflect the higher turnover of these compounds in an area where monoamine oxidase activity is most intense. The posterior neural lobe area of the lamprey neurohypophysis consists of an inner zone containing the cell bodies of the ependymal cells (tanicytes), a middle zone of neurosecretory fibres and an external layer of ependymal end feet and the enlarged terminals of neurosecretory fibres. In *L. fluviatilis*, three types of nerve fibres have been distinguished; peptidergic A^1 and A^2 fibres with granules of 160-340 nm and 120-220 nm and monoaminergic B fibres with small granules 80-100 nm in diameter. The presence of two types of peptidergic granules is interesting in view of the fact that only a single neurohypophysial hormone, arginine vasotocin has so far been identified in the lamprey neurohypophysis, although it should be noted that the A^1 fibres are said to be rare and the concentrations of their secretions may be very low. The neural elements are separated by a basement membrane from the connective tissue zone that lies between the neural lobe and the meta-adenohypophysis. This connective tissue contains pericapillary spaces in those areas where the wide sinusoidal capillaries of the mantle plexus are found. From the structure of this region of the neurohypophysis several routes for the neurosecretion would appear to be possible. Peptides or monoamines might be passed into the capillaries and thus reach the general circulation, or alternatively they could enter the cerebrospinal fluid of the third ventricle. Belen'kii however, believes that as in other lower vertebrates, the meta-adenohypophysis is under the dual control of peptidergic and monoaminergic neurosecretory fibres from the hypothalamus, whose secretions reach the secretory tissues by diffusing through the connective tissue boundary layer.

Myxinoids

By combining fine structural observations with histochemical techniques attempts have been made to elucidate the significance of regional differences in the organisation of the hagfish neurohypophysis. From this point of view, four areas may be distinguished: the anterior and posterior dorsal walls in front of, and behind the infundibular stem and the corresponding anterior and ventral walls of the infundibular process. At an early stage, the fact that the stainable neurosecretory materials were seen to be concentrated largely in the posterior dorsal wall, led to the designation of this area as the neural lobe, while the existence of the pre-hypophysial capillary plexus, suggested that this anterior region might function as a median eminence. On the other hand, in its topography and fine structure it has been claimed that the anterior ventral region which is adjacent to the adenohypophysis offers a closer parallel to the tetrapod median eminence. In this region three types of neurosecretory axons were distinguished by these authors in *E. burgeri*, with small granules of mean diameters, 65, 80 and 110 nm which were thought to contain monoamines. As in the lamprey

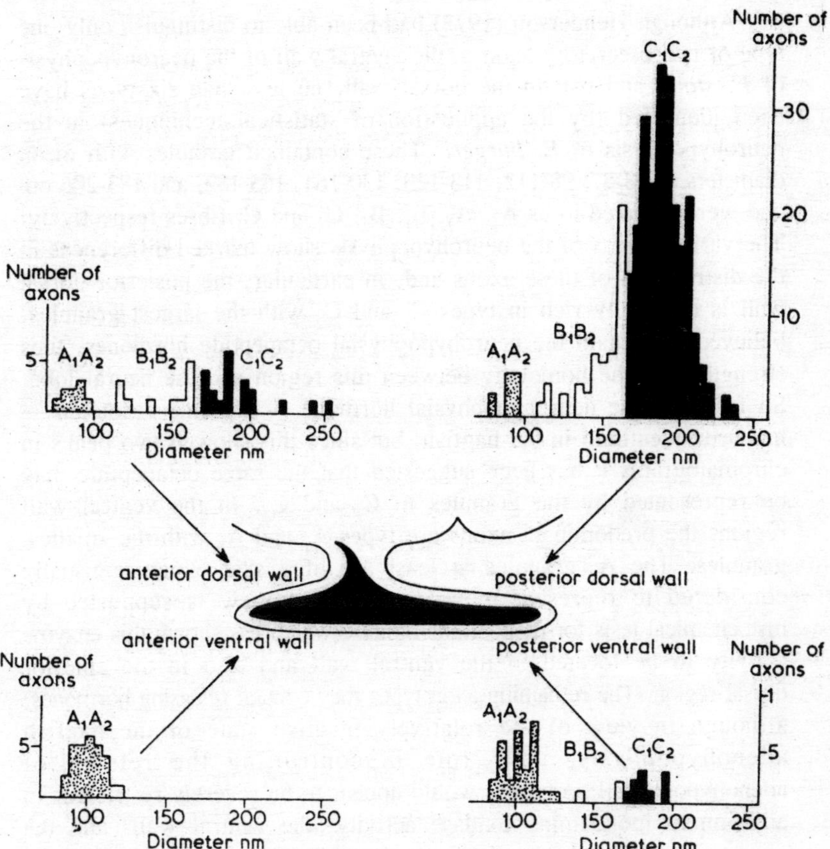

Fig. 15.10. Distribution of various types of neurosecretory axons in the neurohypophysis of the hagfish, Eptatretus burgeri.

neurohypophysis, these type B axons in some instances end on the ependymal cell processes of the outer layer adjacent to the connective tissue separating the neurohypophysis from the adenohypophysis and, although in this species there are said to be some capillaries present in this connective tissue, it has generally been assumed that the transport of neurosecretions takes place by diffusion; a process for which there are many precedents in other cyclostome endocrine tissues. That this type of transport is practicable has been demonstrated by injecting peroxidase into the ventricle and following its passage through the connective tissue between neurohypophysis and adenohypophysis and even as far as the fibrous septa that surround the adenohypophysial cell nests.

Although Henderson (1972) had been able to distinguish only one type of neurosecretory axon in the ventral wall of the neurohypophysis of *E. stouti* and two in the dorsal wall, no less than six types have been identified (by the application of statistical techniques) in the neurohypophysis of *E. burgeri*. These contained granules with mean diameters of 78-87, 98-112, 113-129, 130-164, 165-182, and 183-206 nm and were referred to as A_1, A_2, B_1, B_2, C_1 and C_2 fibres respectively. The various areas of the neurohypophysis show marked differences in the distribution of these axons and, in particular, the posterior dorsal wall is especially rich in types C_1 and C_2 with the largest granules, believed to contain the neurohypophysial octapeptide hormones, thus strengthening the homology between this region and the neural lobe. So far only one neurohypophysial hormone — arginine vasotocin — has been identified in the hagfish, but since this showed two peaks in chromatograms it has been suggested that the same octapeptide may be represented by the granules in C_1 and C_2. In the ventral wall regions the predominant axons are types A_1 and A_2 with the smallest granules. The A_1 granules at least are of a size which is usually considered to represent monoamines. This view is supported by histochemical tests for monoamine oxidase which has shown this enzyme activity to be located in the ventral wall and also in the anterior dorsal region. The remaining axon types may contain releasing hormones, although in view of the relatively inactive state of the hagfish adenohypophysis, their role in controlling the release of adenohypophysial hormones would appear to be severely restricted. In addition to monoamine oxidase activity, the ventral wall, and the anterior dorsal wall have been shown to exhibit acetylcholinesterase activity. In both these respects, a parallel with the tetrapod median eminence, based on histochemical criteria, would have to embrace the dorsal wall anterior to the infundibular stem in addition to the ventral walls, which more especially have been claimed to show the fine structural features of a median eminence. Whether or not these comparisons with the tetrapod median eminence turn out to be justified, there seems little doubt that, in spite of its primitive phylogenetic status, the neurohypophysis of the hagfish shows a surprising degree of organizational complexity and a well-developed neurosecretory repertoire.

Neuroendocrine Functions

Direct experimental evidence for hypothalamic control over the adenohypophysis in the cyclostomes is almost completely lacking.

Because of their relatively stable environment, changes in light or temperature could hardly be expected to play an important role in influencing the activities of hagfishes, although in these respects *E. burgeri* with its breeding season and breeding migration may prove to be the exception. At least in *Myxine*, no cytological changes have been observed when adenohypophysial tissues have been cultured in isolation from their normal connections with the hypothalamus.

On the other hand, in the case of the lamprey, the seasonal incidence of many episodes in the life cycle provide at least a priori grounds for believing that some of these events may be triggered off by environmental factors and mediated through the hypothalamic-hypophysial system. Thus, the onset of metamorphosis occurs in all species of lampreys at the same season of the year and within a comparatively restricted period, the timing of the upstream migration is seasonal and the onset and duration of the spawning season is certainly affected by temperature changes. In spite of this, experimental evidence for hypothalamic control is still lacking. No marked differences in sexual maturation have been observed in *L. fluviatilis* after the pro- and meso-adenohypophysis have been transplanted to sites away from their normal association with the neurohypophysis (heterotopic transplants) as compared with control animals in which the pituitary tissue had been transplanted to its normal site (orthotransplants). Neither were there any obvious differences in the cytology of the pituitary tissues in the two groups. In view of these negative findings it is of interest to note that in the non-parasitic species, *L. planeri*, a peptidergic neurosecretory system has been described in the torus semicircularis of the mid-brain, producing cysteine-rich protein secretions. Significantly, this system has been found only in the female and even there it does not appear until metamorphosis and after the eruption of the paired eyes. Both the numbers of neurosecretory cells and the intensity of their granulation increase with the progress of sexual maturation, suggesting that this system may be involved in the control of gonadal development. Some indication that the hypothalamo-hypophysial system may also be involved in the control of gonadal processes has come from the inhibition of the neurosecretory system accompanied by degranulation of the presumptive gonadotrophs of the adenohypophysis following the administration of chlorpromazine.

There are also other observations that suggest that adenohypophysial tissue may be influenced by the proximity of the neurohypophysis. For example, the pro-adenohypophysis which lies immediately below the 'median eminencelike' area of the anterior neurohypophysis is the first region of the adenohypophysis to differentiate in the larval lamprey and the basophils of this area are the first to show signs of secretory activity. In addition, as several authors have pointed out, pro-adenohypophysial cells in the dorsal regions adjacent to the neurohypophysis may show cytological differences from those in more ventral locations. Thus, in *L. tridentata*, the most deeply staining basophils are those at the most extreme tip of the pro-adenohypophysis and, in *L. fluviatilis*, the dorsal cells are larger and more active in those regions that are the first to differentiate in the ammocoete.

The presence of a common capillary bed between the neural lobe of the lamprey and the meta-adenohypophysis would at least make possible a direct influence of the neurohypophysial hormones on the adenohypophysial tissues. However, apart from such local effects, these hormones may enter the general circulation and exert a wider systemic effect on the organs concerned with salt and water metabolism. The earliest observations that were made on the effects on mammals of cyclostome pituitary extracts, demonstrated the existence of both oxytocic and vasopressor activity and such effects could be important in relation to the role of the gills in ionic regulation. In the hagfish, the endogenous neurohypophysial hormone, arginine vasotocin, produces contractions in the smooth muscles of the ventral aorta, even in very low concentrations, but the corresponding tissues of the dorsal aorta are much less sensitive. This has prompted the interesting suggestion that this hormone could be concerned in controlling the gill circulation. In the sea lamprey, injections of arginine vasotocin have resulted in elevations of the plasma free fatty acid levels, which could be an important factor in the mobilization of the lipid stores of the spawning migrant.

Adenohypophysial Functions

Some aspects of pituitary function are discussed in the chapters dealing with reproduction, osmotic regulation and the various peripheral endocrine tissues. In addition, the reader may be referred to recent reviews of the lamprey pituitary by Larsen and Rothwell (1972) and the more general account of cyclostome endocrinology by Falkmer *et al.* (1974).

Reproduction

The control and integration of reproductive processes is probably the oldest and original function of the vertebrate pituitary. This control is exercised through the gonadotrophic hormones, acting on both germinal and somatic tissues of the gonads. Through these hormones, the adenohypophysis is able to influence gametogenesis, culminating in the liberation of sperms and eggs, as well as controlling the production of steroid sex hormones by the somatic gonadal tissues and thus regulating the development of secondary sex characters and specific forms of breeding behaviour.

The characteristic development of the basophils of the pre- and mesoadenohypophysis of the lamprey during sexual maturation suggests that some or all of these may be gonadotrophs, but the results of hypophysectomy are not easy to reconcile with the cytological observations. In the higher vertebrates, pituitary ablation is usually followed by atrophy of the testis and atresia in the ovary, while castration results in characteristic changes in the pituitary gonadotrophs (castration cells), due to the interference with the normal feedback between gonadal steroids, hypothalamus and adenohypophysis. None of these effects have been seen in hypophysectomised lampreys, but it should be remembered that pituitary removal has so far only been practised on the sexually maturing spawning migrant stages. In the male lamprey, hypophysectomy early in the migratory period does not inhibit spermatogenesis, although the production of mature sperm is generally delayed. On the other hand, spermiation, involving the breakdown of the testis lobule and the liberation of the sperms into the body cavity, is usually inhibited. When the operation has been carried out in early autumn or winter, secondary sex characters generally fail to appear and if hypophysectomy is delayed until after these structures have already begun to develop, they may undergo regression.

Early hypophysectomy in the female lamprey retards the growth of the oocytes and inhibits ovulation. The effects on the granulosa cells of the ovarian follicle are particularly marked. These fail to show their normal development and the accumulation of mucopolysaccharide secretions. During normal sexual maturation, variable numbers of oocytes are subject to atresia, but there are no indications that its incidence is increased by removal of the adenohypophysis. In the female as in the male, an intact pituitary is essential for the development of secondary sex characters. What these experiments have shown therefore, is a marked pituitary influence on the gonadal changes that precede the liberation of the gametes and on the output of sex hormones from the somatic tissues of the gonads.

Earlier work involving partial hypophysectomy suggested that the mesoadenohypophysis was alone responsible for these influences on sexual maturation; a conclusion difficult to reconcile with the obvious development of the pro-adenohypophysical basophils during the spawning migration. However, subsequent work involving carefully controlled partial hypophysectomy has implicated the pro-adenohypophysis as well as the meso-adenohypophysis in sexual maturation. For example, extirpation of the pro-adenohypophysis alone did not prevent the attainment of sexual maturity, but removal of the meso-adenohypophysis gave variable results, from complete inhibition to a lack of effect. After examining the pituitary regions of the experimental animals, it was concluded that complete inhibition of sexual maturity only occurred when pro-adenohypophysial tissue had disappeared after operation and that an intact pro- and meso-adenohypophysis are required for normal reproductive development. The involvement of both lobes would be consistent with the view that these may contain distinct types of basophil, producing two types of gonadotrophin with different target organs.

So far we have no conclusive evidence that hagfish pituitaries produce a gonadotrophic factor, or that they contain cells with the usual cytological characteristics of vertebrate gonadotrophs. Complete hypophysectomy in *E. stouti*, involving removal of both neurohypophysis and adenohypophysis has not provided any clear evidence for pituitary gonadotrophic activity. In the operated animals the number of abnormal testis follicles was greater than in controls, but the progress of spermatogenesis from spermatogonia to spermatozoa was not obviously impeded, neither were any regressive changes noted in the somatic tissues of the gonads. Analyses of plasma testosterone concentration showed lower mean values in the operated animals, but the differences between these and the control values were not statistically significant. In the ovaries, although some abnormal features were noted in the mesovarium, oogenesis appeared to proceed normally in the hypophysectomised hagfish and the proportion of atretic follicles was not increased. No effect of hypophysectomy on oestradiol concentrations was detected. On the other hand, in *E. burgeri*, the only hagfish known to have a definite breeding season, there have been some indications that gonadal growth and spermatogenesis may be retarded after hypophysectomy. However, in spite of earlier reports that hagfish pituitary extracts showed slight gonadotrophic activity when tested in a mouse bioassay, these experiments fail to show any absolute dependence of the gonads on pituitary hormones, although the possibility of a more general and less obligatory influence has not been entirely excluded.

Experiments involving the administration of sex steroids to cyclostomes have so far failed to provide any convincing evidence for the existence of feedback mechanisms operating through the pituitary-gonadal axis. Sex hormones applied to castrated river lampreys, *L. fluviatilis* produced no decisive changes in the cytology of the adenohypophysis, although some increase was noted in the intensity of staining of the pro-adenohypophysial basophils in animals that had been treated with testosterone. In similar experiments carried out on intact hagfishes, *E. burgeri*, oestradiol treatment resulted in a marked degeneration of large oocytes. However there was no indication that the secretory activity of any of the granular cell types in the adenohypophysis had been inhibited and the effect of the hormone on the oocytes was thought to be pharmacological. Perhaps as a result of a direct stimulatory influence on spermatogenesis, testosterone produced a high proportion of mature sperm in the treated males and although there appeared to be some accumulation of granules in the type 1 adenohypophysial cells, there was too great an individual variability in this character in the control animals to justify the conclusion that a feedback control system exists.

Metabolism

Investigations of pituitary influences on metabolism have so far been limited to the migratory period in the life cycle of the lamprey; a time when the animals are no longer feeding and when their main source of energy is the lipid accumulated, either during the larval stage (in non-parasitic forms) or in the phase of parasitic feeding. Moreover, at this same period, the picture is further complicated by the diversion of metabolites to the rapidly maturing gonads. These conditions are reflected in the atrophy of the intestine, reduction in body length and weight, the depletion of lipids in the body wall musculature and the decrease in the liver stores of glycogen.

Hypophysectomy appears to have a retarding effect on many of these changes and diminishes the rate at which energy reserves are depleted. Thus, atrophy of the intestine occurs more slowly after early hypophysectomy and the process may even be reversed when the gonads are removed. In normal animals the reduction that occurs in body length throughout this migratory period can be equated with the reduction in the tissues of the body wall, providing a rough indication of metabolic depletion. Throughout the early migratory period in autumn and winter, there is a slow reduction in body length which is replaced by a rapid shrinkage with the approach of sexual maturation in the spring. Hypophysectomy in the early stages does not affect the total length reduction that the animal has suffered by the time that it dies, but because the operated animals survive longer, the rate of length reduction is reduced. Hypophysectomy carried out in the spring has tended to prevent the final phase of very rapid shrinkage and in many instances has resulted in very prolonged survival. However, the precise mechanism whereby the removal of the pituitary reduces the mobilization of the tissues remains obscure. One possibility is that this effect might be related to a reduction in gonadotrophin output and thus indirectly to retarded gonadal growth. Alternatively, a more general effect on the metabolic rate may be involved. The latter possibility seems to be supported by experiments involving the transplantation of the pro- and meso-adenohypophysis to heterotopic sites. In some instances, these animals became sexually mature, but their survival was prolonged and the metabolic effects were similar to those seen in hypophysectomised animals. This might indicate that it is the separation of the pituitary from the CNS that is responsible for a decreased rate of tissue mobilization.

One of the most puzzling and striking aspects of hypophysectomy is its effect on the life span of the lamprey. These animals normally die very soon after spawning, but operated animals may survive for months after reaching sexual maturity and one lamprey lived for nearly a year beyond its normal breeding season. Since this prolonged survival may occur after the animals have attained full sexual maturity, we may infer that the postponement of natural death is not simply a consequence of the removal of gonadotrophin secretion, but is rather related to the reduced rate of metabolism.

Neither in lampreys nor hagfishes has there been decisive experimental evidence of a growth-hormone-like effect on blood sugar concentrations. Unlike the condition in the tetrapods, hypophysectomised lampreys or myxinoids show no trends towards reduced blood sugar levels, although in operated lampreys these concentrations may be less susceptible to elevation in certain forms of stress.

Metamorphosis

In view of the morphological and cytological changes in the pituitary at the time of metamorphosis, there have been suggestions that the adenohypophysis might, in some way, be involved in transformation, although there is as yet no very convincing evidence to support this belief. The rapid development and differentiation in the pro- and meso-adenohypophysis at the beginning of metamorphosis could be significant, but there is no way of knowing whether these are the result of the morphogenetic changes rather than a factor in their initiation.

An interesting feature of metamorphic development is the apparent correlation between the morphogenetic changes in the nasohypophysial stalk of the ammocoete and the growth and differentiation of the pro-adenohypophysis. In large ammocoetes of *L. planeri*, cavitation of the nasohypophysial cord appears to begin well in advance of the external metamorphic changes and at least in this species, proliferation from the cells of the nasohypophysial epithelium appears to contribute to the tissue of the pro-adenohypophysis. Thus, in some way the developments that occur at metamorphosis in the various parts of the nasohypophysial complex appear to be co-ordinated. This recalls the isolated cases of neoteny that were observed by Zanandrea (1956, 1958) in a local population of the North Italian brook lamprey, *L. zanandreai*, where a few female ammocoetes were observed with well-developed ovaries and mature eggs. Unfortunately we have no information on the condition of the adenohypophysis in these animals but it may be significant that a well-developed nasohypophysial sac was present.

Pigmentary Control

Diurnal changes in the skin colour of lampreys are brought about by the dispersion or concentration of melanin granules in the pigment cells (melanophores) of the skin. After total hypophysectomy, involving the extirpation of the meta-adenohypophysis, the animals remain in a permanent state of pallor. Lamprey pituitary extracts are capable of causing melanophore dispersion in frog skin and injections of mammalian intermedin produce similar effects in the ammocoete. The cytological, topographical and histochemical parallels between the meta-adenohypophysis of lampreys and the pars intermedia of higher vertebrates leaves little room for doubt that this region produces a hormone of the MSH type.

Evolutionary Perspectives

The quite remarkable parallels between the nasohypophysial complex of lampreys and cephalaspids almost compel us to believe that this correspondence would probably have extended to the pituitary itself. The differentiation, already apparent in the cephalaspids, between the openings into the nasal and hypophysial parts of this complex, emphasizes the confusion so often introduced into comparative discussions by referring to the cyclostome nasopharyngeal opening and tract as a nasal opening and nasal tract. In the past, where doubts have been raised on a homology between the nasohypophysial tract and Rathke's pouch these have usually been based on the view that the tubular and persistent hypophysial canal of the cyclostome should be regarded simply as an extension of the nasal sac. In addition, such views have been influenced by misconceptions over the role of the nasohypophysial complex in relation to the origin of the adenohypophysis, which was thought to be derived from the nasohypophysial cord only in embryonic or early larval life. In fact, further contributions are made to the glandular tissue from the epithelium of the naso-hypophysial canal during metamorphosis and it is possible that these processes may extend even further into adult life. While it is true that aspiration of the olfactory sac is a function of the extended nasohypophysial sac of the adult lamprey, this is almost certainly a secondary development and there are good grounds for the belief that this function could not have been carried out by the nasohypophysial complex of the cephalaspids. Finally, the presence in the ammocoete of a solid cell cord rather than a hollow structure is hardly relevant to the question of its homology, since among fishes it is only the embryonic elasmobranchs that show a typical hollow Rathke's pouch and in teleosts, the adenohypophysial tissue proliferates from a cellular cord in which cavities subsequently develop as schizocoeles.

The close association between olfactory and hypophysial components that we see in the myxinoids and lampreys may be reflection of relationships that were established at very early stages in the evolution of the vertebrate pituitary, although the precise manner in which their functions are now carried out may bear little relation to the ancestral condition. If we accept an homology between the nasohypophysial cord or canal of the cyclostomes and Rathke's pouch, what is the explanation for this close association with the olfactory organ? Perhaps, as has been suggested, the early precursor of the pituitary may have been concerned with reproductive control and coordination, responding to the presence in the water of sexual products by secreting hormonal substances to effect the liberation of the gametes. If, in the ancestral condition, the pituitary was primarily concerned with reproductive processes, an association between chemosensory and glandular functions would hardly be surprising. It has been widely assumed that the adenohypophysis originated in a patch of glandular epithelium on the roof of the stomodaeum or immediate pre-oral region, which was subsequently invaginated to form an exocrine gland secreting through a duct into the mouth cavity. This is in harmony with the presence in the primitive polypteroids (and in the earlier ontogenetic stages of other actinopterygian fishes) of a persistent hypophysial duct from the pro-adenohypophysis (rostral pars distalis) to the oral cavity. This may be compared with the widespread occurrence in the cyclostomes of cyst-like cavities in the adenohypophysis, some of which at least may communicate with the lumen of the nasohypophysial tract through duct-like openings. What is more, as seems to be the case in the follicular cavities of the actinopterygian pars distalis described by Olssen (1968), the cells associated with these cysts are similar both structurally and in the nature of their secretions, to the mucoid cells of the nasohypophysial or oral epithelium.

At this early duct stage in the evolution of the hypophysis, Olssen considered that the gland would have been concerned with the production of a 'skin function' prolactin-type hormone, involved in the chemical and mechanical protection of the gut epithelium. This hypothesis may be in line with the localization of prolactin cells in the rostral pars distalis of the actinopterygian adenohypophysis, but fails to take account of the absence of any evidence for prolactin-like hormones or lactotrophs in the cyclostome pituitary. The apparent dominance in the pro-adenohypophysis of the lampreys, of glycoprotein-secreting elements believed to be gonadotrophs, suggests that reproductive control may have been the main function of the pituitary in the primitive vertebrates.

On the other hand a quite different view has been adopted by Fernholm (1972a) in relation to the hagfish pituitary. Here, he considered it likely that the two granular cell types produce hormones different to those associated with the gnathostome pituitary and perhaps akin to the ancestral molecules that might have existed in the Cambrian or Pre-Cambrian common vertebrate ancestor. For example, the type 1 cell with the smallest granules might be related to the primitive adenohypophysial cell, from which in the course of vertebrate evolution were differentiated the cells producing ACTH and MSH, while the type 2 cell could be the forerunner of the differentiated cells eventually producing growth hormone and prolactin respectively. Since the mucous materials produced by the PAS-positive cells were not considered to have any hormonal significance, this interpretation would exclude the existence of any cells corresponding to gonadotrophs in the early vertebrates.

In the absence of any division into separate lobes, the lack of a differentiated meta-adenohypophysis, the scarcity of granular cells and the sparseness of their secretory granules, the hagfish pituitary contrasts sharply with that of the lamprey. Were these conditions to be regarded as truly primitive, it would be difficult to imagine what selective advantages could accrue from the development of a structure with such apparently limited functional capacity. From this point of view, we are almost compelled to believe that this poor differentiation of the hagfish adenohypophysis is a secondary condition, resulting from its restricted habitat and unadventurous mode of life, more especially the general lack of seasonally controlled reproductive rhythms. However, these differences in the adenohypophysis of the two groups of cyclostomes are not parallelled in the structure of the neurohypophysis and indeed, in this respect, the simple infundibular sac of the lamprey may be considered to be more primitive than the well-developed infundibular process of the myxinoids. The fact that the hagfish neurohypophysis has not followed the apparent regression of the adenohypophysis could be related to the more basic metabolic functions of the hypothalamus and neurohypophysis. The widespread occurrence among the invertebrates of neurosecretory systems tends to suggest that the hypothalamic-neurohypophysial system of the vertebrates may be of greater phylogenetic antiquity than the adenohypophysis and for this reason might have been less susceptible to the kind of regressive changes that appear to have overtaken the myxinoid adenohypophysis.

Peripheral Endocrine Tissues

Thyroid and Endostyle

Throughout the vertebrates the thyroid is a derivative of the mid-ventral floor of the embryonic pharynx. The cyclostome pattern differs from that of the gnathostomes only in that the thyroid tissue develops along almost the entire length of the pharynx rather than from a relatively short anterior region. In the earliest myxinoid embryos that have been examined, the thyroid appears to have arisen through the closing off of a ventral pharyngeal groove and not from an endostyle-like structure of the ammocoete type. From this embryonic rudiment, a continuous chain of cell cords is produced, passing upwards towards the oesophagus in the adipose tissue surrounding the gill pouches. These cell groups, at first solid, eventually hollow out to form thyroid follicles. In lampreys, the definitive thyroid tissue does not appear until metamorphosis, when it develops from certain epithelial elements that appear to survive the otherwise complete breakdown of the larval endostyle.

Endostyle

The large number of morphological and physiological investigations that have been carried out on the endostyle are an indication of the significance that has been attributed to this organ in relation to the evolution of thyroidal functions. Like the tunicates and cephalochordates, the ammocoete is a microphagous feeder, using its mucous secretions to trap minute food particles suspended in the water and, in spite of its much greater complexity, there can be little doubt that the specialized, closed endostyle of the ammocoete is morphologically homologous with the simpler organ of the protochordates. The significance of this correspondence was of course greatly heightened by the disclosure of remarkable parallels in the iodine metabolism of the protochordate and lamprey endostyles. Both share a common capacity for iodine concentration and iodine binding, with the formation of the usual range of iodo-tyrosine derivatives — mono- and di-iodotyrosine (MIT, DIT), thyronine (T_3) and thyroxine (T_4). These biosynthetic activities are concentrated within the endostyle, although at least in the protochordates, perhaps not exclusively so.

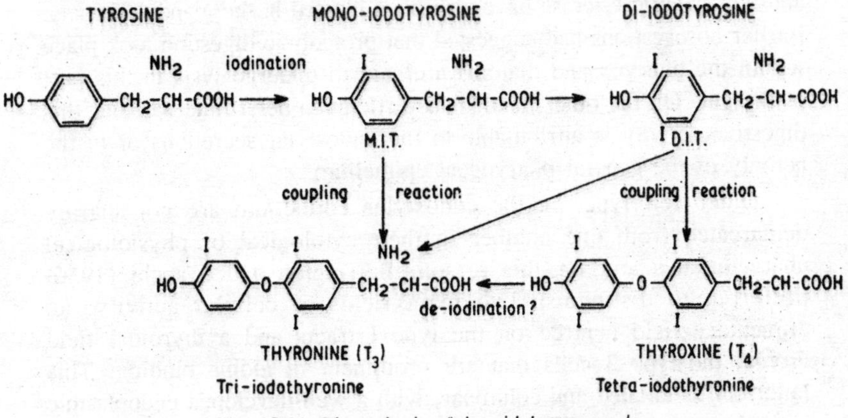

TYROSINE MONO-IODOTYROSINE DI-IODOTYROSINE

iodination

M.I.T. D.I.T.

coupling reaction coupling reaction

de-iodination ?

THYRONINE (T_3) THYROXINE (T_4)

Tri-iodothyronine Tetra-iodothyronine

Fig. 15.11. Biosynthesis of thyroidal compounds.

The ammocoete endostyle consists essentially of paired, tubular lateral chambers, which in their anterior sections are completely separated by a median septum. In the posterior regions this septum is incomplete and the two chambers are in communication with one another. These tubes are also completely cut off from the pharyngeal lumen except at one point. Here, the endostylar duct opens into the pharynx dorsally, while ventrally it leads into a median chamber directed backwards from the point where the duct enters. This median chamber is coiled upwards in its posterior sections. In addition to the endostyle, the pharyngeal floor has a deep ciliated groove and, close to the exit of the endostylar duct, this is joined by two extensions of the lateral chambers, running forwards and upwards along the hyoid arches as ciliated, pseudobranchial grooves. This arrangement presumably serves to spread out the mucous sheets immediately behind the velum, where they would be in a position to trap food particles entering in the respiratory water current.

Within each chamber of the endostyle are four large tracts of glandular type I cells. These groups are approximately triangular in section and the cells show ultrastructural and histochemical resemblances to the mucous cells of the protochordate endostyle. These type I cells do not take part in iodine binding, but their highly developed granular endoplasmic reticulum intimately associated with the Golgi, indicates that they are concerned in the synthesis and transport of protein materials. From the histochemical evidence it is assumed that the secretions of these cells contribute to the mucous feeding mechanisms. A protease is present in endostylar extracts and phosphomonoesterase and non-specific esterase have also been located in the glandular tracts. Earlier observations had suggested that proteolytic digestion took place within the pharynx and diatoms are said to undergo lysis in this part of the gut. On the other hand, it is difficult to determine whether this digestive activity is attributable to the endostylar secretions or to the activity of the general pharyngeal epithelium.

Other cell types in the endostylar epithelium are not sharply demarcated from one another in their cytological or physiological characteristics and for this reason, Barrington and Franchi (1956) preferred to distinguish two main fields of cellular activity; an alimentary field centred on the type I tracts and a thyroidal field around the type 3 cells that are prominent in iodine binding. This latter cell is ciliated and columnar, with a well-developed endoplasmic reticulum and vesicular Golgi, within which are electron-dense materials, tending to concentrate towards the apex of the cell forming multivesicular bodies. These have been interpreted as storage sites for iodinated products. Granules originating in the smooth endoplasmic reticulum are thought to be lysosomes, containing non-specific phosphatase and esterase, whose distribution coincides with that of bound iodine in autoradiographs, after injections of ^{131}I.

Type 2 cells have been further subdivided into types 2a, 2b and 2c; the latter adjacent to the type 3 zone. Both types 2 and 3 are ciliated, with microvilli on their apical surfaces but, in type 2c, the rough endoplasmic reticulum and Golgi are better developed and the lysosomes more conspicuous. Although bound iodine can be demonstrated in all three cell types, it is much more conspicuous at the surface of the type 2c cell. Type 4 cells have no granular endoplasmic reticulum and ribosomes are scanty. Traces of bound iodine have been seen in these cells, but they are considered to have little ability to bind this element or to synthesize protein. Type 5 cells, lining the endostylar lumen show active pinocytosis and are regarded as mainly absorptive elements, although their response to goitrogens might be regarded as evidence for some synthetic capacity.

Metamorphosis and the Lamprey Thyroid

In the adult lamprey, the thyroid lies beneath the tongue musculature and extends from the second to the sixth gill pouch. It consists of follicles of varying diameter, in which the cells vary considerably in height and width. Both ciliated and non-ciliated cells are present and these show the usual hallmarks of protein-secreting cells — a well-developed granular endoplasmic reticulum and Golgi — although neither of these organelles is as well-developed as in the thyroid cells of higher vertebrates. Small vesicles of varying degrees of electron density are present in the apical regions, together with lysosomal bodies that may be derived directly from the endostylar epithelium. In addition to the normal follicles, which may contain variable amounts of a loose-textured, PAS-positive colloid, parafollicles have been distinguished in which the epithelial cells are of a squamous type and in which the follicular lumen is devoid of colloid. Cell nests with no central lumen may also be present between the follicles. Compared to the thyroid of higher vertebrates, the lamprey gland is poorly vascularised and the capillaries lack fenestrations.

During metamorphosis, the structure of the endostyle is completely broken down by a combination of cytolysis and phagocytosis. These processes have made it very difficult to follow the fate of different parts of the endostylar epithelium at the critical periods when the adult thyroid follicles are being formed. In view of these difficulties, it is not surprising that there should still be uncertainty as to which cell types in the endostyle contribute to the adult follicles. Sterba (1953) distinguished two types of thyroid follicles; primary follicles containing a remnant of the original endostylar lumen and secondary follicles in which the lumen was only formed during their metamorphic development. The persistence of the endostylar lumen has also been suggested by Wright and Youson (1976), who observed a proliferation of type 2 and type 5 cells at the angle of the endostylar lumen during the metamorphosis of *P. marinus*. Most of the earlier authors who studied metamorphic changes in the endostyle believed that the greatest contribution to the adult thyroid follicles was made by type 4 cells, although the participation of other cell types has not been excluded. Sterba for example, suggested that the primary follicles were formed mainly by the type 4 cells, but that the secondary follicles originated in type 3 cells. Gorbunova (1975) considered that types 4 and 5 are the least specialized of the endostylar elements and that these persisted throughout metamorphosis, subsequently differentiating to form the definitive thyroid follicles. Wright and Youson found that iodine binding continued throughout metamorphosis, first in the transforming endostyle and later in the reconstructed thyroid follicles. Since the type 4 cells play little or no part in iodine metabolism they considered that other cells, such as types 2 or 3 which are more active in iodine binding, must participate in the development of the adult thyroid follicles. Subsequently, Wright *et al.* (1978b) have been able to detect the presence of thyroglobulin in cell types 2c and 3, as well as in some type 5 cells, by the use of immunocytochemical techniques.

Thyroid Gland of the Myxinoids

The hagfish thyroid consists of diffuse tissue extending over a wide area of the pharyngeal floor, along the course of the ventral aorta as far back as the last branchial pouch. The follicles, which may occur either singly or in groups, are embedded in connective tissue and show great variability in size. For example, in *Myxine* their diameters range from 0.18-1.0 mm. As in lamprey thyroids, the cells vary in height, but are more uniform within an individual follicle and, perhaps because of the absence of an endostylar stage, no ciliated cells are present. The follicular lumen may contain a coarsely granular material, quite unlike the colloid of higher vertebrates. Capillaries are relatively rare and like those of the lamprey are without fenestrations, but pinocytotic vesicles have been described within the endothelial cytoplasm. As in other poorly vascularised cyclostome endocrine tissues, it is presumed that secretion occurs through the basal lamina and the connective tissues.

In their ultrastructure there are distinct parallels between the thyroid cells of hagfishes and lampreys. In the myxinoid thyroid, only one cell type has been described with short apical microvilli, rough endoplasmic reticulum and Golgi field, although like the lamprey thyroid cell, these are not as well developed as in the thyroid epithelia of higher vertebrates. Like the endostylar and thyroid cells of lampreys, the hagfish cells contain dense bodies which have been regarded as secondary lysosomes containing condensed colloid.

Interpretations of Thyroidal Mechanisms

In all vertebrates, thyroidal activity involves the same basic sequence of biosynthetic processes:

1. The accumulation of iodine from the environment and its oxidation
2. Iodination of tyrosine and the formation of MIT and DIT.
3. The coupling of these compounds to form T_3 and T_4 in thyroglobulins
4. The hydrolysis of thyroglobulin and the liberation of the active hormones, T_3 and T_4.

After the injection of radio-iodine ^{125}I or ^{131}I or its addition to the ambient water, it is taken up by the thyroid or endostyle cell, but in comparison with higher vertebrates the rate of accumulation in the cyclostome is slow, perhaps because of the relatively small volume of the tissue and its poor vascularisation. In the lamprey, even greater concentrations of iodine may occur in the notochord or gonad. For example in *L. fluviatilis*, only 0.2-1.8% of the injected dose of ^{131}I appeared in the thyroid, compared to 10% in the notochord and about 3% in the ovary. However, in these extra-thyroidal tissues, the iodine exists mainly in the inorganic form as iodide, whereas in the endostyle or thyroid tissue, a higher proportion is organically bound to protein.

By combining the ultrastructural and autoradiographic evidence, it is possible to outline some of the basic cellular processes that may be involved in the thyroidal activity of cyclostome follicle cells. Within a short period of its administration, radio-iodine is concentrated around the apical surfaces of the cells and, to a more variable extent, within the follicular lumen, suggesting that these are the sites where iodination mainly occurs. Thyroglobulin or its protein precursors, are presumably synthesized within the rough endoplasmic reticulum and processed within the Golgi, where secretory vesicles are formed. These then pass towards the cell surface, where their contents are extruded into the lumen and iodination occurs. Judging by the relative intensity of the luminal contents as seen in autoradiographs, the storage functions of the colloid are much less important in the cyclostomes than in higher vertebrates, although in the sea lamprey, *P. marinus*, thyroglobulin has been detected immunocytochemically both in the luminal colloid and in the colloid droplets within the thyroid cells. The iodinated products are re-absorbed into the cells by pinocytosis and the pinocytotic vesicles are thought to associate with lysosomes in the dense bodies. In both lamprey and hagfish thyroid cells, X-ray microanalysis has shown that these structures contain relatively high levels of iodine although, in the case of the hagfish, the relative concentrations in cytoplasm and luminal

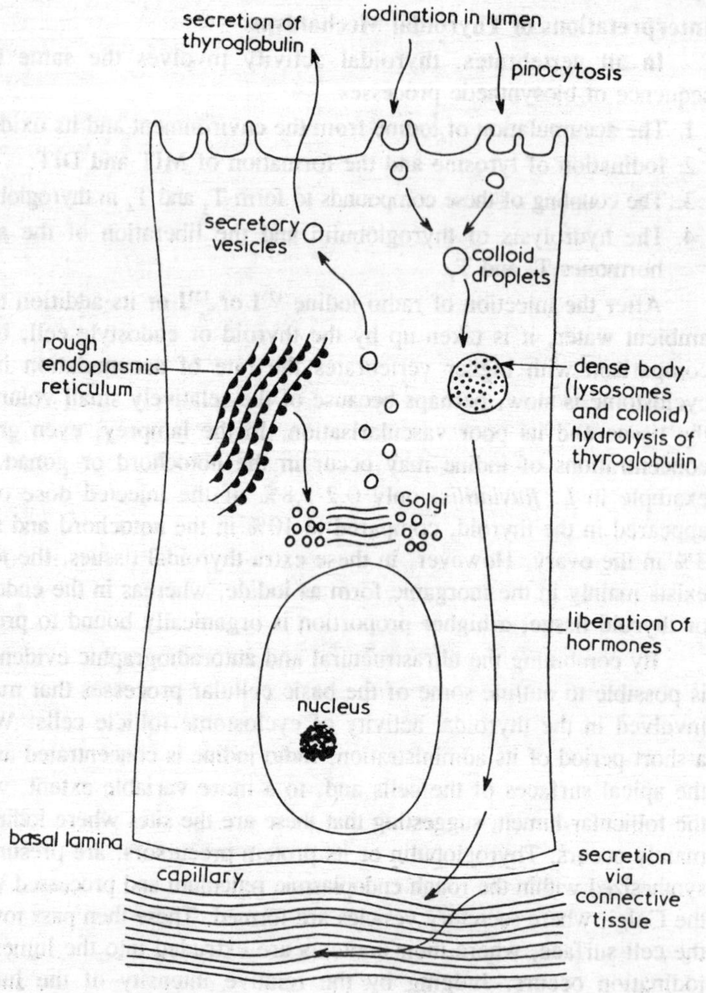

Fig. 15.12. Ultrastructural features of the cyclostome thyroid cell and interpretations of the biosynthetic sequence.

colloid were too small to be measured by this technique. Presumably, hydrolysis takes place within the secondary lysosomes, containing the highly condensed thyroglobulin and the hormonal products are liberated to pass through the basal lamina.

In the case of the endostyle, the secretory processes must differ from those of the thyroid gland, in so far as there can be no extracellular storage of the iodinated products. In autoradiographs, bound iodine has been seen on the cell surfaces of type 2c and 3 cells within

30 minutes of the administration of radio-iodine. After 2 hours, labelling was seen in vesicles at the cell apices and after a further 2 hours it was evident in the multi vesicular bodies. These organelles were considered to be intracellular storage sites for the iodinated products. The distribution of the lysosomes appears to correspond to those regions where iodine binding is most intense, suggesting that these organelles may either be involved in biosynthesis or in the liberation by hydrolysis of the hormonal products. At the same time we cannot exclude the possibility that hydrolysis may be extracellular and that hormonal materials may be liberated by digestion in the gut or, more probably, through the agency of the protease that has been identified in endostylar extracts.

Where analyses have been made on endostylar or thyroid tissue following the administration of radio-iodine, the amounts of the isotope incorporated into thyroxine has usually represented only a small fraction of the total radioactivity, most of which is present as iodide, with smaller amounts in the form of MIT and DIT. Thyronine T_3, which appears to be metabolized more rapidly, has either not been detected or has been present in only very low concentrations. The distribution within the iodoproteins of the various iodinated products is quite similar in cyclostomes with DIT as the major constituent, followed by MIT. T_3 was not detected in either sample in this instance and T_4 represented only 2.5% in the hagfish and 2.9% in the lamprey. Quite similar results have also been obtained for the ammocoete and adult of *L. planeri*, where the iodinated products represented only 0.002% of the total protein. Together, T_3 and T_4 made up 3-5% and there were larger amounts of MIT and DIT. Compared to the thyroglobulin of the dogfish, the most significant features of cyclostome iodoproteins are therefore their low iodine content and lower hormonal levels.

Although it has been said that the cyclostomes are characterized by their low levels of circulating thyroid hormones, this is hardly borne out by the information. This shows that the concentration of T_4 in cyclostome serum is of the same order as in two teleost species reported by Refetoff *et al.* (1970) (4.3-4.5 μg 100 ml^{-1}) and generally higher than in the trout, where a serum concentration of 1.1 μg 100 ml^{-1} has been recorded. Even lower values have been reported in a number of freshwater and marine teleosts by Leloup and Hardy (1976). Whereas in freshwater forms, T_4 levels (0.085-1.5 μg 100 ml^{-1} were always higher than those of T_3 the reverse was true in sea water, where values for the latter (0.36-3.6 μg 100 ml^{-1}) were said to be higher than those of mammalian species.

For the sea lamprey we have information on T_4 levels at all stages of the life cycle with the exception of the parasitic phase. Surprisingly, thyroxine concentrations in the blood of the ammocoete are very much higher than in the upstream migrant stages of either *P. marinus* or *L. tridentata* and are comparable with the exceptionally high concentrations in the blood of spawning female sea lampreys. During the early stages of metamorphosis, there is a dramatic fall in hormone levels and this decline continues right through to the macrophthalmia stage at the completion of transformation. Unfortunately, we have no information on T_4 levels in the period of adult life that intervenes between the end of metamorphosis and their entry into the rivers on their upstream migration, but since the values are somewhat similar in macrophthalmia and early upstream migrants it is quite possible that these low levels may be characteristic of the parasitic phase. Suggestions that the low levels of T_4 in early migrants may be due to the cessation of feeding have not been borne out in the ammocoete, where starvation did not result in decreased hormone levels. The pattern of events in the metamorphosis of lampreys is obviously in direct contrast to the situation in the frog tadpole, where an initially inactive thyroid begins to release increasing amounts of thyroid hormones during metamorphosis. On the contrary, the ammocoete endostyle shows the maximum biosynthetic activity followed by a rapid decline during transformation as this organ degenerates and is replaced by the

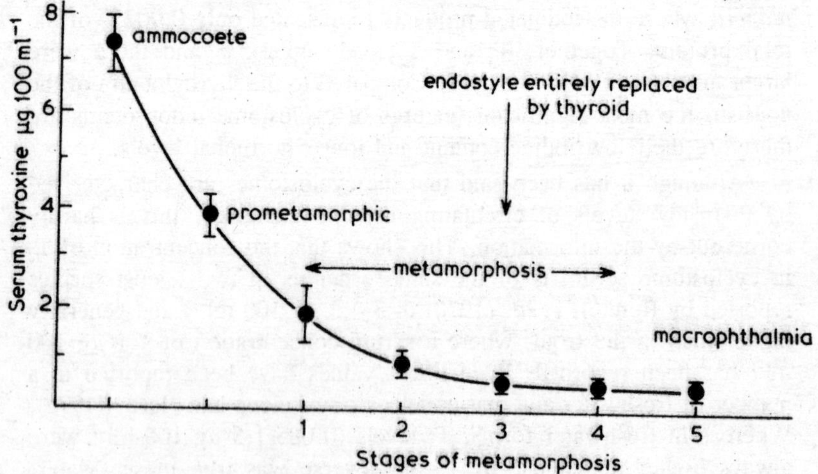

Fig. 15.13. Serum levels of thyroxine (T_4) during the metamorphosis of the sea lamprey, Petromyzon marinus.

developing adult thyroid follicles. This is parallelled by the activity of endostylar oxidative enzymes, which is at its maximum in the earliest stages of metamorphosis, decreasing when the first thyroid follicles appear. Although it would be premature to postulate a causal relationship between the declining T_4 levels and metamorphosis in the ammocoete, it may be recalled that some metamorphic changes have been observed in larval lampreys after treatment with potassium perchlorate; an agent that inhibits thyroidal biosynthesis. Thyronine (T_3), although in only low concentrations, has been detected throughout the migratory period of the sea lamprey, but T_4 was only present in measurable amounts when the animals were already spawning. Indications of an inverse relationship between T_3 and T_4 concentrations suggest that as may be the case in higher animals, the former may be the more active hormonal compound. The sharp increase in T_4 levels, that seems to occur in the spawning sea lamprey, may appear to conflict with observations made by Pickering (1972, 1976a) on upstream migrant river lampreys (*L. fluviatilis*). In these animals, the rate of uptake of ^{125}I by isolated thyroid tissue showed a consistently downward trend from the beginning of the migration in August at least up to November. Moreover, as judged by the decreasing height of the thyroid epithelial cells, the gland was considered to be less active as the animals approached sexual maturity, when compared to those that had just embarked on their upstream movement in the autumn. Nevertheless, a declining level of thyroidal activity during the migratory period would not necessarily be incompatible with a rise in blood hormone concentrations in the brief final period when the lampreys are spawning. Indeed, in view of the well known effects of thyroid hormones on the vertebrate central nervous system, it is conceivable that such an increase might help to explain the marked changes in metabolic rhythms and behaviour patterns that occur in breeding lampreys, when restless and continuous daytime activity replaces the nocturnal habits of the upstream migrants. However, speculation apart, it must be admitted that we are totally ignorant of any possible role for thyroid hormones in the life of the cyclostomes. Based on analogies with the amphibians, attempts to accelerate the transformation of the ammocoete by the administration of thyroxine proved completely unsuccessful and the removal of the endostyle has not inhibited metamorphosis. A metabolic effect of thyroxine comparable with that of higher vertebrates has not been demonstrated and, as we have seen, oxygen consumption rises continuously throughout metamorphosis at a period when T_4 concentrations are undergoing a continuous and marked decline.

Pituitary Regulation of Thyroid Activity

In higher vertebrates, the activity of the thyroid is influenced by thyroid-stimulating hormone (TSH) secreted by basophils in the adenohypophysis. Earlier studies on the lamprey pituitary had suggested that the comparatively sparse basophils of the mesoadenohypophysis might represent such thyrotrophs. Some support for this idea came from experiments involving prolonged treatment with the goitrogen, thiouracil, which resulted in cytological changes in these mesoadenohypophysial basophils, although it is possible that these could have been due to the toxic qualities of this compound. In *L. ftuviatilis*, surgical removal of the pituitary has failed to produce any changes in iodine uptake, thyroxine levels or in the cytology of the thyroid epithelium. Similar conclusions were also reached by Pickering (1976a) who cultured thyroid tissue from the same species together with lamprey pituitary extracts or mammalian TSH. Neither treatment produced significant changes in the uptake of ^{125}I by the gland tissue or in the distribution of the radioactivity in the various iodine compounds recovered from the culture medium. Pickering also reported similar negative results after measuring the iodine uptake of isolated thyroid tissue from animals that had been hypophysectomised two months previously.

In ammocoetes, removal of the pituitary has not produced significantcytological changes in the endostyle, nor has this operation affected the rate of organic iodine binding. On the other hand, a number of earlier authors had reported cytological changes or alterations in iodine metabolism following the administration of mammallian TSH to the ammocoete, although such experiments cannot prove that the animal produces its own TSH. In this connection, experiments involving the application of goitrogens are of some interest. These compounds inhibit iodine binding and by so doing reduce the levels of circulating thyroid hormones. The existence in the higher vertebrates of a negative feedback system between the thyroid and the pituitary thus leads to an increased output of TSH from the adenohypophysis following the administration of goitrogens. After such treatments, ammocoetes show certain changes in the endostylar epithelium that would be consistent with hypersecretory activity, although curiously, these changes were not confined to those areas that are known to be primarily concerned in iodine metabolism and even affected the type I glandular cells. Even more importantly, these hypersecretory responses could be elicited in hypophysectomised ammocoetes, thus excluding the possibility that they are due to a pituitary-endostylar feedback system. These authors were inclined to regard these paradoxical results as due to a feedback mechanism (at a purely intracellular level) on those cellular systems normally involved in the synthesis of proteins materials participating in iodine binding.

If the situation regarding pituitary control of the lamprey thyroid is equivocal, it has been equally difficult to find decisive evidence for pituitary thyrotrophins in the myxinoids from some of the earlier experimental work involving hypophysectomy. These failed to show significant changes either in the cytology of the thyroid tissue or in the plasma thyroxine levels of the operated animals. On the other hand, some evidence for the existence of a thyrotrophic factor in hagfish pituitary extracts has been claimed as a result of bioassays and more recently the uptake of ^{131}I by *E. stouti* was found to be nearly doubled after the administration of mammalian TSH. The treated animals also showed a lower percentage of labelled MIT and higher proportions of DIT and T_4 compared with the control animals. Matty *et al.*, (1976) have also found that transplantations of hagfish pituitary tissue into the cranial sub-cutaneous sinuses of *E. stouti* resulted in elevated T_4 levels and a similar result was obtained after the *in vitro* culture of pituitary and thyroid tissue of the same species. It seems therefore, that there is rather better evidence for a thyrotrophin in the hagfish than in the lamprey and this is all the more surprising when we recall the undifferentiated condition of the myxinoid adenohypophysis and the rarity of granular or basophilic elements.

In view of the uncertainty surrounding the existence in the lamprey pituitary of a TSH factor, it is interesting to find that thyrotrophin releasing factor (TRH) has been identified in the brain tissue of the adult lamprey, *P. marinus*, in the brain and pituitary of the ammocoete and also in the head region of *Amphioxus*. This has led to a suggestion that this tripeptide—Glu-His-Pro— has an ancient chordate lineage and may have been first developed as a central neurotransmitter (perhaps in connection with the olfactory system), before it became adapted as part of the thyroid control system. Certainly, it does not appear to have this latter function in the hagfish, since injections of synthetic TRH have had no effects on the adenohypophysis or thyroid tissues, nor on serum T_4 levels of *E. burgeri*. Furthermore, TRH has not increased the production of T_4 by cultured pituitary-thyroid tissue in *E. stouti* although the pituitary tissue itself increased the output of T_4 from the thyroid tissue by over 50%. TRH is also present in high concentrations in the amphibian hypothalamus, but here again is said to have no effects on pituitary TSH. On the other hand, synthetic TRH has been shown to stimulate the release of α-MSH from the neuro-intermediate lobe of the frog pituitary, suggesting that its normal function may be to act as an MSH-releasing hormone.

Cyclostomes and the Evolution of the Thyroid

That the vertebrate thyroid gland is morphologically homologous with the endostyle of ammocoetes and protochordates has rarely been challenged by zoologists, although there is no direct evidence that this latter structure was present in the fossil agnathans. Many of these animals are thought to have been microphagous feeders, but the manner of their mucus production and the pattern of their ciliary feeding tracts is completely unknown. Undoubtedly, the transformation of the larval endostyle to the closed thyroid follicles within the life cycle of the lamprey has great intellectual attraction, in portraying a simple picture of the phylogenetic history of the gland as originating in the mucus feeding mechanism of some protochordate-like vertebrate ancestor. On the other hand, when the origin and significance of iodine metabolism are considered, the picture is less clear, especially in view of the widespread occurrence of iodine binding in both protostome and deuterostome invertebrates. Furthermore, there is some evidence to suggest that iodine uptake may be an inherent property of the entire vertebrate endoderm, but that this potency is suppressed in the course of ontogeny, except in those special areas from which the thyroid gland is subsequently developed. Thus, iodine incorporation has been demonstrated in the salivary glands, bronchi and even in the duodenum of the embryo rat.

An interpretation of the evolution of thyroidal activity that has gained wide currency is the one originally proposed by Gorbman (1958). This draws attention to the presence of iodoproteins in the exoskeleton and pharyngeal teeth of invertebrates, suggesting that at some point in time these compounds assumed a metabolic role and that, as in the protochordate endostyle, the pharyngeal site became dominant. At this stage, hydrolysis of the protein would have occurred in the gut, leading to the liberation of the active iodinated compounds. In *Amphioxus* and in the ammocoete, the iodoproteins are no longer skeletal proteins and specialized cells have appeared in the endostyle that are capable of synthesizing T_3 and T_4. However, these compounds may still be liberated in the gut by hydrolysis after being conveyed there in the mucus stream. Finally, with the replacement of microphagy by the macrophagous feeding habits of the vertebrates and the loss of the mucus carrier, the iodinated proteins were confined to the closed thyroid follicles, where hydrolysis occurred, setting free the hormonal compounds to enter the circulation.

In what may appear to traditional zoologists as a somewhat heretical approach to the problems of thyroid evolution, Etkin and Goa (1974) have questioned the hallowed position that the endostyle has hitherto occupied in the phylogeny of the vertebrate thyroid gland. In this connection it is interesting to note that thirty years earlier, similar doubts had been expressed by Leach (1944) who remarked that 'the usual statement that the endostyle of ammocoetes is phylogenetically intermediate between the acraniate endostyle and the thyroid of true vertebrates is to be seriously questioned.' Accepting the widespread and 'accidental' nature of iodine binding as a consequence of the chemical affinity between iodine and tyrosine, Etkin and Goa suggest that in the early stages of thyroidal evolution, the iodinated proteins may have had some structural or enzymatic role and that iodine metabolism may have been developed by vertebrates invading freshwater as a way of sequestering and storing iodine. The presence in blood or skeletal tissues of proteins capable of binding iodine would have facilitated the coupling of iodotyrosines to form T_4, but at the stage which might be represented by the protochordates, the presence of such compounds should not be assumed to have any hormonal significance. Although in the vertebrate ancestors, iodine-metabolizing cells may have tended to concentrate towards the floor of the pharyngeal region, it would be unnecessary to assume that they were associated with endostylar function. As these authors point out, iodine metabolism in the protochordates is not confined to the endostyle and even in the ammocoete the glandular tracts have no thyroidal activity. The presence of cells showing such activity in the lamprey endostyle could be interpreted as a result of the suppression of thyroid differentiation consequent on the introduction of a prolonged larval stage. Such an interpretation is in harmony with the views expressed elsewhere in this volume on the place of the larval stage in the evolution of lampreys. The presence of an embryonic thyroid primordium or 'rest' whose final differentiation has been delayed by the extension of larval life has its counterpart in other groups of animals with specialized larval forms, such as the heterometabolous insects or the amphibians.

As explained earlier, we are still uncertain as to the exact contribution made to the adult thyroid follicles by the various cell types that have been identified within the larval endostyle, although several authors have considered that this may involve the type 4 cell, which plays little or no part in iodine binding. In this connection it is interesting to note a comment made by Barrington and Sage (1972) who, in reference to the contribution made by this cell type to the adult thyroid, remarked 'that this could be readily understood if they were genetically programmed for activity in the adult'.

HORMONES OF PANCREAS AND GUT

Organization of the Exocrine and Endocrine Pancreas

Unlike the higher vertebrates, the cyclostomes have no discrete pancreatic organ lying outside the gut and connected to it by pancreatic ducts. Nevertheless, both components of the vertebrate pancreas can be recognized; exocrine zymogen cells secreting their digestive enzymes directly into the gut lumen and endocrine tissue representing the islets of Langerhans of the higher vertebrates.

The distribution of zymogen cells in the anterior intestine and intestinal caecae of ammocoetes and adult lampreys have already been described and, at metamorphosis, there are important changes in this region of the gut. The most posterior segment of the ammocoete oesophagus persists in the adult, but the remainder of the definitive oesophagus is a neoformation, developed from a dorsal lamina on the roof of the larval pharynx. In Northern Hemisphere lampreys, the adult intestine forms a blind caecum over the posterior end of the oesophagus, where it breaks up into a complex of branching diverticula surrounded by islet tissue. In the Southern genera, the larval caecae are reduced in size at metamorphosis, but in *Mordacia* a small structure persists ventral rather than dorsal to the gut as in the Northern forms. This difference like those between the ammocoete caecae of the various genera need not necessarily be of profound phylogenetic significance and both are probably a result of variations in the torsion that takes place in the intestine during early larval and metamorphosing stages. In so far as they may foreshadow the evolutionary history of the gnathostome exocrine pancreas, these intestinal evaginations are of considerble interest, pointing towards a closer affinity between the petromyzonts and the gnathostomes.

Equally significant in relation to the evolution of the vertebrate pancreas is the condition of the cyclostome endocrine tissue. First described by Langerhans himself in 1873 and later to be known as the follicles of Langerhans, the islet tissue of the ammocoete is formed by the proliferation of cells at the base of the intestinal submucosa in

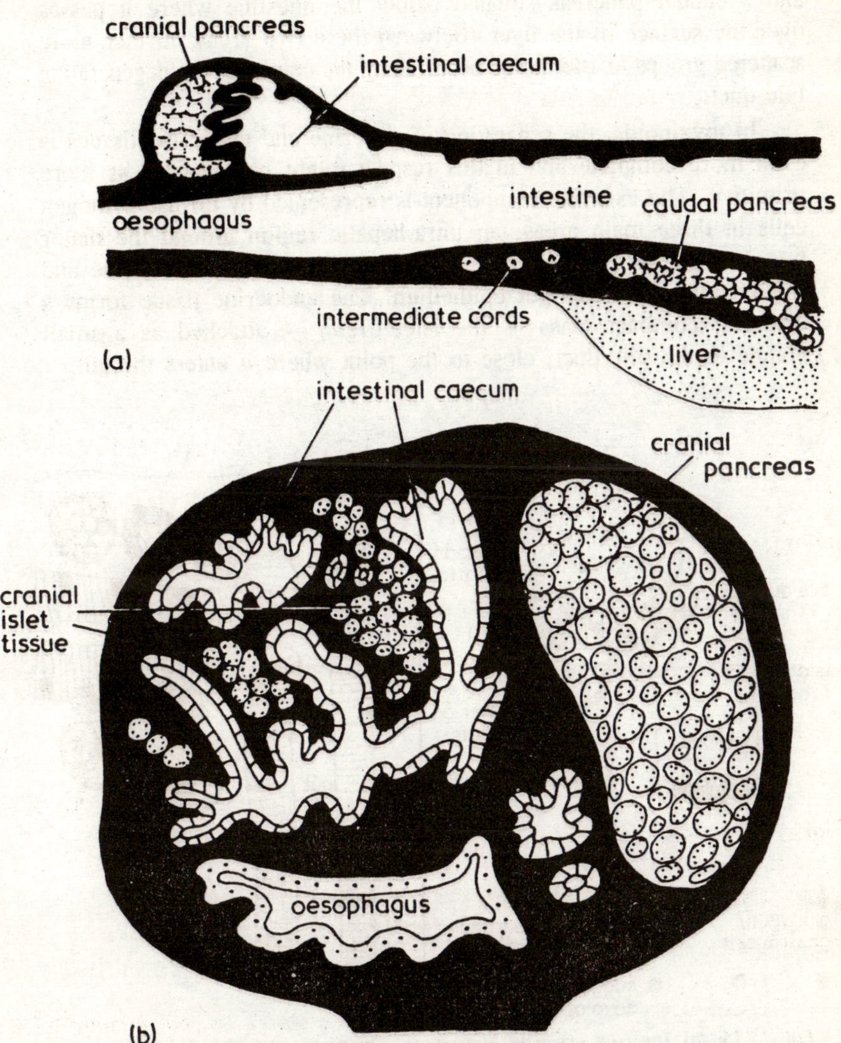

Fig. 15.14. The islet organ of the adult lamprey (a) Sagittal section through the junction of oesophagus and intestine. (b) Transverse section through the region of the cranial pancreas.

the region around the junction of the oesophagus and intestine. These cells were later shown to be involved in the control of blood sugar concentrations and therefore comparable with the b cells of the pancreatic islet tissue. In adult lampreys, this tissue is concentrated in two main areas; a cranial pancreas, lying above the intestinal caecae and a caudal pancreas situated below the intestine where it passes over the surface of the liver. Between these two areas, further more scattered groups of islet tissue occur along the course of the degenerating bile duct.

In myxinoids, the separation of exocrine and endocrine tissues is even more complete and in this respect might be regarded as more primitive. The exocrine component is represented by diffuse zymogen cells in three main areas; an intra-hepatic region around the major liver vessels, in the mesenteries between the liver and intestine and finally within the mid-gut epithelium. The endocrine tissue forms a discrete, compact mass — the islet organ — attached as a small nodule to the bile duct, close to the point where it enters the gut.

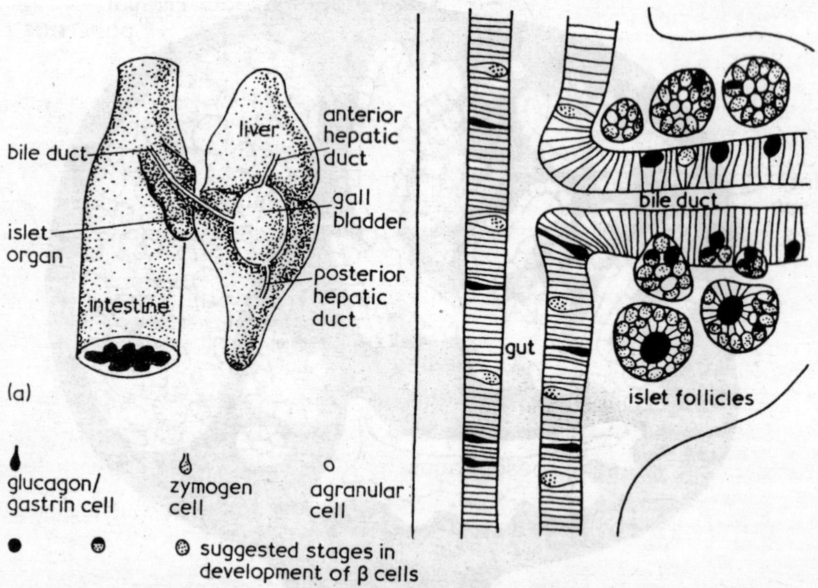

Fig. 15.15. (a) The islet organ of Myxine. (b) Distribution of endocrine cells in the intestine and bile duct of Myxine and the formation of islet follicles.

Islet Tissues

Both in their general structure and in their cellular components, there are some quite striking parallels between the islet tissues of lampreys and hagfishes and these show significant divergencies from the typical gnathostome pattern. The islet lobules of *Myxine* tend to form hollow follicle-like structures, containing cavities 25-100 μm in diameter but these are less frequent in the islet tissues of *Eptatretus* species. In *Myxine*, these follicles sometimes enlarge to form cyst-like tumours (hamartomas) which may occupy almost the entire islet organ. The functional significance of the normal follicular structure of the islet organ of *Myxine* is not understood but, at least in some cases, the lining cells show microvilli and coated vesicles, suggesting that they may be involved in the absorption of protein materials from the luminal contents. The islet tissue of larval lampreys often contains follicular cell groups, but in the adult the islet cords or lobules are more usually solid structures. However, in the cranial pancreas of upstream migrant river lampreys, *L. fluviatilis*, follicular structures and islet cell tumours closely resembling those of the *Myxine* islets, have been found in a considerable proportion of the animals examined and appear to increase in frequency towards the end of the migratory period.

In both groups of cyclostomes, a feature of the islet tissue is its relatively poor vascularisation and this is most marked in the hagfish, where capillaries are said to be absent. As in the thyroid gland, secretion is believed to take place by emiocytosis into the surrounding connective tissues. Judging by their cytochemical and immunological reactions it is generally believed that apart from some agranular cells and connective tissue elements, the islet organs of cyclostomes consist only of β cells, reacting to antisera raised against mammalian insulin. In the islet tissue of *Myxine*, a rare additional type of cell has been described, which in its ultrastructure appears to show some of the characteristics normally associated with A cells. The same cell type has been found in the bile duct epithelium where it has been regarded as a precursor of the functional β cell. The islet tissues of lampreys contain both light and dark cells, probably representing phases in the secretory cycle of the β cell. More difficult to evaluate are the four types of granular cell described by Brinn and Epple (1976) in the islet tissue of *P. marinus* and although these may be derived directly or indirectly from the β cells, these authors did not exclude the possibility that they are functionally distinct cell types.

Endocrine Cells of the Gut and Bile Duct

The vertebrate gut and its various appendages, produces a whole spectrum of polypeptide hormones concerned in the integration of the digestive processes, including the secretory activities of the glands and the passage of food through the tract. Amongst the better known of these hormones are glucagon, produced by the A cells of pancreatic islets, gastrin secreted in the pyloric glands of the stomach, secretin and cholecystokinin-pancreozymin (CCK).

Investigations into the endocrine cells of the cyclostome gut have relied mainly on immunological techniques, based on antibodies raised against mammalian polypeptides. These have been supplemented by electron microscopy and by studies of the effects of cyclostome gut extracts on mammalian tissues. In the bile duct epithelium of the hagfish *Myxine*, there are scattered cells which react to mammalian insulin antisera, in addition to the rare granular cell already referred to. These latter cells have been seen to proliferate within the bile duct mucosa and are believed to contribute to the formation of fresh islet tissue. Eventually, after acquiring an envelope of connective tissue, groups of these cells separate from the bile duct epithelium and enter the lobules of the islet organ. Although these accounts have established the immediate origin of the β cells from the bile duct mucosa of the adult hagfish, they do not of course resolve the problem of their more remote embryological history.

A number of detailed studies have been made on the ontogeny of the islet tissues of larval lampreys. These larval follicles arise from cells lying at the base of the oesophageal epithelium at its junction with the intestine, but further contributions are also made by the epithelium of the intestine and the bile duct. From these parent cells, follicles are formed which then pass from the epithelium to the underlying sub-mucosa, where they are surrounded by connective tissues and blood vessels. After a central lumen has developed within the follicles, the contents of the granular cells, together with cell debris may be discharged into the lumen, which may also contain blood cells from the surrounding capillaries. Throughout larval life, further follicles continue to develop, both by proliferation of the epithelium as well as by amitotic divisions of existing follicle cells.

The cranial pancreas of the adult lamprey which develops at the posterior end of the persistent larval oesophagus, is produced directly from existing larval follicles in the same region. These tend to lose their central lumen to form the more compact cords and lobules of the adult islet tissue. However, with the degeneration of the bile duct during metamorphosis, further cords are proliferated from its epithelium to form the caudal pancreas. Thus, the larval islet tissue originates mainly from the gut epithelium, but as in the hagfish, some of the endocrine tissue of the adult is derived from the bile duct epithelium.

The intestinal epithelium of *Myxine* contains scattered endocrine cells, which judged from the distribution of their secretory granules and the fact that these are not passed into the gut lumen, have been described as 'open' basal granulated cells. These elements, with their base resting on the epithelial basal lamina, extend throughout the entire depth of the gut epithelium, with their apices bearing microvilli, projecting into the intestinal lumen. From these morphological characteristics it is clear that these cells pass their secretions through the basal lamina into the underlying connective tissue of the gut wall. In animals that have been starved it has been noticed that the numbers of secretory granules increases, suggesting that their liberation is linked to feeding and digestion. Despite their uniform ultrastructure, these cells nevertheless show immunofluorescent reactions to antisera to porcine glucagon and also (although less intensely) to antisera against gastrin, the terminal C-peptide of gastrin (pentagastrin) and to caerulein, a peptide with biological activity characteristic of cholecystokinin-pancreozymin (CCK). These cells have shown no reactivity to either insulin or secretin antisera.

Using similar techniques, parallel investigations have also been carried out on the pancreas and gut of larval and adult lampreys. In the ammocoete, endocrine cells in the intestinal epithelium are scattered along the length of the gut in the larger animals, although they tended to occur more frequently in the anterior third. Similar cells in the adult lamprey (*L. fluviatilis*), were more numerous and larger than those of the ammocoete and occurred either singly or in groups throughout the entire length of the intestine from the liver to the cloaca although, as in the larval stage, they were more abundant in the anterior sections. These endocrine cells are oval or flask-shaped, with a wider base resting on the basal lamina and a slender process extending upwards to reach the luminal surface of the epithelium.

No immunoreactivity to glucagon or secretin antisera has been observed in the islet tissue of larval or adult lampreys and the islet cells respond only to anti-insulin sera, supporting the view that the β cells are the only functional cell type in the endocrine pancreas. On the other hand, in the intestine, cells believed to be identical with those seen in electron micrographs were found to be reactive to glucagon, gastrin and caerulein antisera. These reactions were always more intense in adults than in ammocoetes and are believed to occur within the same cell. There is thus a general agreement in the immunological characteristics of intestinal endocrine cells in lampreys and hagfishes.

Intestinal extracts from both lampreys and hagfishes exert a CCK-like activity when tested on the mammalian pancreas. The same authors also found some evidence for secretin-like activity in their intestinal extracts, in spite of the apparent absence of cells with secretin-like immunoreactivity in either group of cyclostomes. Glucagon immunoreactivity has also been observed in lamprey intestinal extracts as well as in extracts from the cranial pancreas, although in the latter case the response is probably due to glucagon-secreting cells in the intestinal diverticula which penetrate the islet tissue.

In view of the possibility of cross-reaction and the absence of specificity, results based on immunological responses to mammalian antisera must be interpreted with caution. What these observations show is that the intestinal endocrine cells of the cyclostomes contain molecules sharing certain immunological properties and reactive sites with the hormones used in the preparation of the antisera.

Physiology of Cyclostome Islet Tissues

Early investigations on the islet tissues of lampreys were directed mainly towards structural and functional comparisons with the endocrine pancreas of higher vertebrates. Historically, the presence in cyclostome islet tissue of insulin-like activity and its essential role in blood sugar regulation was first demonstrated in larval lampreys, where destruction of the follicles of Langerhans by cautery was followed by an elevation of blood sugar concentrations. Conversely, after glucose loading, the vacuolisation of the islet tissue gave clear indications of hypersecretion. Subsequent studies on the adult lamprey have confirmed that as in other vertebrates, the insulin produced by the β cells is involved in the maintenance of relatively constant levels of circulating glucose and in the utilization and storage of carbohydrate reserves.

Surgical removal of the islet tissue of the adult lamprey (pancreotectomy) has been followed by decreased levels of insulin-like activity in the serum, associated with an increase in blood sugar levels to concentrations some five times higher than those of normal animals. Comparisons of their glucose tolerance curves showed that the operated animals had completely lost their ability to regulate blood sugar concentrations following a single glucose injection.

Earlier experiments on *Myxine* gave no evidence of hyperglycaemia in isletectomised animals, even after force feeding, and their glucose tolerance curves showed no indications that their capacity for blood sugar homeostasis had been impaired. Furthermore, in contrast to the lamprey, where anti-insulin sera have provoked hyperglycaemia and decreased glycogen concentrations in liver and heart muscle, this procedure did not result in elevations of blood sugar levels in *Myxine*.

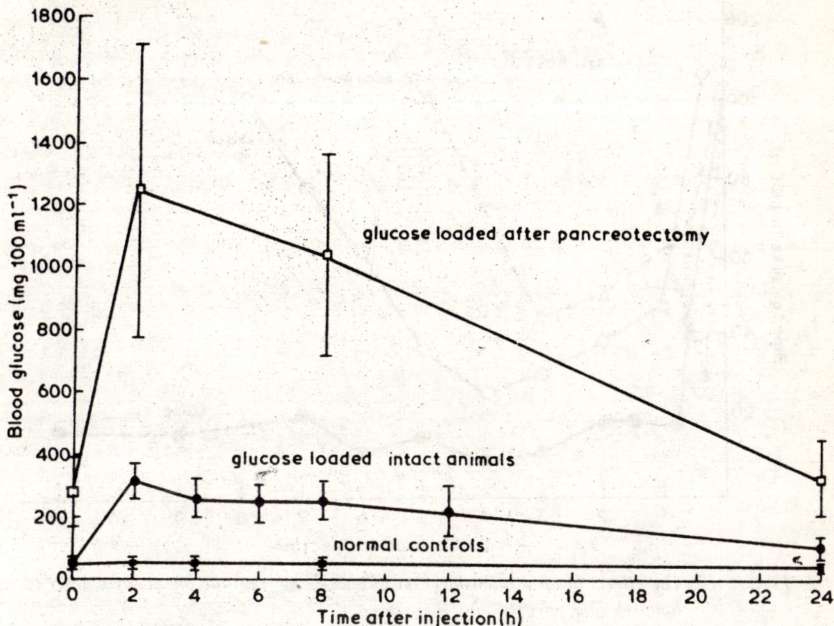

Fig. 15.16. Effect of removal of the islet tissue on the blood sugar homeostasis of the river lamprey, Lampetra fluviatilis.

More recently however, some evidence of hyperglycaemia has been found in isletectomised *E. stouti*, subjected to 10 weeks starvation, but in this as in other divergencies between the myxinoid species in their carbohydrate metabolism and its hormonal control, it is difficult to decide whether these should be attributed to differences in handling or experimental methods rather than to real species differences. For example, Falkmer and Matty (1966) had commented that *Myxine* was able to withstand prolonged starvation without displaying a reduction in blood sugar levels, whereas in *E. stouti* these declined from 28 mg 100 ml^{-1} in freshly caught specimens to 11 mg 100 ml^{-1} after 10 weeks without food.

A common characteristic of both lampreys and hagfishes is the slow rate at which a load of injected insulin is cleared from the circulation and the delay in the establishment of normal blood sugar levels after glucose loading. For example in *Myxine*, blood sugar levels following glucose injection, showed a rapid rise within the first few hours, succeeded by a slow return to normal values over a period of 1-2 days. Similarly slow responses have been observed in lampreys and hagfishes afterinjections of mammalian insulins. This hormone

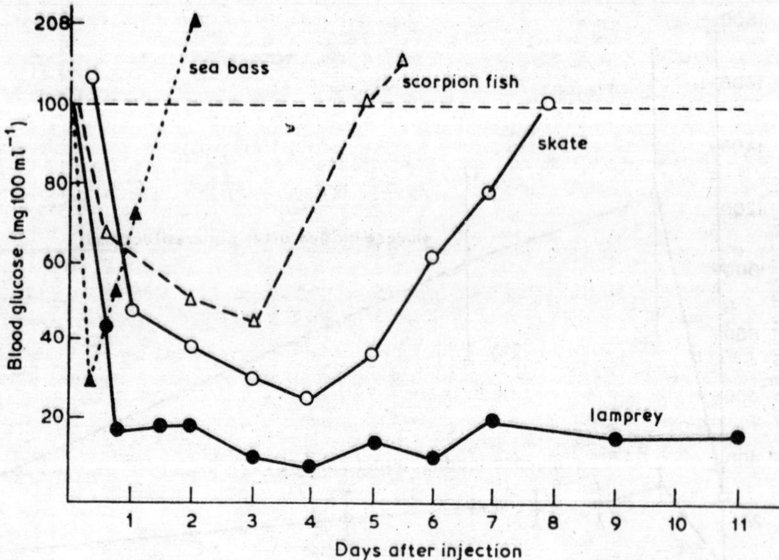

Fig. 15.17. The effects of an insulin injection (30-60 IU kg⁻¹) on the blood sugar level of lamprey and teleosts.

produces a long-lasting hypoglycaemia, which may take many days to return to baseline levels. While similar slow responses are characteristic of the lower vertebrates, they are certainly more marked in the lamprey than in teleosts. The Pacific hagfish, *E. stouti* in contrast to *Myxine* requires only small dosages of bovine insulin (0.1-0.5 IU kg^{-1}) to induce hypoglycaemia and its sensitivity to the exogenous hormone is said to be some 1000 times greater than that of the Atlantic species. After a dose of 10 IU kg^{-1}, *E. stouti* still showed depressed blood sugar concentrations 15 days after the injection. Convulsions have rarely been seen in ammocoetes or adult lampreys, and never in the hagfish, after the administration of insulin and this has been related to the high glycogen content of the brain and its low levels of glucose-6-phosphatase activity. When administered to *Myxine*, mammalian insulin has been reported to increase the glycogen content of the muscles and decrease carbohydrate levels in the liver, but these effects were not observed in *E. stouti*. However, in the latter species, insulin appears to be involved in protein synthesis, since its administration has been followed by depression of plasma nitrogen and increased incorporation into muscle protein of ^{14}C-labelled glycine. Exogenous insulin appears to have no effect on the glycogen content of the body wall muscles of the lamprey and reports of its influence on liver glycogen have been unconvincing or even conflicting.

Lampreys are quite remarkable in the extent to which they are able to maintain their blood sugar concentrations during the long periods of fasting that are a normal feature of their life cycles; the first occurring during and for some time after metamorphosis, and the second during the upstream spawning migration. Coinciding with these changes from feeding to fasting, there are considerable changes in the orientation of their metabolism. Thus, in spite of the cessation of feeding, the metamorphosing sea lamprey, *P. marinus* shows a marked rise of blood sugar levels, with indications that the glycogen content of the muscles and liver diminish. These changes could indicate a drop in insulin output at a period when the islet tissue is undergoing its transition from the larval to the adult condition.

In the upstream migration of the river lamprey, *L. fluviatilis*, blood sugar concentrations appear to remain relatively constant throughout a period of 6-8 months without feeding, although a slight rise has been observed at spawning time. During the winter and early spring, the glycogen content of the liver and heart muscle decrease, whereas the carbohydrate reserves of the muscles of the body wall remains unchanged and are only exhausted during the final burst of activity that accompanies spawning. This remarkable ability of the lamprey to maintain its blood sugar levels and the carbohydrate content of the somatic muscles throughout long periods of starvation, argues for the existence of efficient means for synthesizing carbohydrates from amino acids and above all, from the extensive lipid reserves that were built up during the preceding period of parasitic feeding. Both the somatic and heart muscles have been shown to contain highly active glycerol-3-phosphate dehydrogenase and glycerol kinase and rapidly incorporate ^{14}C-labelled glycerol. On the other hand, in the liver this gluconeogenic activity was very weak, confirming that the muscles are the main site of glycogen synthesis during the upstream migration.

In lamprey muscle it has been shown that insulin increases the activity ofglycogen-synthesizing enzymes. This has led to the view that this hormone serves to maintain the carbohydrate content of the muscles throughout the period of winter starvation; an idea that is strengthened by the effect on the lamprey of antisera to mammalian insulin. This has resulted in hyperglycaemia and in a decrease in the glycogen concentrations of heart muscle and liver; parallelling the changes observed in the upstream migrant.

Changes in glycogen levels during the migratory period have been correlated with the levels of insulin reactivity in the blood. When the first migrants enter the rivers in October, these insulin levels are high, but throughout the months following until February, they decline in parallel with the decreasing glycogen content of the liver and heart muscle. A rise in insulin levels occurs in the spring as the animals are approaching sexual maturity, followed by a dramatic fall at spawning time. The exhaustion of the carbohydrate reserves of certain organs has been linked by Plisetskaya to chronic insulin deficiency, developing at a time when the animals are no longer feeding. A similar fall is also said to occur in the anadromous sturgeon, although in this case the carbohydrate levels are restored when feeding is resumed.

Little is known of carbohydrate metabolism in the hagfish, but it appears that the liver is a relatively unimportant glycogen store and that, as in the lamprey, the muscular tissues represent the major proportion of the body carbohydrate. Unlike the situation in higher vertebrates, where removal of the liver is followed rapidly by hypo glycaemia and death, the hepatectomised hagfish has been found to survive for periods up to 30 days, without showing reductions in its blood sugar levels. Judging by their relative glucose-6-phosphatase activity, the kidney and intestine are probably major factors in the maintenance of blood sugar concentrations, and enzyme activity in the latter organ is about half that of the liver tissue. Similar conditions have also been observed in the lamprey, where hepatectomy has had little effect on blood sugar homeostasis. On the other hand, the importance of the liver in the elaboration of yolk proteins during sexual maturation was shown by the almost complete inhibition of vitellogenesis in the female lamprey.

Evolution of Islet Tissue and Gut Hormones

The mode of origin of cyclostome islet cells from the epithelium of the gut, or its derivative the bile duct, points towards an ancestral condition in which there may have been a widespread and diffuse system of insulin-secreting cells within the digestive tract. This is apparently true of many invertebrates, where insulin immunoreactive cells have been detected throughout the digestive tract and its associated glands. Evidence for such 'insulin cells' has now been found amongst molluscs, crustaceans, echinoderms and tunicates, whereas the existence of glucagon-like materials has yet to be established decisively. On the other hand, immunoreactivity to glucagon as well as to gastrin and insulin antisera has been reported in the gut of *Amphioxus*. In this protochordate, no secretin-like activity has been detected, but, as in the cyclostomes, both the glucagon and gastrin-like activities appear to be located within the same cell type.

While the immediate derivation of vertebrate islet tissue from the gut endoderm is not in question, the ultimate and early embryological source of these cells is still disputed. A recent hypothesis has postulated the origin of islet cells together with other polypeptide secreting cells of the thyroid, adenohypophysis and gastro-intestinal tract from the neural crest or neurectoderm of the vertebrate embryo. In this respect, this view has grouped together in one family, a wide range of endocrine elements, including the corticotrophs of the adenohypophysis, the C-cells of the thyroid, the chromaffin cells and the endocrine cells of the gut; all of which share a number of cytochemical features in common, above all the ability to take up and decarboxylate certain precursors of biogenic amines, believed to be concerned in the processing and secretion of cell granules. For this reason this family of endocrine cells has been termed the APUD series (amine precursor uptake and decarboxylation). However, there is now some doubt whether the origin of all these cells can in fact be traced back in ontogeny to the neural crest. For example, in bird embryos, transplanted primordia of the neural crest were found to give rise to parasympathetic ganglia, but not to cells with the cytochemical characteristics of pancreatic endocrine elements and in rat embryos from which neural crest precursors had been removed, functional β cells were subsequently identified.

At what evolutionary stage, or in what sequence, the close association of exocrine and endocrine pancreatic tissues occurred is open to a number of alternative interpretations. In the lamprey, and to a lesser extent in the hagfish, the islet tissues have already left their ancestral site within the intestinal epithelium to form masses of endocrine tissue, separated from the zymogen cells which still remain within the gut mucosa. Is this condition representative of an intermediate stage in the phylogeny of the gnathostome pancreas? If so, the condition in the lampreys with their so-called 'protopancreas' would be regarded as less primitive than that of the hagfish. At the same time, the segregation of the islet tissue from the gut epithelium in both groups might indicate that this migration preceded phylogenetically, the extramural concentration of exocrine elements in the history of the gnathostome pancreas. Such a view is not without its difficulties. The absence in cyclostome islets of A cells or of other differentiated elements of the gnathostome endocrine pancreas could only be explained on the assumption that the movement of the β cells out of the gut epithelium preceded that of the A cells and that the latter entered the islet tissue later, at a time when a distinct exocrine pancreas was already developing. An alternative hypothesis would involve the simultaneous migration from the intestinal wall of both endocrine and exocrine elements, but this would imply either a secondary regression of the exocrine pancreas in the cyclostomes, which seems highly improbable, or an independent evolution in agnathans and gnathostomes of distinct types of pancreatic organization. Finally, a third possibility would involve the origin of exocrine and endocrine tissues from a common anlage, which at first produced only exocrine tissue, but from which, at a later stage, endocrine elements were differentiated. This hypothesis, which has no support from our knowledge of the relation between zymogen and endocrine cells in cyclostomes would, like the second alternative, place the condition in these animals outside the phylogenetic sequence followed in the gnathostomes.

The first indications that the cyclostomes might possess gut hormones with some of the properties of those of higher vertebrates, came from observations on the physiological effects in mammals of intestinal extracts from lampreys and hagfishes. When tested on rats, intestinal extracts from the lamprey increased both the flow and the enzyme content of the pancreatic juice and in rabbits produced contraction of the gall bladder. In mammals, these physiological activities are controlled by gastro-intestinal hormones; secretin promoting the flow of pancreatic secretion and cholecystokinin-pancreozymin, the release of pancreatic enzymes and stimulation of the gall bladder. Secretin-like activity has also been detected in intestinal extracts of *Myxine* and immunoreactivity to glucagon in similar extracts from lampreys.

How can these findings be related to the ultrastructural and immunoreactive characteristics of the endocrine cells in the intestinal mucosa of the cyclostomes? Neither in the hagfish nor in the lamprey has secretin-like immunoreactivity been observed, but glucagon reactivity has been obtained from both hagfish and lamprey intestinal extracts and from their endocrine cells. The presence of secretin-like biological activity in intestinal extracts can probably be explained by the existence of a glucagon-type polypeptide, whose molecular structure is sufficiently like that of secretin for it to exert the appropriate physiological effects on the mammalian pancreas. In view of the absence in the cyclostomes of a discrete pancreas, it is difficult to visualise the necessity for secretin-like control of the diffuse zymogen cells, which are themselves in direct contact with the contents of the gut. This interpretation is in line with the close similarities that exist in the molecular structure of glucagon and secretin and which has led to the suggestion that both have been evolved from a common parent molecule.

How are we to regard the evidence for cholecystokinin-pancreozymin (CCK) activity in the intestinal extracts? The endocrine cells of the hagfish intestine, in addition to their immunoreactivity to glucagon antisera, also reacted less strongly with antisera to gastrin, pentagastrin and to caerulein and similar reactions occur in the endocrine cells of lampreys. Caerulein, a decapeptide isolated from frog skin and a more potent stimulator of gastric acid secretion than gastrin, shares with the latter and with CCK an identical C-terminal pentapeptide and even its last eight amino acid residues differ from those of CCK at only one position. This explains why antisera to caerulein can cross-react with CCK and gastrin in radioimmunoassays. The presence in both lamprey and hagfish intestinal extracts of CCK biological activity could thus equally well be attributed to the presence of a gastrin-like polypeptide. This has led to a suggestion that the same cyclostome endocrine cell may produce two separate polypeptides; one belonging to the secretin-glucagon family and the other to the gastrin-CCK group, each perhaps released in response to different conditions in the intestinal environment. Many of the organs or tissues which in the higher vertebrates are targets for their differentiated gastro-intestinal hormones, are either absent altogether in cyclostomes or are present in only a less developed form. Thus, there is no stomach or discrete exocrine pancreas and, in adult lampreys, even the gall bladder and bile duct have disappeared. One way out of these difficulties would be to assume that the target organs for the cyclostome hormones have been modified during the evolution of the higher vertebrates; a process for which the history of vertebrate endocrine systems provides ample precedents. In addition, the biological properties of the undifferentiated cyclostome polypeptides may differ considerably from those of the related mammalian hormones. For example, it has been suggested that hagfish endocrine cells may be involved in controlling the islet tissues, since in mammals both CCK and enteric glucagon are said to act as signals for insulin release.

Unfortunately, one of the limitations inherent in the immunological techniques that have been used in the study of cyclostome polypeptides is that they do not always enable us to decide whether we are dealing with separate hormones or with different manifestations of a single larger molecule. Thus, those who cling to the traditional doctrine of 'one cell one hormone' may prefer to believe that the intestinal endocrine cell of the cyclostome is in reality producing a single type of polypeptide. For example, Weinstein (1972) had proposed that the gastro-intestinal hormones of higher vertebrates were originally derived from a primitive ancestral molecule which he termed 'prosecgastrin' with sequences at one end of the chain resembling secretin and at the other end, like gastrin. Elaborating on this concept, van Noorden and Pearse (1974) suggested on the basis of their observations on the lamprey intestine, that a corresponding parent molecule 'proglucgastrin' might exist, having as one part of the chain sequences similar to both glucagon or secretin and another region with a gastrin-like sequence. Such a molecule might then be expected to show both the secretin and CCK-type of biological activity that has been found in cyclostome gut extracts. If indeed the cyclostomes should be found to possess such a common parent molecule, the splitting up of this polypeptide might have been brought about by successive gene duplications producing first a glucagon-secretin and a gastrin-CCK peptide, followed later by the evolution of the separate hormones and accompanied by the increasing differentiation and the alimentary tract and its associated glands. With regard to the timing of these events, Weinstein (1968) has calculated from a comparison of the sequences of secretin and glucagon, that these two polypeptides were separated by gene duplication during the Mesozoic, some 200 million years ago; a figure that agrees broadly with the distribution of glucagon in amphibians, birds and mammals. Judging from the distribution in various vertebrate groups of separate gastrin and CCK polypeptides, it would appear that these may have been separated somewhat more recently than the secretin-glucagon gene duplication. Thus, in mammals birds and in reptiles it has been found that gastrin-like and CCK-like molecules occur within separate cell systems whereas, in amphibians and in teleosts, the same cell appears to be responsible for the production of both types of activity. These authors believe that gastrin and CCK have both evolved from a common caerulein-like ancestral molecule, which could in its turn have shown the kind of glucagon-like immunoreactivity that is found in the cyclostomes. Although it is possible that some of our views on the evolution of these molecules may have to be modified when the position of the protochordate intestinal polypeptides has been clarified, there seem little doubt that the cyclostomes occupy a primitive and key position in relation to the phylogeny of the vertebrate enteric hormones.

INTERRENAL AND CHROMAFFIN TISSUES

The mammalian adrenal consists of two distinct functional components, differing fundamentally in their embryological origins and in the type of hormonal compounds that they produce; the adrenal cortex elaborating a range of steroid hormones and the medulla producing catecholamines. In the cyclostomes, these two tissues although recognizable, are not combined to form a discrete organ, but are loosely distributed over wide areas of the body. The tissue that is regarded as homologous with the adrenocortical tissue of the higher vertebrates is commonly referred to as the interrenal, while the counterpart of the adrenal medulla is the system of chromaffin cells.

Localization and Structure of the Interrenal

First recognized in the lamprey embryo shortly before hatching, the origin of the interrenal has been traced to a proliferation and thickening of the coelomic epithelium at the angle between the mesentery and the somatopleur at the level of the second pronephric tubule. After separating from the epithelium, the interrenal cells migrate dorsally away from the coelomic surface, eventually concentrating around the ventral sides of the great vessels in the vicinity of the proaephric tubules.

This tissue was first described in *L. planeri*, where it appeared that the largest masses of interrenal cells remained within the pronephric region, although it was known that smaller groups subsequently extended posteriorly throughout the greater part of the trunk. However, more recent studies on ammocoetes and adult sea lampreys *P. marinus*, have shown that much greater numbers of interrenal cells may occur in association with the mesonephric kidney. In the pericardial regions of *Lampetra* larvae, small groups of interrenal cells are found on the ventral and lateral surfaces of the dorsal aorta, around the cardinal veins and among the pronephric tubules, but during the later stages of metamorphosis and in the adult, the largest groups are usually to be found at the base of the persistent pronephric funnels. In the sea lamprey, the interrenal cells of the trunk region are mainly associated with the arterial supply of the kidney where they have presumably been carried in the walls of the renal arteries during the development of the mesonephros. Entering the kidney along these arteries, the interrenal cells occur along the course of both efferent and afferent arterioles of the glomus as well as within the renal capsule itself,

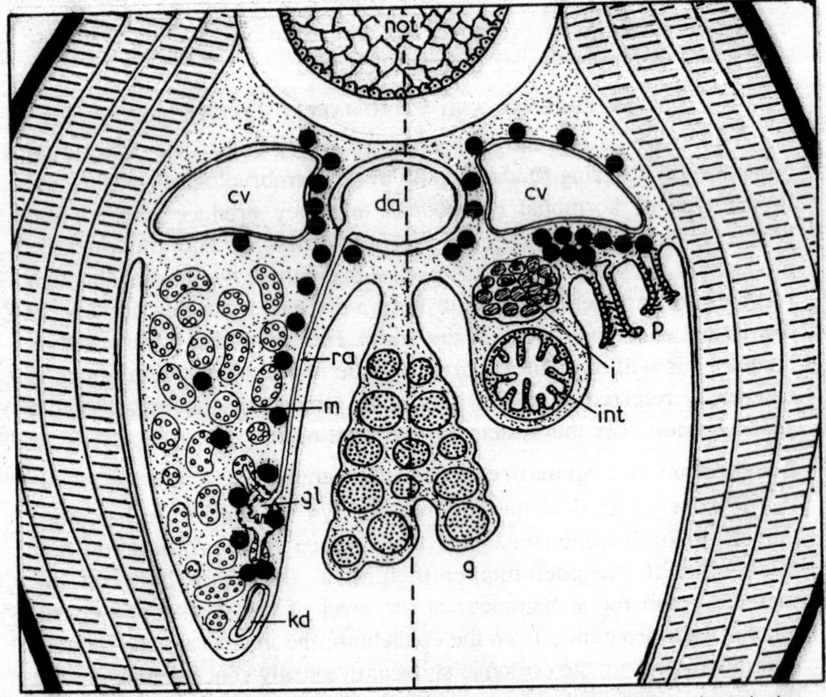

Fig. 15.18. The distribution of interrenal tissue in a transverse section through the pericardial region of an adult lamprey.

where they are separated only by the basement membrane from the capillaries and visceral epithelium. Other groups of cells are found between the kidney tubules. In *L. planeri*, the total mass of interrenal tissue has been put at about 0.04 mg 100 g^{-1} body weigh^{-1} ; a minute figure by mammalian standards, where for several species the adrenal tissue represents from 10-126 mg 100 g^{-1}.

The larger groups of interrenal cells are often arranged in cords or follicles, divided internally by thin connective tissue septa. Characteristic of these cells are the large liposomes, often exceeding in diameter the size of the nucleus and containing lipid and cholesterol. Ultrastructurally, they have all the hallmarks of other vertebrate steroid-secreting tissues. The smooth endoplasmic reticulum is well-developed, the mitochondria have tubular or vesicular cristae and the cytoplasm contains large numbers of free ribosomes or polysomes. Furthermore, the relationships between the mitochondria, lipid vacuoles and endoplasmic reticulum parallel the patterns seen in other steroidogenic cells.

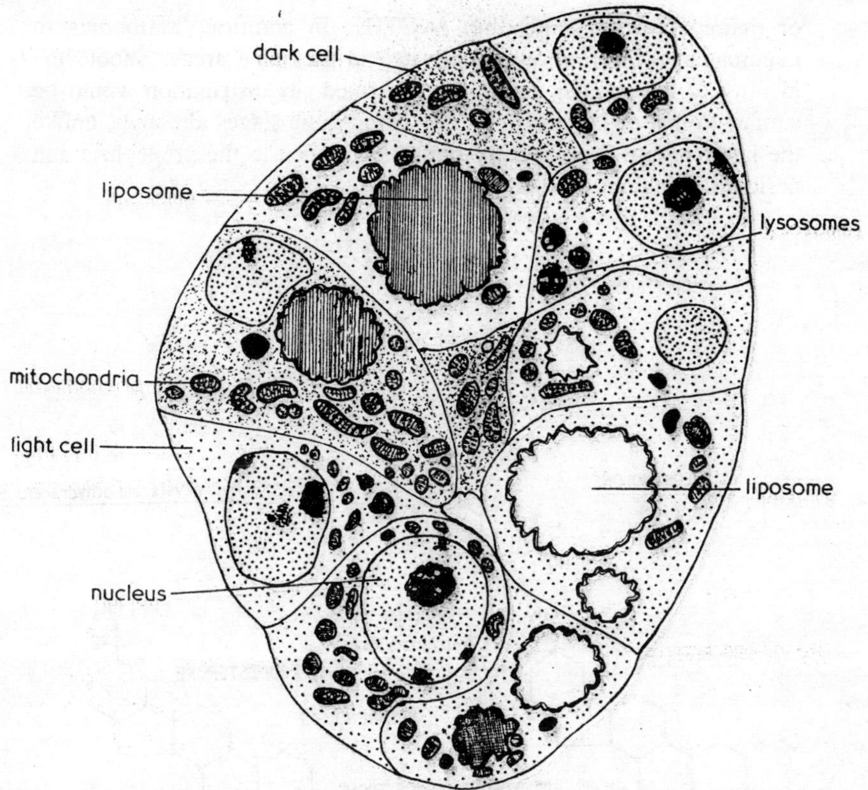

Fig. 15.19. A group of interrenal cells from the lamprey.

Until recently, the interrenal tissue of myxinoids had not been identified. By using ^{14}C-labelled isoaxole, an inhibitor of the steroid enzyme Δ^5, 3β-hydroxy steroid dehydrogenase, it has been possible to locate sites where the enzyme may be concentrated. Since this is a key enzyme in steroid biosynthesis, its presence should be regarded as evidence that the tissue concerned is involved in steroid secretion. In *Myxine*, the activity of the enzyme inhibitor was found to be concentrated in the pericardial and pronephric regions, but there was no evidence for its presence in either the mesonephros or posterior cardinal veins. Histological examination of the pronephros revealed cells with some of the characteristics of interrenal cells, situated around the bases of the pronephric tubules and in the central mass. Equally significant was the hypertrophy and hyperplasia that occurred in the central mass of the pronephros after injections of saline solutions or of mammalian corticotrophin (ACTH). In addition, antibodies to mammalian ACTH were concentrated in the same areas. Should the identity of this hagfish tissue be confirmed, its distribution would be similar to that of lampreys in early ontogenetic stages although, unlike the latter group, it apparently remains confined to the pronephric and pericardial regions.

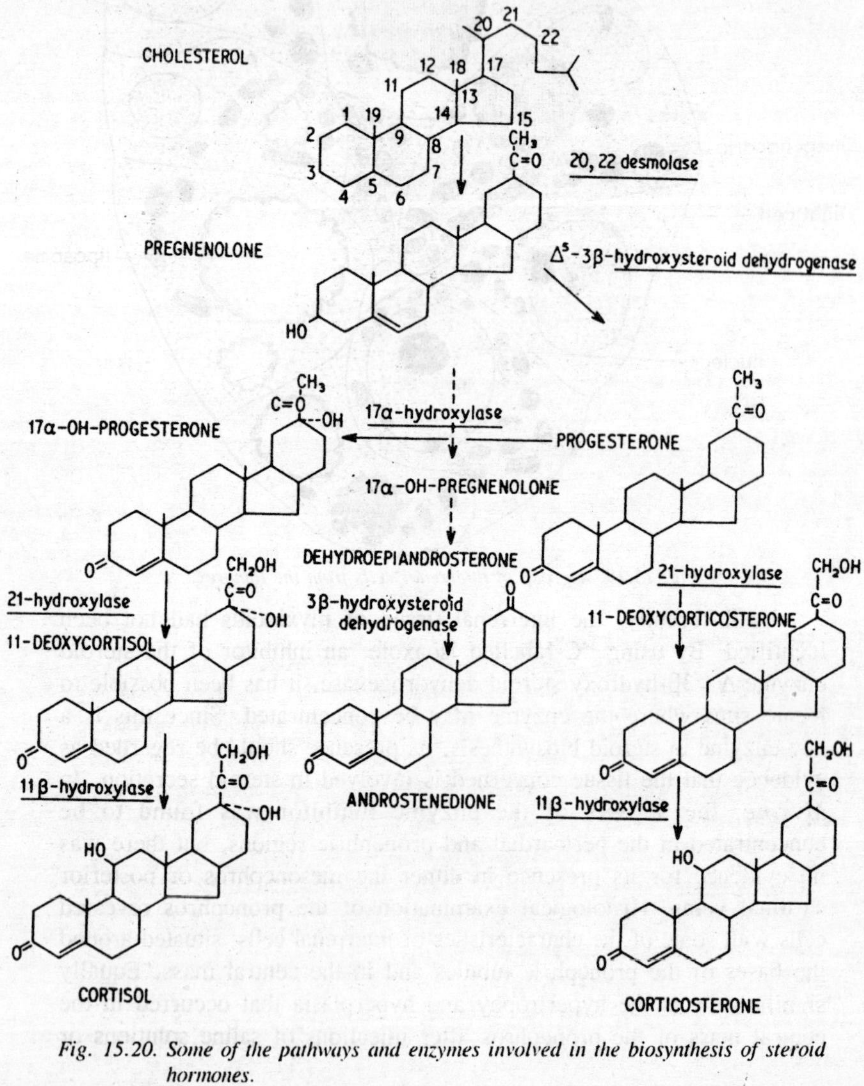

Fig. 15.20. Some of the pathways and enzymes involved in the biosynthesis of steroid hormones.

Biosynthetic Activity and Pituitary Control

Although the cytological, histochemical and embryological evidence points very clearly towards an homology between the lamprey interrenals and the adrenocortical tissues of higher vertebrates, the nature and extent of their biosynthetic activities are still obscure. On the other hand, in the myxinoids, where the plasma contains some of the usual vertebrate corticosteroids, their cellular source remains to be firmly established. In lampreys, the minute volumes of active interrenal tissue and the consequently low concentrations of hormones or enzymes might be at least partly responsible for some of the difficulties encountered in biochemical studies, such as the repeated failure to localize the enzyme Δ^5,3b, hydroxysteroid dehydrogenase by histochemical methods, although some evidence of its activity has come from the application of sphectophotometric techniques using tissue homogenates.

Attempts to establish the existence of corticosteroids in lampreys have involved blood analyses as well as *in vitro* investigations on tissue containing interrenal cells. Cholesterol, the parent precursor in steroid metabolism has been shown to be present in the liposomes of interrenal cells and these take up ^{3}H-labelled cholesterol thus strengthening their homology with adrenocortical tissue. However, in spite of the use of highly sensitive methods of analysis, the presence of corticosteroids in lamprey blood has not yet been demonstrated unequivocally and even if such compounds are present, their concentrations must be so low that it would be difficult to imagine that they have any physiological significance. On the other hand, experiments involving the incubation of kidney tissue containing interrenal cells with the appropriate ^{14}C-labelled steroid precursors, indicate that this tissue is capable of forming compounds which are intermediates in the synthesis of vertebrate steroid hormones and that it can produce some of the enzymes involved in steroid metabolism. In spawning migrant sea lampreys, tissue from the kidney was incubated with ^{14}C-labelled cholesterol, 11-deoxycorticosterone, 11deoxycortisol and progesterone. As a result, neither corticosterone, cortisol, cortisone or deoxycorticosterone were formed, but 17α-hydroxyprogesterone was synthesized from cholesterol, indicating the presence of the enzyme 17α-hydroxylase. A further series of experiments involved kidney and interrenal tissue from ammocoetes and feeding stage sea lampreys. After incubation with ^{14}C-progesterone, again no corticosterone, cortisone or cortisol were produced, but in this case 11-deoxycortisol,

17α-hydroxyprogesterone and androstenedione were detected (Weisbart and Youson, 1975). These results indicate the presence of 17α - and 21-hydroxylase activity in addition to 20-desmolase and, when considered in relation to the observations on spawning migrants, they suggest that there may be some changes in the biosynthetic capacities of the interrenal cells when the animals abandon parasitic feeding to embark on their spawning run. The ability of feeding stage animals to form androstenedione is of particular interest in view of the ability of other vertebrate adrenocortical tissue to produce C-19 steroids. Curiously, in vivo experiments on parasitic phase sea lampreys involving the intracardiac administration of ^{3}H-labelled progesterone, have only provided evidence for 21-hydroxylase activity and the formation of 11-deoxycorticosterone.

In the blood of *Myxine*, many of the normal vertebrate corticosteroids have been detected after prior treatment of the animals with mammalian ACTH. In this animal, cortisol, cortisone, corticosterone and 11-deoxycortisol were present and, although their source is not known, the fact that their production was stimulated by ACTH suggests that this may have been interrenal tissue. Moreover, this possibility is enhanced by the fact that, although the gonadal tissue of a hagfish has been shown to be capable of producing a range of steroid compounds *in vitro*, none of identified metabolites (with the exception of 11-deoxycortisol) include the steroids that were found in the blood of *Myxine*. Thus it seems likely that these corticosteroids were synthesized in some tissue other than the gonads and that this tissue, like the adrenocortical tissues of higher vertebrates can be stimulated by the pituitary ACTH.

Evidence for pituitary control of the lamprey interrenal is at present far from conclusive. Although various cell types in the adenohypophysis have been nominated as possible corticotrophs, their identification has not been finally settled. Hypophysectomy does not result in any obvious changes in the morphology of interrenal cells at least at the level of the light microscope but, on the other hand, injections of mammalian ACTH have resulted in nuclear enlargement and hyperplasia of the interrenal cells of both ammocoete and adult lamprey. Much more significant however, are the ultrastructural changes that have been seen in sea lamprey interrenal tissue after ACTH injections. These include alterations in nuclear shape, increased numbers of mitochondria, development of the Golgi and smooth endoplasmic reticulum, reduction in the numbers of lipid droplets and a folding of the plasma membrane adjacent to the perivascular spaces. All of these are reminiscent of the changes that have been seen in the adrenal cortex of the higher vertebrates following the administration of ACTH.

Chromaffin Cells

The chromaffin cells of the adrenal medulla of the higher vertebrates originate from the embryonic neural crest which also gives rise to the sympathetic ganglia. These cells are therefore regarded as modified post-ganglionic neurones, which in the course of their differentiation have lost the characteristics of nerve cells and have assumed glandular functions. Throughout the vertebrates, cells similar in their origins and cytochemical characteristics to those of the adrenal medulla occur in other sites throughout the body and are classified as chromaffin cells. Although we have no direct evidence of their origin from the neural crest, the chromaffin cells of the cyclostomes resemble those of other vertebrates in their cytochemistry, fine structure and the presence of storage granules containing catecholamines.

In lampreys, apart from the chromaffin cells of the heart wall referred to the greatest concentrations of these elements are to be found in the pronephric region. Here they are usually located in the tunica adventitia of the major arteries, particularly the dorsal aorta and the proximal part of the mesenteric artery, and separated from the lumen of the corresponding veins only by a thin endothelial layer. They are especially numerous in the walls of the cardinal veins and the sinus venosus and are often closely associated with nests of interrenal cells. In hagfishes, their distribution has not been described in detail, but they are said to be particularly numerous in the branchial and portal hearts and in the cardinal veins.

In the ammocoete, the chromaffin cells measuring from 12-35 μm in diameter, are often stellate in form with processes extending between neighbouring cells and separated from the underlying connective tissue of the arterial wall by a thin basal lamina. The cytoplasm is filled with granules about 130 nm in diameter, surrounded by a translucent halo and associated with the tubules of the smooth endoplasmic reticulum. Although nerve fibres may be present in the underlying connective tissue, there is no evidence that the chromaffin cells are innervated. From their cytochemical reactions it is believed that the extra-cardiac chromaffin cells of larval lampreys contain a primary catecholamine, either noradrenalin or dopamine.

The chromaffin cells of the lamprey heart are said to contain mainly adrenalin, whereas noradrenalin predominates in the circulating

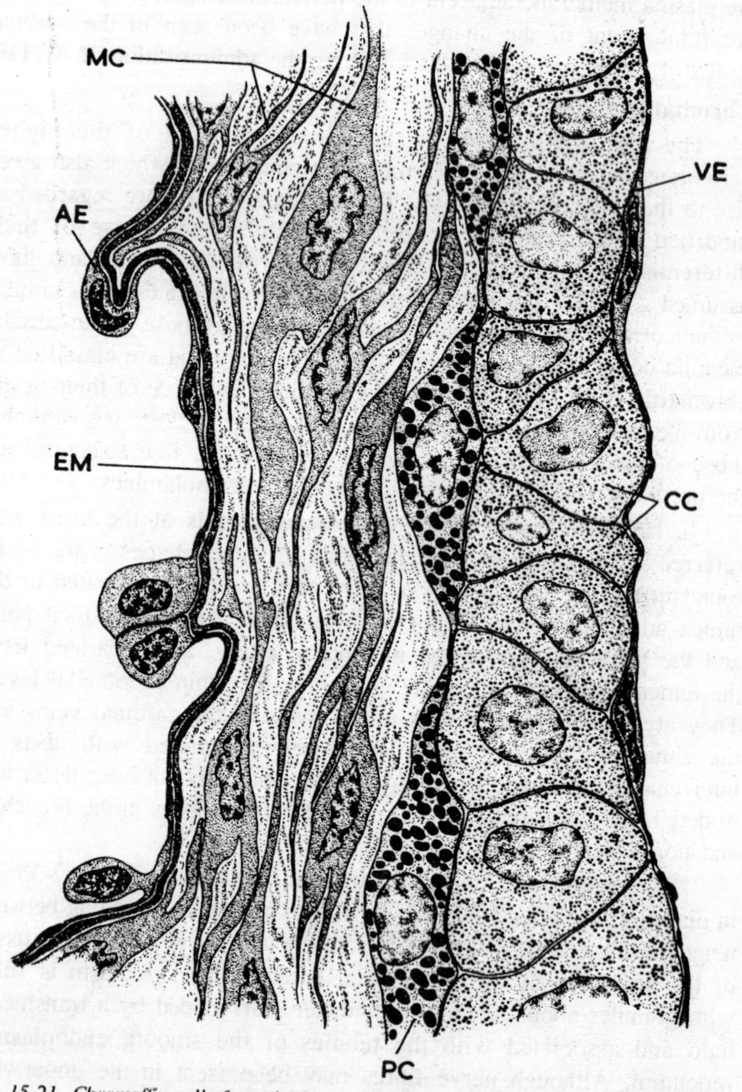

Fig. 15.21. Chromaffin cells from the lamprey. CC–chromaffin cells; PC–pigment cells; MC–smooth muscle; EM–elastic membrane; AE–endothelial cells making up tunica intima of artery together with elastic membrane; VE–single layer of endothelial cells in wall of vein.

blood and is also present in the chromaffin cells of the cardinal veins and the sinus venosus. These differences in the tissue distribution of the two catecholamines are in general agreement with biochemical analyses, which have shown that adrenalin accounts for most of the catecholamine content of the atrium and ventricle, whereas in the exceptionally high catecholamine concentrations of the sinus venosus, the ratio of noradrenalin to adrenalin is about 3:1.

Physiological Role of the Interrenal and Chromaffin Tissues

Activation of the interrenals and/or the chromaffin tissues has been implicated in many of the short-term responses of lampreys to various stresses. These include the diuresis that occurs when larval or adult lampreys are handled and the rise in blood sugar concentrations that is seen when the animals are subjected to reduced oxygen tensions or surgical treatments. Hyperglycaemia has also been observed in *Myxine* in various forms of injection or surgical trauma. Under conditions of hypoxia, hypertrophy of the chromaffin cells occurs in the lamprey and increased levels of blood catecholamines have been observed after animals have been maintained in poorly oxygenated water, or after forced swimming or surgical trauma.

In their responses to adrenalin injections, the two groups of cyclostome show a curious difference. In lampreys, the effect is a marked but transient hyperglycaemia, which has not been observed in *Myxine*. It is difficult to accept the suggestion that this is due to the high levels of endogenous catecholamines in the hagfish, since this must also be true of the lamprey. As in the mammal, adrenalin reduces the glycogen synthetase activity of the liver and muscles of lampreys, thus tending to decrease the rate at which glycogen is produced and stored in these organs. At the same time it promotes the degradation of glycogen to glucose in these tissues by increasing the level of activity of amylolytic enzymes.

In higher vertebrates, stress responses mediated in the first place by catecholamines, are maintained over longer periods by the pituitary-adrenocortical system and the secretion of corticosteroids. Whether a similar mechanism exists in cyclostomes is unknown. A possible intervention of the interrenal of lampreys in certain forms of stress, has been suggested by the hypertrophy and hyperplasia of this tissue, particularly after surgical interventions. Evidence that the interrenal might also be involved in carbohydrate metabolism rests solely on the effects of a single dose of injected cortisol, which is said to increase the liver stores of glycogen and raise blood sugar concentrations.

By analogy with other vertebrates it might be anticipated that the interrenal would be implicated in water and electrolyte metabolism, but the indications we have are conflicting. Some evidence of interrenal hyperplasia and hypertrophy has been reported after saline injections in lampreys but, on the other hand, in young sea lampreys, acclimated to full strength sea water, the nuclear and cell volumes of the interrenal are said to show a significant decrease. The common embryological origin of adrenocortical and nephrogenic tissue from the mesonephric blastema has been advanced as a rationale for the involvement of corticosteroids in water and salt balance and the control of kidney function, and especially in lampreys, the close association between the interrenal cells and the vessels of the kidney or glomus might suggest some functional relationship.

The trend in higher vertebrates towards an increasingly close association of the adrenocortical and chromaffin tissues has yet to receive an entirely satisfactory explanation. In the past, this close proximity of the two tissues has been attributed to the dependence of catecholamine biosynthesis on corticosteroids. Thus, the conversion of noradrenalin to adrenalin involves the transfer of a methyl group catalysed by the enzyme, phenylmethylanolamine-N-methyl transferase (PNMT), whose synthesis and activity are promoted by corticosteroids. Since the levels ofthese hormones required for this activity are thought to be much in excess of their normal concentration in the circulation, a close contact between the two tissues might be physiologically advantageous. However, there are considerable doubts whether this explanation is generally applicable to the lower vertebrates and, in lampreys, there is some evidence that the methylation of noradrenalin may be independent of corticosteroids. Nevertheless, even in these animals these two tissues, although widely dispersed, are frequently closely associated and the two cells types quite often occur within a common connective tissue envelope.

Endocrine Tissues of the Gonads

Endocrine Tissues of the Testis

In teleosts where the organization of the testis is similar to that of lampreys, the steroid-secreting tissues are either located in the walls of the lobules (lobule boundary cells) or, like the Leydig cells of higher vertebrates, in the interstitial tissues between the lobules.

In sexually immature river lampreys, *L. fluviatilis*, scattered undifferentiated cells of a fibroblast type are found in the interstices between the testis lobules and below the fibrous coat covering the surface of the gonad. By the late winter or early spring, these cells have divided to produce small islets of polygonal cells whose cytoplasm now contains numbers of mitochondria and small granular lipid

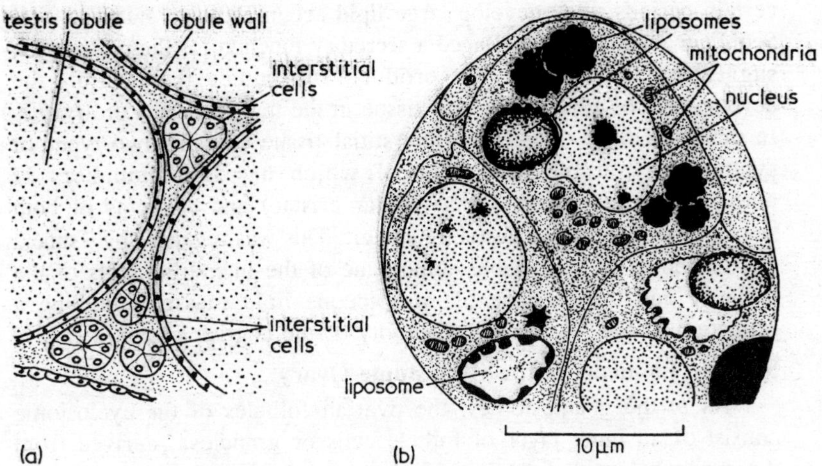

Fig. 15.22. The interstitial cells of the lamprey testis. (a) The interlobular location of the tissue. (b) High power view of a group of active interstitial cells.

inclusions. With further progress towards sexual maturity and the appearance of spermatids and spermatozoa, these interstitial cells show large cholesterol-rich, lipid masses (liposomes) like those of the interrenal cells, the smooth endoplasmic reticulum is well-developed and the mitochondria show a dense matrix and tubular cristae, characteristic of steroid-secreting cells. Furthermore, their implication in steroid biosynthesis has been confirmed by their positive reaction to the techniques used for the histochemical demonstration of 3β-hydroxysteroid dehydrogenase. Shortly before the sperm are set free into the body cavity of the sexually mature animal, significant changes occur in the lobule wall cells, suggestive of their involvement in some form of steroid metabolism. These undergo a rapid and synchronous swelling, lipid masses appear in the cytoplasm, cytoplasmic vesicles increase in size and number and the mitochondria develop structural characteristics similar to those of the interstitial cells.

The lobules of the hagfish testis are surrounded by a continuous layer of somatic cells (Sertoli cells) and in the testis of *E. stouti* these show occasional small lipid droplets, while some of the mitochondria may develop tubular cristae. However, these features were not sufficiently developed to justify a conclusion as to their possible steroidogenic activity. Earlier light microscopists considered that the Sertoli cells of *Myxine* had a nutritive function although it is clear from the latter author's descriptions that they undergo marked changes in the course of the spermatogenic cycle and that, at least at certain periods, they develop large lipid accumulations. Furthermore, Schreiner himself had envisaged a secretory function for these elements, suggesting that they might produce substances that inhibit the development of adjacent ovarian tissue in the mixed gonads of *Myxine*. In the testis of *E. stouti*, the interstitial tissue contains two types of synthetically active cells; one of which has large numbers of mitochondria (a majority with tubular cristae) and elements of both smooth and rough endoplasmic reticulum. This was regarded by Tsuneki and Gorbman as a probable homologue of the interstitial cells of the lamprey and the absence of conspicuous lipid masses could be a reflection of the low synthetic activity of the hagfish gonad.

Somatic Elements of the Cyclostome Ovary

As in the gnathostomes, the ovarian follicles of the cyclostome consist of an inner layer of follicle cells or granulosa, derived from the peritoneal epithelium (germinal epithelium) on the surface of the developing ovary. This is surrounded by a thecal layer derived from modified fibroblasts.

In the ovarian follicles of the lamprey, the granulosa at first consists of fusiform cells with spindle-shaped nuclei but, as vitellogenesis proceeds, these become prismatic and their lateral and apical surfaces

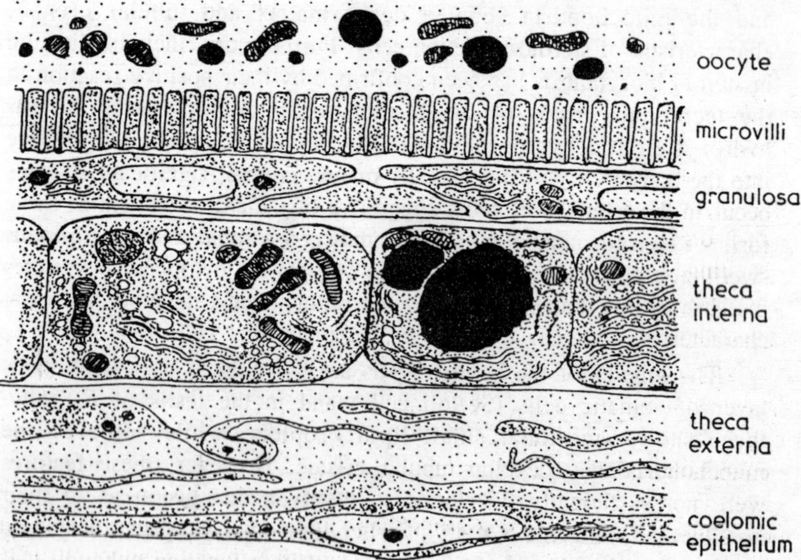

Fig. 15.23. A section through the ovarian follicle of a lamprey at the onset of vitellogenesis.

adjacent to the oocyte develop short microvilli. Initially, the granulosa covers most of the vegetative surface of the oocyte and the cells increase in height, particularly at the vegetal pole. This growth reaches its peak at the spawning season, shortly before ovulation, when the individual cells begin to separate. After ovulation, the remnants of these cells remain attached to the egg, forming a sticky coat that enables it to adhere to the sand grains in the nest. Mucopolysaccharides are an important element in the secretions of the granulosa cells and are thought to playa significant role in the final rupture of the follicle at the time of ovulation. Judging from their ultrastructural features, it seems more likely that the thecal tissues are responsible for the steroid secretions of the ovary and it is significant that these undergo a marked differentiation during the breeding season. At first slender and fusiform in shape, they become polygonal, developing a smooth endoplasmic reticulum and mitochondria with tubular cristae.

In the ovaries of *Myxine*, Fernholm (1972b) was unable to detect the presence of the enzyme 3 β-hydroxysteroid dehydrogenase by histochemical techniques and no cellular elements with the structural characteristics of steroidogenic tissues could be distinguished. In the course of oocyte growth the granulosa cells undergo changes similar to those that occur in the lamprey follicles and in the largest oocytes they form a single layer of high columnar cells with long processes extending to the surface of the oocyte. At this stage, the cytoplasm contains dense whorls of rough endoplasmic reticulum, vesicular smooth endoplasmic reticulum, Golgi, lipid droplets and mitochondria with lamellar or irregular cristae. As the oocyte increases in size the theca, which at first is composed of a single layer of modified fibroblasts, becomes multi-layered and divisible into a theca interna and externa, but although some of these cells may show a well-developed Golgi and rough endoplasmic reticulum, together with some lipid droplets, there are no indications that they are concerned in steroid biosynthesis. Thus, neither the granulosa nor the thecal elements show the typical hallmarks of active steroidogenic tissues and the former layer at least, is no doubt primarily concerned in the transport and processing of metabolites involved in the maturation of the oocyte. At the same time, as conditions in the lamprey gonad indicate, the assumption by somatic cells of the structural differentiations associated with active steroid metabolism may be relatively transient and sudden and in the case of the hagfish these might easily be overlooked without a more intensive investigation, taking into account the likelihood that there may be considerable changes in physiological state depending on precise phases of the spermatogenic or ovarian cycle.

Hormonal Activities of Cyclostome Gonads

It has now been established that the vertebrate sex hormones, oestradiol and testosterone are present in the blood of cyclostomes or in gonadal extracts although in hagfishes the concentrations of these substances is extremely low. In the ovary, the precise cellular site of steroid synthesis has not been firmly established, but in the lamprey testis the involvement of the interstitial cells has been confirmed by the presence in these cells of the enzyme 3β-hydroxysteroid dehydrogenase. The steroidogenic activity of the lamprey gonad has been further demonstrated by the failure of these animals to develop secondary sex characters after castration and the reappearance of these structures when sex hormones are administered to the operated animals. That the sex hormones are specific to the development of the male and female secondary sex characters has been shown by experiments involving the administration of oestradiol to male lampreys and testosterone to females. In some cases this led to the inhibition of the sex characters normal to the particular sex, but in other cases there appeared to be some simulation of the sex characters of the opposite sex. Although the mechanism is not understood, it is interesting to note that oestradiol tended to prolong the life span of the sexually mature lamprey.

Other than their involvement in the development of these sex characters, we have few indications of the wider role of sex steroids in the physiology of lampreys, although it is likely that as in other vertebrates, they may be involved in some aspects of gametogenesis and in the onset of spawning behaviour. In the female lamprey, oestrogens are believed to be implicated in the synthesis of calcium-rich phosphoproteins in the liver, and the administration of this steroid has resulted in increases in ovarian weight and oocyte volumes during the vitellogenic phase. This is in line with the processes of vitellogenesis in other oviparous vertebrates, where under the influence of oestrogens, yolk proteins are synthesized in the liver and transported to the oocytes.

A wider involvement of oestrogens in metabolic processes is also indicated by the effects of these compounds on the intestinal atrophy that occurs in the course of the spawning migration. Castration early in the migratory period of the river lamprey, *L. fluviatilis*, between August and November, prevented this intestinal degeneration. Furthermore, when the operation was delayed until the gut had already atrophied, castration resulted in some restoration of the gut tissues. On the other hand, when pellets of oestradiol or testosterone were implanted into the body cavity, the atrophy of the intestine was stimulated in both castrated and normal animals, but these treatments were ineffective either in normal animals late in the migratory season, or in cases where the gut had already hypertrophied as a result of castration. As the removal of the pituitary early in the migratory period is also capable of preventing or retarding the atrophy of the gut, it may be assumed that this acts by virtue of its control of the secretory activity of the gonadal tissues.

There is some evidence that steroidogenic activity in the gonads of lampreys does not become significant until comparatively late stages of sexual maturation. In the testis, for example, the differentiation of the interstitial cells does not seem to occur until spermatogenesis is well advanced and the concentrations of testosterone in the blood remain at very low levels until the spring, when the secondary sex characters appear. Similarly in the female sea lamprey, oestradiol concentrations in the plasma have been found to rise sharply at sexual maturation, declining even more rapidly after ovulation has occurred. This may explain the apparent failure to demonstrate the production of testosterone by testis tissue *in vitro* after incubation with ^{14}C-labelled progesterone. In these experiments, using the most refined methods of steroid analysis, it was only possible to establish the presence of 11-deoxycorticosterone (and therefore of the enzyme 21-hydroxylase) in gonadal tissue from larval and feeding stage sea lampreys.

Although Fernholm (1972) was unable to find any indications of steroid biosynthesis or hydroxysteroid enzymic activity in ovarian tissues of *Myxine*, clear evidence of steroid secretion has been reported in the ovaries of *E. burgeri*, a species with a seasonal breeding season. In experiments on ovarian tissue from this species, Hirose *et al.* (1975) cultured the tissue with ^{14}C-labelled pregnenolone as substrate and identified among the metabolites, 17 β-hydroxyprogesterone, androstenedione and 11-deoxycortisol. When 17 β-hydroxypregnenolone was used as the substrate, androstenedione was produced, suggesting that as in higher vertebrates, there may be two alternative biosynthetic pathways from pregnenolone to androstenedione. These authors believed the presence of 11-deoxycortisol to be of particular interest, since corticosteroids have been implicated in fish ovulation and 21-hydroxylase has also been found in the ovaries of teleosts and protochordates.

In another member of the same genus, *E. stouti*, measurements have been made of both testosterone and oestradiol concentrations in the blood in hypophysectomised, sham-operated and anaesthetised animals. As mentioned elsewhere there was no evidence that the removal of the pituitary has any effect on the levels of circulating sex hormones and these concentrations appeared to decline the longer the animals were kept in captivity. In the male hagfish, there was no significant correlation between body length and plasma testosterone concentrations, but in the females there was a distinct tendency for hormone concentrations to decrease with increasing body length. If the latter is assumed to be an index of the degree of sexual maturation, this relationship might imply that, in smaller animals, the testicular tissues tend to be better developed (or to have undergone less regression) and therefore may retain a greater capacity for testosterone production. In a very high proportion of the hagfish examined, the oestradiol concentrations were below the sensitivity of the analytical technique and of the individuals with the highest hormone levels, five were sexually mature females and two were large males.

16

Ecology and Behaviour

Habitats

Myxinoids

Characteristically, hagfishes are adapted for life at considerable depths at high salinities and low light intensities, but within the group there are distinct species differences, particularly in the depths to which they penetrate. *Myxine glutinosa* normally occurs between 50-100 m, but has been recorded down to 1100 m off the Norwegian coast and in Japanese waters *M. garmani* is found between 400-800 m. In some inner fiords, *M. glutinosa* may occur in much shallower water at about 30 m, but in these areas there may be marked stratification with the freshwater overlying water of high salinity (Tambs-Lyche, 1969). *Myxine* is apparently less tolerant of reduced salinity than some species and this may explain its absence from the Baltic. From its distribution, it has been inferred that the upper temperature limit for this species is between 10-13°C. When hauled to the surface in a trap from deep water, the animals are stressed by the rapid change in salinity and exude large quantities of mucus.

Some at least of the members of the genus *Eptatretus* are able to tolerate conditions in much shallower water, with higher temperatures and lower salinity. For example, the populations of *E. burgeri* studied by Fernholm (1974) in Japanse waters were living at depths of only 10 m and in other areas the same species has been found at 6-8 m. A similar shallow water habit is also characteristic of the New Zealand hagfish, *E. cirrhatus* which occurs at depths from 4-100 m. Shallow water species such as *E. burgeri* must be able to tolerate considerably higher temperatures than those that limit the distribution of *Myxine*.

Thus, in the area where these animals were studied by Fernholm, the average winter temperature was 13°C rising to 24°C in July. As far as is known this species is unique in showing a seasonal migration to deeper water, where spawning is believed to occur. This takes place in late summer and autumn when the shallow areas reach their maximum temperature (25°C) and the animals return again in late autumn when temperatures are once again beginning to fall. Off the Californian coasts, *E. stouti* has been reported at depths varying from 45-1000 m and a temperature of 10°C is said to be close to its upper limit. However, it is generally regarded as a relatively shallow water species and has been seen swimming near the surface where salinities may be as low as $24^{0}/_{00}$. This would be consistent with the tolerance that this species has shown to low salinity in laboratory experiments. *E. deani*, on the other hand is a deep water species and is reported to be a dominant element in the fish populations of the San Diego Trough 12 km off the California coast at depths of 1200 m. Evidence cited by Strahan and Honma (1960) suggests that other members of the *Eptatretidae* may also be more tolerant to reduced salinities than *Myxine*, since on the Japanese coasts, *E. (Paramyxine) atami* is said to be found around river mouths. Similarly the New Zealand hagfish, *Nemamyxine* has been taken within an estuary.

Myxine displays the most extreme adaptations to deep water life in the almost complete absence of skin pigmentation and in the extent of its optic degeneration. On the other hand, in the shallower water species of the genus *Eptatretus*, the skin is less transparent (except over the eyes), a vitreous body is present in the eye and retinal structures are more highly differentiated. However, the extent of the degeneration in the eye of *Myxine* does not necessarily imply that its deep sea habits are of very great geological antiquity and the existence of similar atrophy in cave-dwelling vertebrates indicates that regression may, in an evolutionary context, take place over relatively short periods.

Myxine inhabits areas with a soft bottom, often on the continental shelf adjacent to estuaries where extensive silt deposits are most likely to occur. This preference for a soft substrate has been illustrated by Cole (1913) who described the movements of a population of hagfishes on the N.E. coast of England several miles inshore, towards an area that had previously had a hard bottom, but which had been covered with mud dropped from the hoppers of dredgers. Given these conditions, local populations of *Myxine* were said to be as numerous as earthworms in a fertile soil. Earlier reports had indicated that some *Eptatretus* species favoured hard or rocky bottoms and this led Strahan (1963a, b) to suggest that these forms were not so firmly committed to the burrowing habit as the members of the genus *Myxine*, where the fusion of the efferent branchial ducts could be regarded as an adaptation to this mode of life. However, observations in the field have shown that both *E. stouti* and *E. burgeri* inhabit areas of soft mud and the burrowing habits of the latter species have been confirmed by skin divers.

Lampreys

The marine phase of the lamprey is almost certainly a secondary introduction into the life cycle of a primarily freshwater group and even today represents no more than a third to a quarter of the total life span. The ultimate success of a lamprey population must therefore depend to a large extent, on the quality of its freshwater environment. This must provide appropriate conditions for the development of the eggs and the survival of the larval stage, together with a quite different set of environmental conditions that are necessary for the completion of spawning. Evolved as a means of exploiting the organic rich deposits of the river systems, the ammocoete phase requires relatively stable silt beds into which the larvae can burrow, together with water of sufficient productivity to supply adequate amounts of algal food. Within a given river system, the distribution of the larval population results from the interaction of the characteristic behaviour of ammocoete and adult; the mainly passive downstream drift of the larva and the rheotactic upstream migration of the spawning adult. Thus, throughout the larval period the ammocoete population will tend to move downstream towards the middle and lower reaches of the river system, but this movement is counteracted each year by the ascent of the spawning adults towards the higher reaches.

A typical ammocoete habitat is an area where the current is slack and so placed that it is protected from major fluctuations in water levels or stream velocities. Such conditions are most likely to be found in eddies or backwaters or at bends in the river bed, where there are likely to be sufficient accumulations of soft silt and sand to provide a suitable substrate for the burrowing ammocoete. In such places, often partially shaded by trees, diatoms may form an incrustation or 'polster' on the interface between the silt and the water and this may be an important factor in the stability of these ammocoete 'beds'. Current velocities over these sites have been put at an average of 0.4-0.5 m s^{-1} and limiting velocities are said to be about 0.6-0.8 m s^{-1}; rates of flow that are of the same order as the maximum swimming speeds of which a large ammocoete is capable.

For spawning lampreys, the important requirements are first and foremost a gravel bed, moderately fast currents and cool, well-oxygenated water. In general, current velocities on the spawning grounds will tend to be greater than the maximum swimming speeds of which lampreys are capable. As a result, once the animals move out of the nest (which protects them from the current) they are only able to maintain their position by anchoring to the substrate with their suctorial discs.

Burrowing Habit

In their anatomical and physiological adaptations to a burrowing habit there are remarkable parallels between larval lampreys and hagfishes. These need not necessarily have any phylogenetic significance, although it is possible that both hagfishes and lampreys have been derived from benthic ancestors with mud-feeding habits. Similar common adaptations to burrowing are to be found in a wide variety of vertebrates, where they have arisen independently by convergent evolution.

Larval Lampreys

When about to burrow, the ammocoete points its head region downwards towards the substrate and once it has made contact, the normal swimming movements are replaced by violent whiplash contractions of the tail, propelling the head and branchial region below

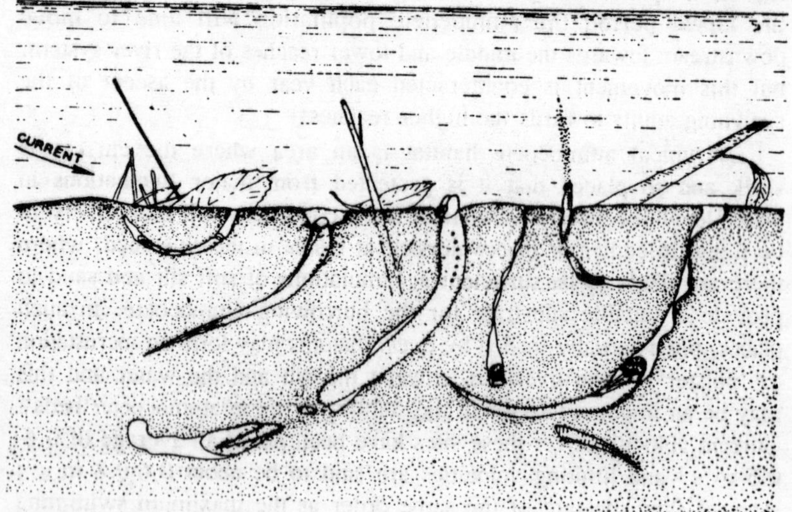

Fig. 16.1. Burrowed ammocoetes of the sea lamprey, Petromyzon marinus.

the surface. At this point, the vigorous contractions of the tail subside and this region is now laid more or less horizontally over the surface. Final submergence is achieved by contractions of the more anterior trunk regions using the expanded oral hood as an anchor below the surface, literally pulling the tail region into the mud. Once buried, the animal assumes an arched posture forming a U-shaped burrow which can be detected by a conical depression on the surface of the mud, with a central hole marking the position of the mouth. These openings are said to lie at an angle of 60-70° to the mud surface in the direction opposite to the current flow. The depth of the burrow varies with the size of the ammocoete and when stimulated mechanically or by an increase in light intensity, they retreat to greater depths, in the case of large ammocoetes, up to 18 cm below the surface. The walls of the burrow are cemented by mucous secretions, particularly from the dorsal columns of the endostyle facilitating the passage of the exhalant current from the gills.

Although they may change their position during the daytime, on each occasion forming a new mucus tube, ammocoetes normally leave their burrows only at night. During the daytime, any disturbance in the water only has the effect of driving them deeper into the burrow or at least causes the withdrawal of the snout, which normally protrudes from the entrance hole. Under simulated field conditions, ammocoetes show a diurnal rhythm of activity, leaving their burrows for periods during the night to swim freely in the surrounding water. Under laboratory conditions, this activity has been found to be most intense in the second hour of a 12 hour dark-light cycle, with a second smaller activity peak during the eighth hour of the dark period. This pattern is quite similar to the diurnal activity rhythm of the adult lamprey. Except in the genus *Mordacia*, ammocoetes also show a diurnal rhythm in the pigmentation of the skin. This darkens during the day through the expansion of the melanophores and pales at night when the pigment cells contract. If the ammocoete is kept in continuous darkness, this rhythm will persist for a time, eventually becoming irregular and finally disappearing to leave the animal permanently dark. Permanent darkening can also be produced by subjecting the animals to constant illumination or by removal of the pineal. Although these pigmentary responses might at first sight appear too trivial to have much adaptive value in an animal that spends nearly all the daytime hours burrowed beneath the mud surface, pigmentary protection from light is fundamental to a wide variety of animals. The importance of this protection can be seen in *Myxine*, where the skin is almost devoid of pigment and quite transparent. When subjected to continuous illumination on a white background, *Myxine* dies within a matter of days, showing every sign of being severely stressed.

In their natural habitat, ammocoetes are likely to be affected by seasonal changes in the oxygen content of the water. When first exposed to low oxygen tensions under laboratory conditions, the ammocoete first tends to protrude its head or even the whole of the branchial region above the surface of the substrate, thus tending to inhale water which would have a higher oxygen and lower CO_2 content than that within the borrow. At even more extreme levels of oxygen depletion approaching lethal concentrations, the larvae may emerge from the burrow completely, displaying restless movements punctuated by bursts of swimming which, in their natural habitats, might be expected to move them towards more favourable conditions.

Physiological adaptations to burrowing include the responses of ammocoetes to light and to contact stimuli. In a partly illuminated aquarium, ammocoetes respond to the light stimulus by random swimming, which ceases when they arrive in the shaded area. Although the spinal cord itself is sensitive to light, the tail region is especially so and contains photoreceptors innervated by the lateral line nerves. Light responses from this area are obviously significant in ensuring the complete withdrawal of the ammocoete below the substrate. In the aquarium, free-swimming ammocoetes usually settle down with their lateral surfaces in contact with the bottom. Equally, they will lie quietly in an artificial substrate of glass beads or inside glass tubes. This response to contact stimuli (thigmokinesis), combined with a reduction in light intensity below the threshold level that induces swimming, are no doubt important factors in reducing the daytime movement of burrowed ammocoetes.

Hagfishes

Like the ammocoete, the hagfish remains for the most part, buried in the mud during the daytime, emerging at night, either to rest on the surface or to engage in short bursts of swimming. However, it is not known whether this pattern of behaviour has an innate rhythmic basis, like the diurnal rhythms of lampreys or is simply due to the animal's photokinetic responses. Even if such rhythms are present they could not be regulated in the same manner that is thought to operate in the lamprey, where the pineal is believed to be involved in the control of metabolic rhythms.

In Japanese waters, the commercial fishery for *E. atami* is normally carried out at night although, where fishing is operated at depths below 200 m (under conditions of very low light intensity), there are apparently no differences in the size of day or night catches. Similar conditions have also been reported from Norway, where daytime catches of *Myxine* in shallow water (30 m) are said to be poor, whereas large numbers are caught in daytime in deep water (120 m). These examples suggest that environmental changes in light intensity may be more important than innate circadian rhythms in determining the levels of activity of hagfishes. In spite of their degenerate eyes, hagfishes, like ammocoetes, will congregate in the shaded part of a partially illuminated aquarium. Again like the larval lamprey, they possess skin photoreceptors, concentrated mainly in the head region and around the cloaca, although these differ from those of the ammocoete in being innervated by spinal nerves rather than by a lateral line nerve. In *Myxine*, these dermal receptors show their greatest spectral sensitivity to light of 500-520 nm, corresponding to the wavelengths of the light that would penetrate into their deep water habitat. Their responses to light are characterized by very long reaction times. These vary from 10 seconds at the maximum intensities employed to between 2 and 5 minutes at threshold intensities, similar to those that the animal would experience in its natural environment.

In its method of burrowing the hagfish does not differ greatly from the ammocoete. With powerful swimming movements it dives head first into the substrate at an angle of 45–90°. According to Fernholm (1974), swimming movements continue until the animal is completely buried, but Strahan (1963) and Foss (1968) have described a gliding movement, rather like that of an ammocoete, which draws the hinder part of the body under the mud, after the head and anterior regions have submerged. Eventually, the head may reappear some distance (50-100 cm) from the point where the animal had first entered. Shaped like a shallow U, the entrance to the burrow may show a conical depression, marking the position of the head and nasohypophysial opening, while a small elevation some distance away, indicates the point where the exhalant current emerges. The head and tentacles normally project from the burrow and during the night many animals have been seen with the anterior third of their body exposed. Although Foss has described a copious secretion of slime immediately after the animal has submerged, the walls of the burrow are said to be unsupported by mucus and collapse once the hagfish has left.

In the Hardanger Fjord there are shallow areas where *Myxine* abounds at depths of only 30 m, making observation by diving practicable. Here the sea bed shows large numbers of hillocks, like miniature volcanoes, with a craterlike hole at the top. In an earlier series of observations, hagfishes were seen to enter these mounds through the top or sides, but later investigations failed to provide direct evidence that they were inhabited by hagfishes, although their structure was suggestive of burrowing activity. Observations by divers on *E. burgeri* have failed to reveal similar structures and their relevance to the normal life of the hagfish remains in doubt.

Laboratory observations had suggested that *E. stouti* was a species that preferred to rest on a hard bottom in a coiled position and because of this behaviour, Strahan believed that the members of this genus were less adapted to burrowing than *Myxine* and that the coiled position was a way of ensuring maximum surface contact on a hard and rocky sea bed. This has now been discounted by bathyscaph observations on the muddy terraces and deep water canyons off the Pacific Coasts of North America that are the habitats of *E. stouti* and *E. deani*. These have shown that both species burrow into the substrate in the same way as *Myxine* and similar observations have been made on *E. burgeri* in its natural environment. Like the ammocoete, *Myxine* normally rests on its side on the hard surface of an aquarium and the same position is often adopted by adult lampreys at times when they are no longer attached by their suctorial oral disc. In the case of *Myxine*, this behaviour has been linked with the peculiar orientation of its labyrinth, so placed that with the animal lying on its side, the two toroidal rings and their maculae would lie one above the other in or near the horizontal plane. Another possibility, considered by the same authors, is the presence in *Myxine* of a specialized skin region over the site of the degenerate eye and innervated by the V nerve, which responds to pulsed mechanical stimuli of a similar frequency range to those that excite the vibration receptors of the lamprey labyrinth. However, in view of the fact that this behaviour is shared by the ammocoete, it seems more likely that it reflects the importance of skin contact stimulation in reducing the activity of a burrowed animal.

Adult Lampreys

Although characteristic of the ammocoete, burrowing responses do not entirely disappear after metamorphosis and the macrophthalmia of parasitic lampreys may remain for considerable periods buried in the bottom of the stream, emerging mainly during the night. It is probably

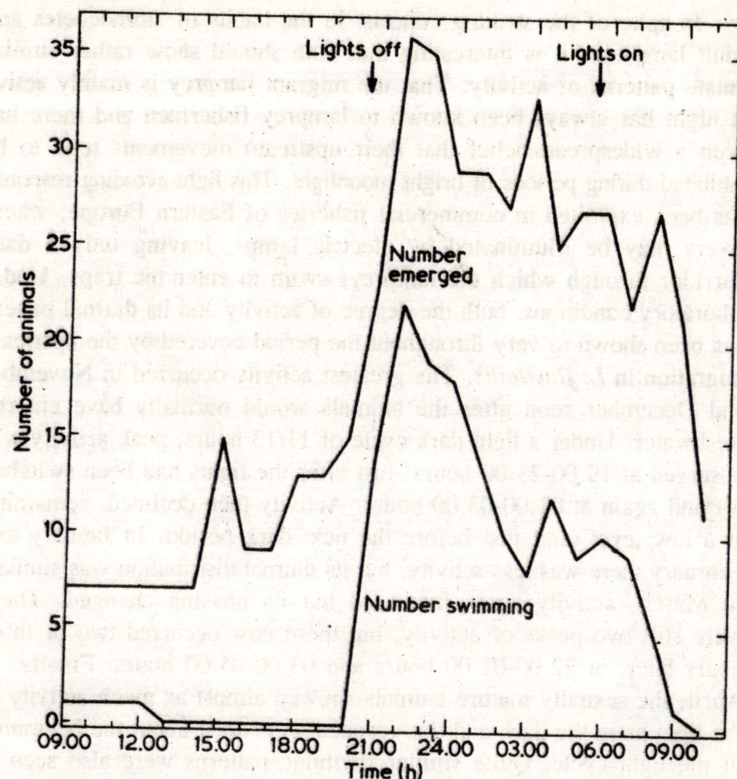

Fig. 16.2. Nocturnal activity of recently metamorphosed river lampreys L. fluviatilis in an aquarium.

during these phases of nocturnal activity that they are carried passively downstream, especially during periods of flood. In non-parasitic species, the burrowing habit and light-avoiding responses are retained for longer periods, only to be lost when spring water temperatures approach the levels at which spawning behaviour is initiated. At these times the adults begin to show a preference for lighted areas and the spawning grounds are often exposed to direct sunlight.

Throughout the earlier stages of the spawning migration, anadromous lampreys avoid the light during the daytime, hiding under rocks or river banks and resuming their upstream movement only during the hours of darkness. These responses are exaggerated in the southern lamprey, *Mordacia mordax*, which is said to burrow deeply into the river bed during its long upstream migration and whose unique dorsolaterally placed eyes are thought to be a specialized adaptation to this mode of life.

In spite of the striking contrast in the habits of ammocoetes and adult lampreys, it is interesting that both should show rather similar innate patterns of activity. That the migrant lamprey is mainly active at night has always been known to lamprey fishermen and there has been a widespread belief that their upstream movements tend to be inhibited during periods of bright moonlight. This light-avoiding response has been exploited in commercial fisheries of Eastern Europe, where rivers may be illuminated by electric lamps, leaving only a dark corridor through which the lampreys swim to enter the traps. Under laboratory conditions, both the degree of activity and its diurnal pattern has been shown to vary throughout the period covered by the upstream migration in *L. fluviatilis*. The greatest activity occurred in November and December soon after the animals would normally have entered freshwater. Under a light-dark cycle of 11/13 hours, peak activity was observed at 19.00-23.00 hours, just after the lights had been switched off and again at 01.00-03.00 hours. Activity then declined, remaining at a low level until just before the next dark period. In January and February there was less activity, but its diurnal distribution was similar. In March, activity again increased but its phasing changed. There were still two peaks of activity, but these now occurred two or three hours later, at 22.00-01.00 hours and 03.00-05.00 hours. Finally, in April, the sexually mature animals showed almost as much activity in the light as in the dark and this reached its peak at about the beginning of the light cycle. Quite similar rhythmic patterns were also seen in heart and breathing rates and these metabolic rhythms correspond quite closely with the behaviour of river lampreys in their natural environment. For example, the greatest swimming activity would normally occur in the early months when the lampreys are making their passage through the estuary towards upstream spawning sites, while in January and February, when activity in the laboratory declined, they might be expected to have reached the areas where breeding occurs. Renewal of activity on a considerable scale in March would coincide with a period when nest-building and pre-spawning activity are in full swing and marked daytime activity appears. This change in activity patterns has been attributed to the accession of additional daytime activity rather than to the loss of the basic circadian rhythm of nocturnal activity. The pineal, already known to be involved in diurnal pigmentary changes, may well be more widely implicated in these metabolic rhythms, although its role has not yet been investigated.

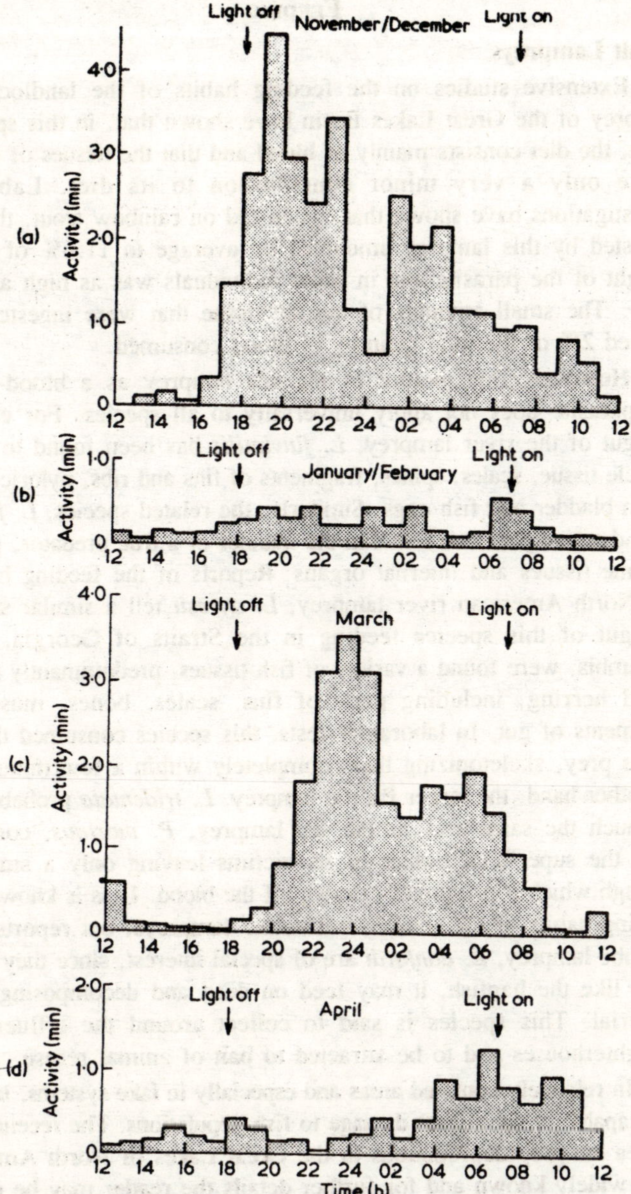

Fig. 16.3. Changes in the circadian activity rhythms of adult river lamprey, L. fluviatilis at various periods of the upstream migratory phase and maintained in a constant light-dark cycle of about 11-13 hours.

FEEDING

Adult Lampreys

Extensive studies on the feeding habits of the landlocked sea lamprey of the Great Lakes Basin have shown that, in this species at least, the diet consists mainly of blood and that the tissues of the prey make only a very minor contribution to its diet. Laboratory investigations have shown that, when fed on rainbow trout, the blood ingested by this lamprey amounted on average to 11.6% of the wet weight of the parasite and in some individuals was as high as nearly 30%. The small amounts of muscle tissue that were ingested never exceed 2% of the total quantity of blood consumed.

However, this picture of the sea lamprey as a blood-sucking ectoparasite does not apply universally to all species. For example, the gut of the river lamprey, *L. fluviatilis* has been found to contain muscle tissue, scales, spines, fragments of fins and ribs, pyloric caecae, swim bladder and fish eggs. Similarly, the related species, *L. japonica* is said actively to attack fish in the manner of a true predator, gnawing out the tissues and internal organs. Reports of the feeding habits of the North American river lamprey, *L. ayresii* tell a similar story. In the gut of this species feeding in the Straits of Georgia, British Columbia, were found a variety of fish tissues, predominantly those of small herring, including parts of fins, scales, bones, muscle and fragments of gut. In laboratory tests, this species consumed the flesh of its prey, skeletonizing them completely within a few minutes. On the other hand, the larger Pacific lamprey, *L. tridentata* probably feeds in much the same way as the sea lamprey, *P. marinus*, consuming only the superficial tissues of its victims leaving only a small hole through which it presumably draws off the blood. Less is known of the feeding habits of the smaller freshwater lampreys, but reports on the Danube lamprey, *E. danfordi* are of special interest, since they suggest that, like the hagfish, it may feed on dead and decomposing animal material. This species is said to collect around the effluent from slaughterhouses and to be attracted to bait of animal refuse.

In relatively confined areas and especially in lake systems, lampreys are capable of significant damage to fish populations. The recent history of sea lamprey depredations in the Great Lakes of North America is now widely known and for further details the reader may be referred to the account given by Smith (1971). This dwarf freshwater derivative of the anadromous sea lamprey of the North Atlantic, *Petromyzon marinus*, has apparently existed in Lake Ontario and other smaller lake systems, probably as far back as the last Glacial period, but was prevented by the Niagara Falls from entering the upper Great Lakes until the construction of the Welland canal in 1829. When these lampreys first penetrated through the canal is not known, but the first specimens were not recorded from Lake Erie until 1921 and although they have never flourished in this lake, by the middle 1930's they were already established in Lakes Huron and Michigan, finally appearing in Lake Superior somewhat later. Once in these upper Lakes the lampreys underwent a veritable population explosion, presumably due to a combination of favourable factors, including the abundance of suitable host fishes, the relative absence of predators and the existence in the Lakes' watersheds of ideal conditions for successful spawning and for the maintenance of their larval stages. The establishment of the sea lamprey was accompanied by a serious decline in the stocks of a number of fish species. This was most important in the case of the lake trout, and led to the collapse of a flourishing commercial fishery. Although it is admitted that overfishing may have been a contributory factor to this decline, there can be no doubt of the destructive potential of the lamprey predator as a major factor in the dwindling populations of many of the more commercially valuable fish species. With the setting up of a Great Lakes Fishery Commission, control measures were instituted on an extensive scale and eventally with the development of selective larvicides, lamprey populations have been reduced to more acceptable levels. As an example of the extent of fish mortality attributed to the lamprey at the height of its invasion, it has been estimated that from 1954 to 1961 the lampreys of Lake Superior alone would have destroyed about 5800 million tons of fish.

Experiments described by Kleerekoper (1972) point to the importance of the olfactory sense in the location of the prey. Kept in dim light, feeding stage sea lampreys maintained for several days a diurnal rhythm of activity, but greatly increased movements occurred when water in which trout had been held, was introduced into the experimental apparatus, at periods when the animals would normally be relatively quiescent. These responses did not occur after the nasohypophysial opening had been blocked. If the olfactory stimulus was diffused throughout the whole body of water the movements of the lampreys was random in direction, but in a compartmentalized apparatus the lampreys moved towards the inlet through which the trout water was introduced, showing that it would be possible for these animals to locate and follow the source of an olfactory stimulus. Chemical isolation of the components of the trout water showed that the effective odoriferous compounds were amines and that one of these, isoleucine methyl ester, was particularly effective not only in lampreys, but also in sharks and some teleosts.

The effectiveness of the lamprey's parasitic habits may be judged by the growth rates that it achieves and the rate at which it is able to accumulate its energy stores of lipid. For example over a parasitic phase of 2 or 3 years, the anadromous sea lamprey increases it body weight from 1-4 g to about 1000 g and its lipid content (as a percentage of wet weight) is increased about eightfold from 1.3% over 10%.

Hagfishes

Lacking the suction apparatus of a lamprey, the hagfish would be incapable of making a successful feeding attack on a mobile living fish and it is now generally accepted that they attack only dead or dying animals. Before the widespread adoption of trawling, hagfishes were indeed a menace to line fishermen, attacking and totally destroying their bait within the space of one or two hours, as well as making inroads into the catch itself. For example, Cole (1913) quoted one example of a cod which, when hauled to the surface, had no fewer than 120 *Myxine* attached to it. As a result of his experience in trapping these animals. Cole asserted that they showed a distinct preference for recently dead or even dying fish, rather than for stale fish. Entering through the gills, it uses its slime secretions to block the movements of the operculum of a dying fish, only entering the body cavity when these movements had ceased. During its initial efforts at penetration, the hagfish is said to retain a vertical burrowing posture, all the time making vigorous swimming movements with the hinder part of the body. According to Cole, the liver is eaten first, then the gut and heart and finally the muscle tissues between the skin and backbone, working forwards from the posterior end. In underwater observations on *E. burgeri*, Fernholm (1974) found that an anchovy could be completely demolished within 12 minutes leaving behind only the skin and the backbone. Throughout their feeding, hagfishes show continuous swimming movements and in their efforts to tear off fragments of food they frequently resort to knotting behaviour to give them a better purchase. A number of observers have noted the secretion of slime around the food and although this may be a by-product of their strenuous activity, it could conceivably help to deter other competitive scavengers in the deep sea community from sharing the food resources of the hagfish.

The fact that hagfishes attack fish when these are used as bait does not of course, necessarily imply that this is their main source of food in their natural environment. As Strahan (1963) has pointed out, it seems unlikely that fish would normally be present in sufficient numbers to maintain the dense populations of hagfishes that are known to occur in certain areas. Some support for this view came from his analyses of gut and faeces. In one series, the gut contained only polychaete remains, but a larger number of faecal analyses made on *Myxine* showed parts of ribs, vertebrae and pectoral girdles of small herring, as well as hermit crabs, shrimps, priapuloids, remains of polychaetes, small lamellibranchs and gastropods. The fish and crustaceans were presumed to be dead when they were ingested, but Strahan believed that other organisms such as the small molluscs were probably ingested accidentally while the hagfishes were searching for food under the surface of the mud. Some experiments have also been carried out in the field involving the presentation of a variety of food materials. Underwater television observation showed that a polychaete and an anomuran were accepted by the hagfishes, but not a sea anemone. When herring was offered together with invertebrates, this was taken preferentially, as was decomposing rather than fresh fish.

Recent investigations have shown that the deep sea benthic community contains a surprisingly large variety of scavenging species and Jensen (1965) has remarked on the richness of the vertebrate and invertebrate fauna in the habitats of the hagfish off the San Diego coast, as revealed by bathyscaph observations. No doubt, within such competing communities, speed in the location and consumption of animal remains is at a premium. According to the stability-time hypothesis, the long-term stability of the deep sea environment has encouraged the development of diverse communities of competing species which have reached an equilibrium through refined niche differentiation, particularly in regard to food specialization. On the other hand, all our information tends to point to the hagfish as an indiscriminate feeder, taking from the sea bed any animal material that comes its way. As Dayton and Hessler (1972) have maintained, in order to survive, deep sea benthic species need to be flexible and opportunist in their choice of food, if they are to make full use of the available resources. This has therefore tended to blur a distinction between predators and deposit feeders, resulting in a community of 'croppers' which consume both live or dead and decaying organic materials. In the case of the hagfish, any preferences that it may appear to show are probably based on the ease with which a particular food can be located and the intensity of the olfactory stimulus that it generates. As a nocturnal feeder, depending entirely on its olfactory sense, the discovery of food, unless aided by currents, has been considered a random and inefficient process. This has been illustrated by aquarium experiments which have shown delays of 18-24 minutes between the introduction of bait and the first responses, such as the protrusion of the head from the burrow and the initiation of swimming. On the other hand, experiments in the field using intact and crushed fish as bait have shown that the responses may take place very rapidly when the olfactory stimulus is more intense. For example, when whole fish were used, the time that elapsed before the first appearance of the hagfishes was 6-16 minutes, whereas when crushed fish were presented, they appeared within half a minute.

Hagfishes are believed to feed only at infrequent intervals and, under aquarium conditions, are said to neglect food when this is offered more than once a week. However, this is compensated for by the voracity of their feeding and the amount they consume on a single occasion. Within minutes of the start of feeding, fragments of undigested material are passed from the anus as faecal tubes, surrounded by the mucous envelope secreted by the intestinal epithelium. Feeding normally occurs during their periods of nocturnal activity, although for a hungry hagfish, the olfactory stimulation may override the normal photokinetic responses. In underwater observations, *Myxine* has been seen to continue its search for bait even under the lighting used for underwater photography and once feeding began, it had to be touched before it reacted. A characteristic of laboratory-maintained hagfishes is their ability to survive for months without food. This is no doubt explained by their low metabolic rates and the extensive lipid deposits in the subcutaneous tissues and intestinal submucosa.

Larval Lampreys

Although superficially resembling the protochordates in its filter feeding mechanisms, the presence in the ammocoete of a velum, capable of boosting the nutritive and respiratory currents, must have increased the rate of food intake and thus the body size that the ammocoete has been able to attain. While this feeding mechanism must put an upper limit on the size of food particle that can be accepted, it is doubtful whether it is otherwise selective as to the kind of organisms that are used as food. In general, the micro-organisms and detritus particles that have been found in the ammocoete gut are representative of those that occur in the surrounding water or on the surface of the silt substrate in the vicinity of the burrows. Some of the Chlorophyta and diatom species may be more abundant in the water than in the gut, but this may be due to their tendency to form filaments too large to pass through the sieve-like mechanism of the buccal cirrhi. A similar selection based on size is also indicated for several genera of episammic diatoms. As might be expected, the size and composition of the diatoms found in the gut varies with the age of the ammocoete. Thus, the mean particle size in 0+ class of ammocoetes was found to be 35 μm compared with 100 μm in four-year-old animals. In sea lamprey ammocoetes, there were three times as many diatoms in the gut in the summer as compared with the winter. Surprisingly, at winter temperatures, up to 90% of the diatoms were not completely digested and the surviving chloroplasts were capable of reproduction in suitable culture media, whereas at summer temperatures only 45% were

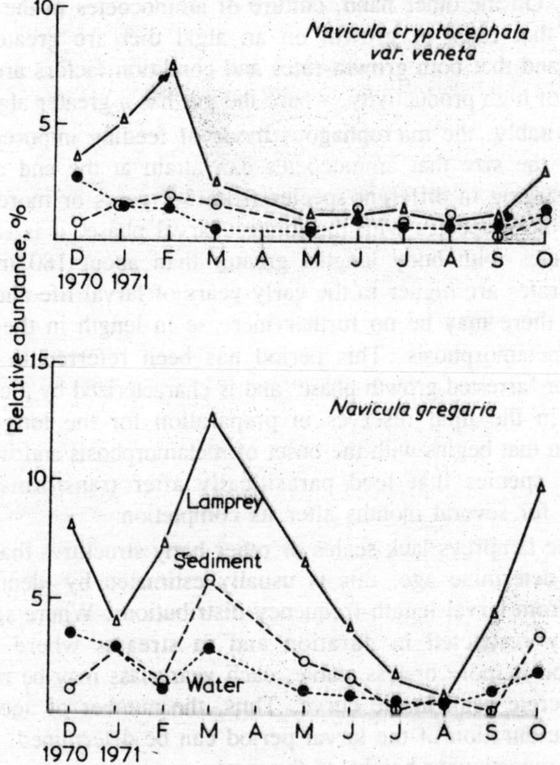

Fig. 16.4. Comparisons of the relative abundance of two species of diatoms in the water and in the sediments or the midgut of ammocoetes of the sea lamprey (Petromyzon marinus).

undigested. Consistent with these figures is the confirmation that the assimilation efficiency of the ammocoete is low and temperature-related.

Although diatoms form an important part of the ammocoete diet, their resistance to extraction techniques has probably tended to exaggerate their importance. Although making up a relatively small proportion of all the organisms in the gut a wide variety of algal groups have been identified, including Chlorophyta, Chrysophyta, Cyanophyta and Euglenophyta. Other organisms identified in the gut include ciliates, rhizopods, rotifers, cladocerans, ostracods and copepods.

In addition to micro-organisms, ammocoetes ingest considerable quantities of detritus, more especially in winter and in streams where productivity is low. Indeed, in one English stream, algae accounted for only 0.01-0.2% of the gut contents in winter and 0.1-0.3% in summer. On the other hand, culture of ammocoetes in the laboratory showed that rates of growth on an algal diet are greater than on detritus and that both growth rates and condition factors are higher in streams of high productivity, where the gut has a greater algal content.

Inevitably, the microphagous mode of feeding imposes an upper limit on the size that ammocoetes can attain at the end of a larval period varying in different species from 3-7 years or more, although even in those species with the longest larval phase, it is rare to find ammocoetes with body lengths greater than about 180 millimetres. Growth rates are higher in the early years of larval life and, in many species, there may be no further increase in length in the final year before metamorphosis. This period has been referred to as a 'rest period' or 'arrested growth phase' and is characterized by a conspicuous buildup in the lipid reserves in preparation for the long period of starvation that begins with the onset of metamorphosis and which (even in those species that feed parasitically after transformation) may continue for several months after its completion.

Since lampreys lack scales or other bony structures that might be used to determine age, this is usually estimated by identifying age classes from larval length-frequency distributions. Where spawning is relatively restricted in duration and in streams where the larval population is more or less stable, each year class may be represented by a discrete peak in the curve. Thus, the number of age class and hence the duration of the larval period can be determined. As judged by the proportionate height of the peaks representing successive age classes, it appears that annual mortality is low and relatively constant. In addition to the protection afforded by their concealed habits, it is thought that ammocoetes may be distasteful to predators such as eels

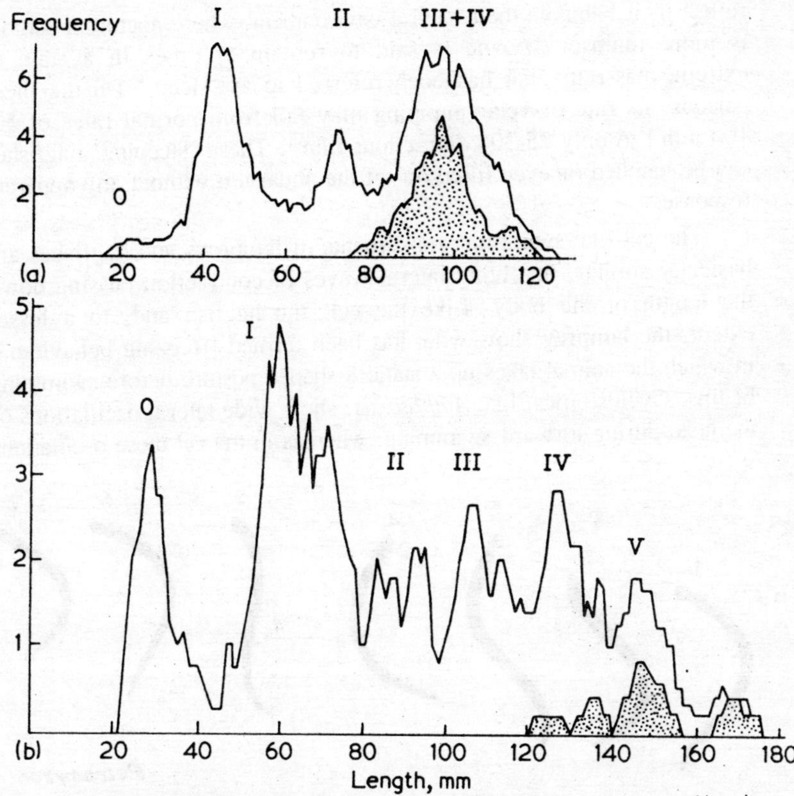

Fig. 16.5. The determination of age classes in ammocoete populations by the use of length-frequency distribution. (a) Length-frequency curves for a population of ammocoetes of the parasitic lamprey, Lampetra fluviatilis. (b) Length-frequency curves for population of ammocoetes of non-parasitic lamprey, Lampetra planeri.

because of their skin secretions. These are produced from mucous cells and club and granular cells, elaborating proteins and peptides. As suggested by the expression 'a surfeit of lampreys' these animals have acquired a traditional reputation for producing digestive disturbances or even toxic effects and they are usually skinned or otherwise treated to remove the mucus before they are cooked. To some extent these unpleasant effects could be due to the presence of a biologically active peptide allied to, but not identical with, bradykinin that has been isolated from their skin secretions.

Movements and Swimming Ability

In the laboratory, hagfishes exhibit long periods of inactivity, punctuated by only intermittent movements; but this may not necessarily reflect their habits in their natural environment, where nocturnal activity is more intense. *Myxine* is said to remain at times in a state of extreme passivity that has been referred to as 'sleep.' During these periods, its rate of velar pumping may fall from normal rates of 50-100 min^{-1} to only 25-30 contractions min^{-1}. These 'sleeping' hagfishes can be handled or even lifted out of the aquarium without any apparent response.

The eel-like swimming movements of lampreys and hagfishes are basically similar, involving lateral waves of contraction passing down the length of the body. Like the eel, the hagfish and, to a lesser extent, the lamprey show what has been termed 'freezing behaviour', in which the animal takes up a static S-shaped posture before swimming begins. Cyclostomes, like *Amphioxus*, show wide lateral oscillations of the head during forward swimming, whereas in the eel these oscillations

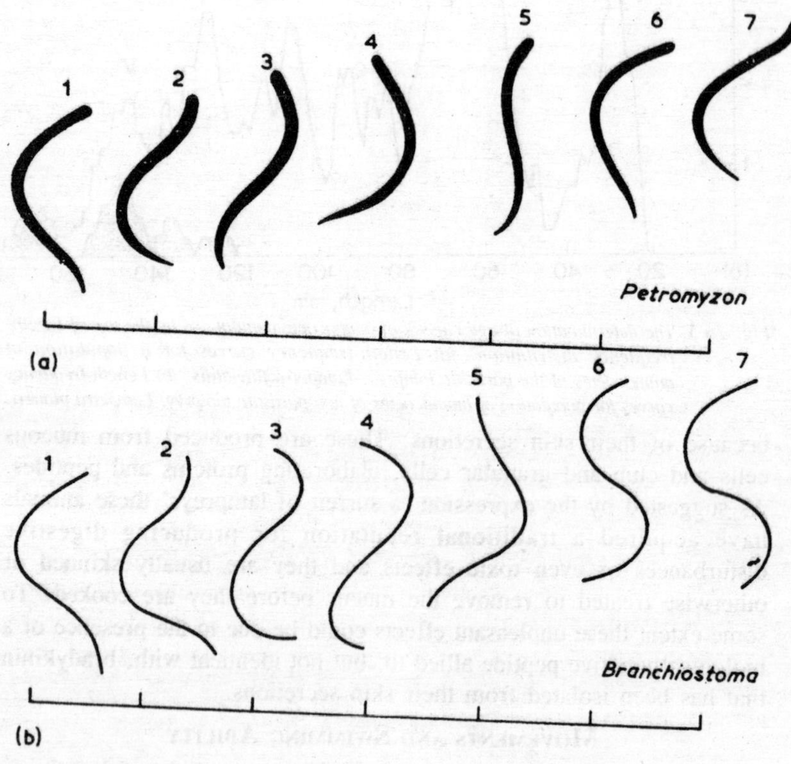

Fig. 16.6. Swimming movements of a 7 mm ammocoete of Petromyzon (a) and of a 45 mm adult of Branchiostoma (b) traced from cine film records.

are much more heavily damped in front of the animal's centre of mass, where the amplitude is usually at its minimum. Behind the centre of mass, amplitude increases reaching a maximum at the tip of the tail, which makes the widest lateral excursions.

Because of the absence of a backward flexure in their myotomes, Nursall (1956) had incorrectly assumed that cyclostomes would be incapable of reverse swimming, involving the propagation of undulatory waves in a tail to head direction. On the contrary, hagfishes often swim backwards; a facility that they share with cephalochordates, eels and lampreys. In an intact hagfish, waves of undulation starting near the head and progressing backwards can be produced by stroking the tail, but after the spinal cord has been divided, these waves begin immediately behind the point of section. Stroking the gill region, on the other hand results in reverse waves, which after spinal section, begin just anterior to the point of operation. Thus, all segments of the cord are able to initiate undulatory waves in either direction.

In *Amphioxus*, a major factor in the production of either forward or backward waves of contraction may be a longitudinal gradient in internal bending resistance resulting from changes in the flexibility of the notochord. Contractile elements like those of the cephalochordate notochord have not been recognized in the hagfish, but the presence of microtubules and membrane arrays surrounding the central vacuole of its notochordal cells at least raises the possibility that similar mechanisms may exist. Lampreys, when faced by obstacles or by noxious stimuli may retreat backwards for a short distance, before turning and swimming away. However, as Blight (1977) points out, backward swimming in a lamprey is not achieved by a simple reversal of the processes involved in forward progression, since the head is held almost as straight as in forward swimming and does not therefore function as a substitute tail.

Hagfishes are described as graceful, elegant swimmers, but perhaps poor spatial orientation due to the degeneration of the eyes, may explain why they normally swim close to the bottom. Observations by divers have shown that they rarely swim more than 1.5 m from the sea bed and that they maintain this position by keeping the head above the horizontal, with the rest of the body sloping obliquely downwards. The fact that they often swim in an upside-down position could be related to their simple, toroidal labyrinth, which is said to be unable to monitor rolling movements as efficiently as pitch or yaw. On the other hand, Whiting (1972) has suggested that the ability of lampreys to maintain the dorsal side uppermost might be assisted by the fatty tissue of the fat column or 'protovertebral arch' above the nerve cord, acting as a buoyancy organ especially helpful to an animal without paired fins. However, while a similar body of adipose tissue is absent in the myxinoids, it is equally possible that it might be of adaptive value to the lamprey as a cushion above the nerve cord which (in the absence of complete vertebral arches), could offer some protection against the attacks of predators during periods when the animals are exposed in relatively shallow water in daylight; a protection that would be less vital to a hagfish that normally lies completely buried during the daytime.

After destruction of the brain, the spinal hagfish completely loses its equilibratory reflexes and rolls over continuously whilst swimming, although in other respects its swimming movements are executed in a normal manner.

Because of the different arrangement of their parietal muscles, the body of the hagfish is much more flexible than that of a lamprey. An example of this flexibility is the way that they are able to perform complex contortions, described as 'figure of eight' and 'knotting'. Both forms of behaviour are based on local responses to contact stimulation, seen in its simplest form in the type of movement called 'gliding'. This involves a lateral wave of contraction immediately behind the point of stimulation proceeding in a forward direction. An example of this type of movement is seen when a hagfish escapes tail first over the edge of a bucket. In the 'figure-of-eight' movement, the tail is first coiled so as to touch the body. This then glides over the body and the body under the tail, each contracting at or just behind the first points of contact. In 'knotting' behaviour, which may take one of two forms, the head may pass through the loop formed by the tail around the body, or in a second form described by Strahan, the tail is also passed through the first loop formed by the body. These responses may be used to escape from entanglement in slime or when the animals are held in the hand. They have also been seen in animals attempting to tear off pieces of food when feeding in their natural environment.

Swimming Speeds

Estimates of the rate of movement of sea lampreys migrating upstream have put their average rate of progress at about 5 cm s^{-1} or 0.18 km h^{-1} although it should be remembered that they are probably inactive for most of the daylight hours. Another estimate for the same species gave an average speed of 3.2 km h^{-1} against a slow current

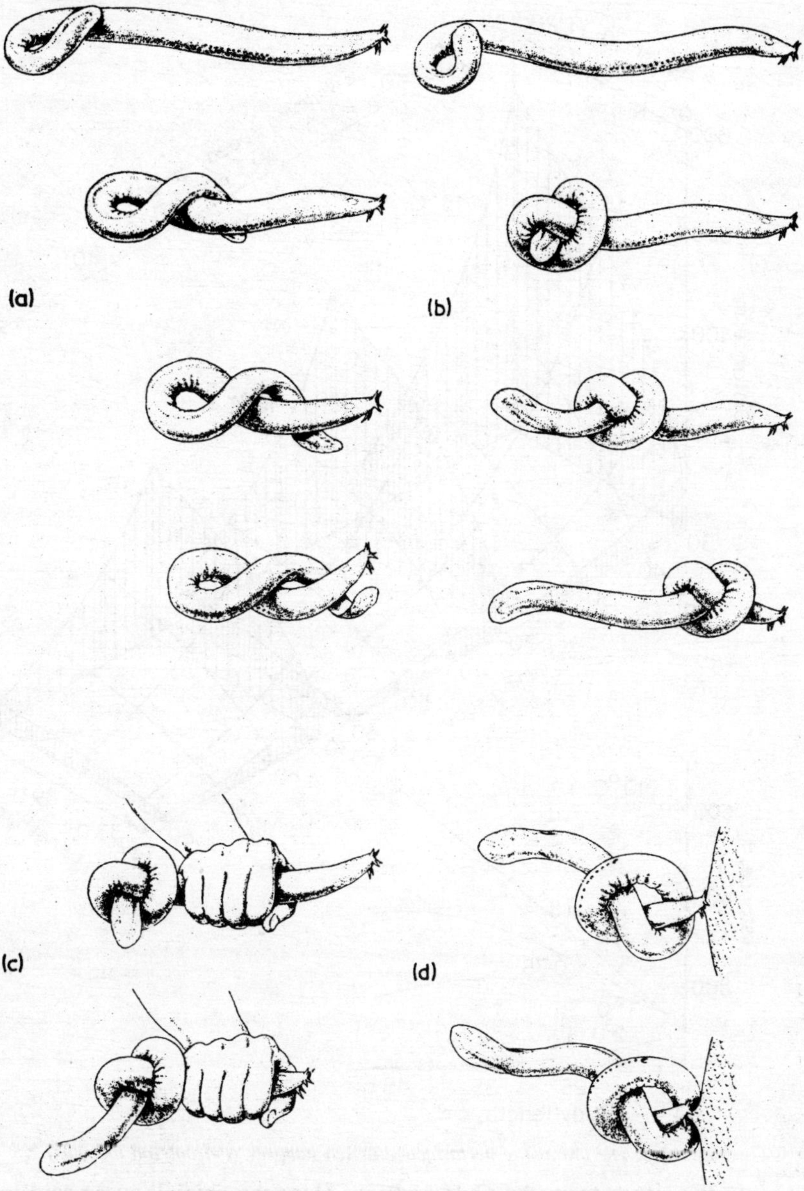

Fig. 16.7. Figure-of-eight (a) and knotting movements of a hagfish (b, c, d).

and 0.5 km h^{-1} in faster flowing water further upstream. Higher estimates have been given for the European river lamprey migrating

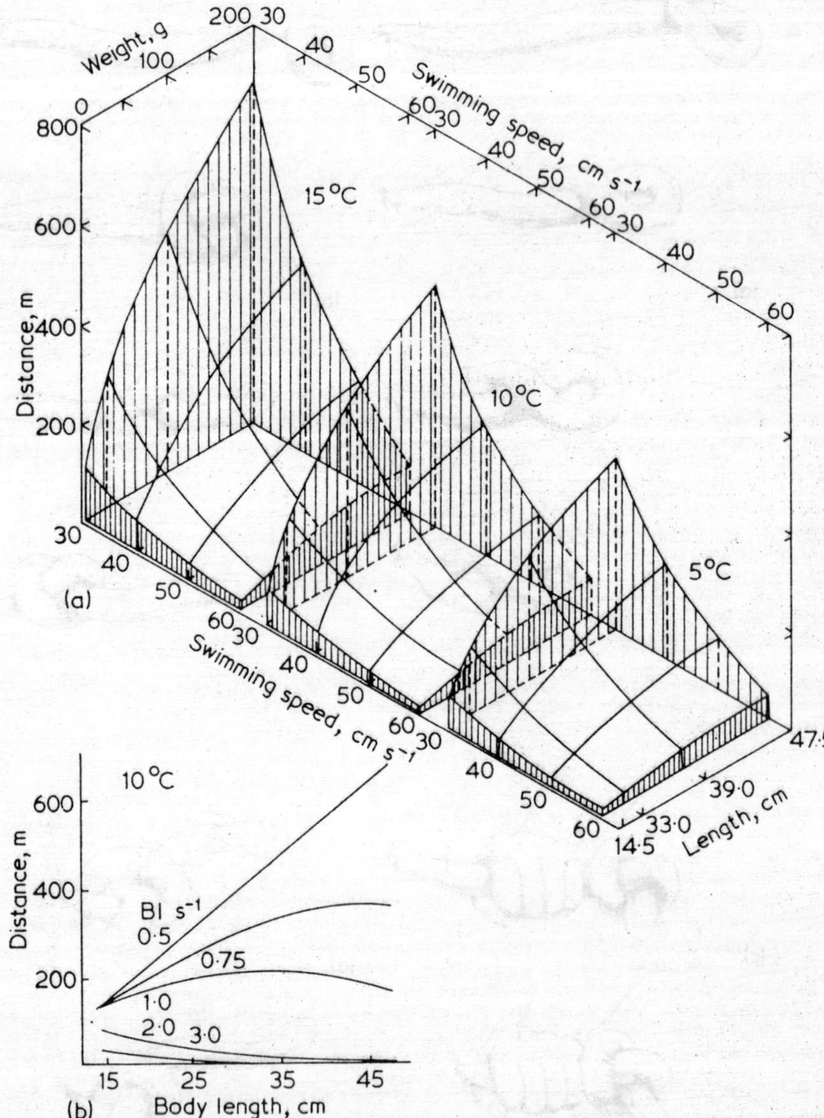

Fig. 16.8. The swimming performance of sea lamprey (Petromyzon marinus).

into the rivers from the Gulf of Riga. Here the animals were said to travel up to 8-13 km in a single night and the average velocity in the early stages of the migration was estimated at 1-4 km h^{-1}. Indirect calculations based on the rate of depletion of energy reserves have put the average swimming speed of upstream migrating anadromous sea lampreys at about 6-7 cm s^{-1} or 200-250 m h^{-1}. For large ammocoetes of the same lamprey, velocities of 0.11 and 0.26 m s^{-1} were recorded at 4-7°C and 0.45 m s^{-1} at temperatures of 20°C, but it is likely that speeds of this order would be maintained only for very brief periods. Laboratory experiments on feeding-stage sea lampreys showed that they were able to swim at their maximum speed of 35.5 cm s^{-1} for about 15 minutes, covering a distance of about 315 m. On the other hand, the sexually mature migrant gave a vastly inferior performance and was able to sustain the same speed for only a minute, corresponding to a distance of 21.3 m. This deterioration in swimming endurance might as the author points out be related to the cessation of feeding and the fact that at this stage the lamprey is relying on energy reserves in the form of lipids, built up during its parasitic phase.

Although there is no precise information of the swimming ability of hagfishes, observations made by skin divers indicate that they are capable of maximum speeds comparable with those of lampreys and that they are able to swim as fast as a diver is able to follow them. Strahan (1963) gives a maximum speed for *Myxine* of 25 cm s^{-1}, but other estimates both for this species and for *E. burgeri* have mentioned figures of the order of 1 m s^{-1}. This represents velocities of about three body lengths s^{-1}; considerably better than the performance of the sea lamprey.

Beamish has drawn attention to the fact that the swimming performance of the lamprey compares very unfavourably with that of teleosts. For example, the salmon is said to be able to maintain speeds about twice as great as the lamprey for periods up to at least an hour. The reason for this relatively poor swimming ability is obscure and there is certainly nothing in the respiratory or cardiovascular physiology of these animals that can be held responsible. For this reason, Beamish considered that the answer might lie in hydrodynamic factors and perhaps in the relatively small caudal fin of the lamprey, which he considered to be poorly adapted to delivering a propulsive thrust. This seems all the more plausible, when it is borne in mind that the caudal fins of teleosts are said to be responsible for between 45 and 84% of the propulsive thrust.

17

BIOCHEMISTRY AND IMMUNOLOGY

Discussion of the phylogenetic status of the cyclostomes has so far been very largely confined to morphological considerations, with little regard to comparative biochemistry and physiology. To a certain extent, the responsibility for this situation can be laid at the door of the biochemists themselves who have often shown some reluctance to follow the logic of their own data and have adopted an attitude reflected in the words of Florkin (1966) that 'we must adopt the methodological rule to be led by knowledge of phylogeny in our search for biochemical evolution, rather than be brought by biochemistry to the discovery of new aspects of phylogeny'. We must of course, recognize that when taken in isolation, the distribution of a particular molecule within or between groups of animals may be of little phylogenetic significance, although such information may become more important where we have more detailed information on biosynthetic pathways and enzyme systems. An example is the erratic distribution of creatine even within animals belonging to the same taxonomic group. This nitrogenous compound has been found in the tissues of cephalochordates, myxinoids and lampreys, although only the latter animals possess the necessary enzymes to synthesize it. In the other two chordate groups, as in some invertebrates, its presence may be due to the ability of these animals, either to recover it from their food or to absorb it from their aquatic environment. However, as Barrington (1974) has pointed out, even where we have information on biosynthetic pathways, their presence in different organisms or groups of animals may be open to several alternative explanations. While it may be evidence of a common ancestry, it is also possible that it is a result of parallel evolution or reflects biosynthetic abilities that are widely distributed among diverse animal groups, as exemplified in the occurrence of steroids or polypeptides such as insulin in a broad spectrum of animal groups.

These reservations, however, apply with much less force to the rapidly advancing techniques that are being developed in the field of molecular evolution. The increasing numbers of protein sequences now becoming available, which because they relate to the fine structure of the genome, constitute a rich source of genetic and evolutionary information which can supplement or even perhaps, in a few cases, supplant the more subjective judgements of the morphologist or palaeontologist.

Considerable precision can now be obtained from protein sequence data by the application of computer techniques for the construction of phylogenetic trees, in which the branch lengths are proportional to sequence differences, taking account of the probabilities of superimposed or back mutations. Providing the geological or palaeontological evidence is adequate to determine the age of the major points of divergence (nodal points) on such trees, it is then possible to estimate the rates of protein evolution, either in terms of accepted point mutation (PAM) 100 residues^{-1} 100 million years^{-1} or as nucleotide replacements (NR) 100 codons^{-1} 100 million years^{-1}. Although it may be true that we cannot at present interpret morphology in terms of molecular structure and that point to point information transfer from DNA stops at the level of the polypeptide to a quite surprising extent, phylogenetic trees for the best-known proteins tend to show agreement with the views of conventional taxonomists. This suggests that evolution at the level of the gene adequately reflects changes in the whole organism at the morphological level.

Unfortunately, sequence data for cyclostome proteins is still very incomplete and for none of the molecules discussed in this chapter do we have comparable information for both myxinoids and lampreys. For this reason, we are bound to rely on comparisons of amino acid analyses, although at best these can provide only crude approximations to the sequence changes that may have occurred. From time to time a number of statistical techniques, of varying degrees of sophistication, have been proposed for estimating the degree of homology between proteins from comparisons of amino acid composition. One of the simplest of these used extensively in this Chapter and devised by Marchalonis and Wellman is the difference index $S\Delta Q = (X_{i,j} - X_{k,j})^2$ where $X_{i,j}$ and $X_{k,j}$ represent the proportions of an amino acid j in the two proteins i and k. This index has been found to correlate quite closely with differences in amino acid sequences in cytochrome *c*, haemoglobin or immunoglobulin. For example, unrelated proteins show a much greater dispersion of SΔQ values than homologous proteins; the median value for the former being 300 units, and for about 12% of the proteins examined the index was over 1000 units. On the other hand in comparisons of related proteins the median SΔQ values were only 80 for the haemoglobins, 30 for immunoglobulins and 20 for cytochrome *c*.

Cyclostome Haemoglobins

Structure of Lamprey and Hagfish Haemoglobins

The complete primary structure of one of the haemoglobins of the sea lamprey, *P. marinus*, is now known together with one of the two major haemoglobins of the river lamprey, *L. fluviatilis*. In addition, X-ray diffraction studies have demonstrated the essential similarity in the three-dimensional structures of lamprey and other vertebrate haemoglobin or myoglobin chains. Unfortunately, sequence data for the hagfish haemoglobin is limited to the first 30 residues at the N-terminal in one of the major components of *E. stouti*, but amino acid analyses are available for five haemoglobin components of this species, as well as for four haemoglobin species of *E. burgeri* and three components in *Myxine*.

For the two species of lampreys belonging to the genera *Lampetra* and *Petromyzon*, the haemoglobin sequences show substitutions at only 12 positions. Five of these would involve single, and seven double nucleotide replacements. At positions 8-10 and 136-139 there are reversals of the sequence which would require two nucleotide replacements. In addition, the methionine at 141 in *fluviatilis* is absent in *marinus* and the latter also has an alanine at 119 which appears to be missing in *fluviatilis*.

The partial sequence for the hagfish shows some striking differences from the lamprey haemoglobins with no less than 17 substitutions out of the 30 residues. In 14 cases these would require single base changes and three would involve double nucleotide replacements. Both hagfish and lamprey share a unique additional segment of nine residues at the N-terminus and in this sequence five of the positions are occupied by the same amino acids in both groups. As Li and Riggs (1972) have pointed out, the fact that this agnathan segment has been preserved through the long period of evolutionary separation, suggests that it may have some functional significance peculiar to this group of vertebrates.

The tertiary structure of the lamprey haemoglobin shows the same predominantly α-helical structure of other globins, accounting for about 79% of all residues. These helices are arranged in basically the same pattern as in the monomeric myoglobin molecule, making it possible to use the same nomenclature that is applied to other globin chains. As in other haemoglobins, the haem group is situated in a haem pocket between the E and F helices, lined by hydrophobic residues and linked to the two histidine residues at E7 and F8.

In the absence of complete sequence data for hagfish haemoglobin, it is instructive to compare its amino acid composition with that of the lamprey and mammalian haemoglobins. Although there have been some discrepancies in the molecular weights of hagfish haemoglobins as reported by different authors, it is generally agreed that these are higher than in the lamprey molecule and the difference is especially marked in the case of *Myxine* haemoglobin. It has been suggested that this could be explained by assuming that hagfish haemoglobin possesses the extra six residues that are found at the carboxyl terminal of the myoglobin chain, as well as the nine residues that are missing from the lamprey chain between positions 130 and 131.

Analysis of the difference indices for haemoglobins of the hagfish and lamprey in table 17.1 discloses a number of interesting features:

1. Whereas the haemoglobins of *L. fluviatilis* and *L. japonica* are identical in their amino acid composition and show only relatively minor differences from haemoglobin V of *Petromyzon marinus*, there are very marked divergerrcies in the haemoglobins of the three hagfish species, where the comparisons between the individual haemoglobins, yield mean values for SΔQ of 45-65.
2. Both in comparisons within and between species, the various individual haemoglobins of the hagfish show a very wide range of variability in amino acid composition. For example, components A, Band C of *E. stouti* are virtually identical (SΔQ < 10) and closely resemble F4 of *E. burgeri* (SΔQ < 10). There is also a considerable measure of agreement between these *Eptatretus* haemoglobins and the M2 component of *Myxine*. In *E. burgeri*, F1 and F2 are almost identical, but differ widely from F3 and more especially, F4 (SΔQ: 116, 126).
3. These intraspecific and interspecific divergencies in amino acid composition are apparently matched by differences in the physiological properties of the haemoglobin components. For example, components F1 and F2 have low oxygen affinities

Table 17.1. Values of difference index SΔQ for comparisons between lamprey and hagfish haemoglobins and human α and β chains.

			Lampreys		Hagfishes												Human	
					E. stouti					*E. burgeri*				*Myxine*				
			LF	PM	A	B	C	D	E	F1	F2	F3	F4	M1	M2	M3	α	β
Human		β	220	185	154	155	124	245	203	348	357	237	129	258	178	209	88	0
		α	153	130	170	189	164	226	273	398	427	291	178	321	207	202	0	
Hagfishes	*Myxine*	M3	93	93	54	38	54	42	60	72	82	69	60	42	46	0		
		M2	81	71	23	15	27	43	44	100	100	62	30	57	0			
		M1	167	169	61	52	73	37	52	45	47	41	66	0				
	E. burgeri	F4	72	62	8	10	7	51	60	116	126	67	0					
		F3	165	166	61	52	66	25	63	55	54	0						
		F2	215	223	109	82	114	56	63	2	0							
		F1	200	206	101	75	105	48	62	0								
	E. stouti	E	135	141	63	31	50	48	0									
		D	115	114	34	25	40	0										
		C	57	51	5	7	0											
		B	68	60	7	0												
		A	55	48	0													
Lamprey		PM	10	0														
		LF	0															

comparable with lamprey haemoglobins, whereas F3 and F4 have the high oxygen affinities characteristic of myxinoids. Likewise, in *E. stouti* the similarity in amino acid composition of A, Band C is also reflected by similar oxygen affinities, even higher than those of the haemoglobins F3 and F4 of *E. burgeri*.

4. Although comparisons between the lamprey haemoglobins and all the 12 hagfish components gives an overall mean SΔQ value of about 120, interpretation is complicated by the variability in the individual hagfish haemoglobins. The closest agreement is between A, Band C of *E. stouti* (SΔQ: 48—68), F4 of *E. burgeri* (62-72) and M2 and M3 of *Myxine* (71-93). These values may be compared with indices of 200-223 for comparisons between the lamprey and Fl and F2 of *E. burgeri* and 114-141 in the case of components D and E of *E. stouti*.
5. Similar difficulties arise in comparing cyclostome haemoglobins with human α- and β-chains. Here again, the closest correspondence between the various hagfish components and human haemoglobins is between A, B and C of *E. stouti* (SΔQ: 124-266), F3 and F4 of *E. burgeri* (SΔQ: 129-291) and M2 and M3 of *Myxine* (SΔQ: 178-202). In contrast there is much greater disparity in the case of D and E of *E. stouti* (203-273) and more especially, Fl and F2 of *E. burgeri* (348-427). For comparisons between lamprey and human chains the difference indices are of a similar order to the smallest values for some of the hagfish components (130-220).
6. In these comparisons with human α- and β-chains the two cyclostome groups display a striking divergence. Both lamprey haemoglobins show a distinctly closer approach to the human a-chain and a similar relationship has also been established in comparisons with the α and β-chains of the rhesus monkey (Marchalonis, 1977). On the other hand, the position is reversed in the hagfishes where, with the sole exception of M3, everyone of the 12 haemoglobins shows a closer similarity to the human β-chain. At least in the case of the lamprey, this situation can be confirmed from the known sequences. These indicate that 73 substitutions have occurred in relation to the human chain, compared to 78 in the β-chain.

Any attempt to draw phylogenetic conclusions from this haemoglobin data must obviously be highly speculative, bearing in mind the extent of the variability in the amino acid composition of the various hagfish haemoglobins, where differences both within and between the species are often as great or even greater than the divergence between hagfish and lamprey molecules. Unfortunately, much less information is available in the case of the lampreys, although the existence of only minor differences between three species representing two genera, seems to indicate that haemoglobin variability may be much less marked than it is in the myxinoids. In relation to the hagfishes, it may be significant that most of the observed variability in the two species of *Eptatretus* relates to F1 and F2 of *burgeri*, which together make up only 30 % of the total haemoglobin and to D and E, which are only minor components in *stouti*. It might therefore, be argued that because these minor components make a less significant contribution to the physiological properties of the blood, they have been under less stringent selective constraints and have therefore, been able to accept mutation rates much higher than those of the major haemoglobins whose physiological properties are more closely adapted to the mode of life and environment of the hagfish. If this is the case, we should therefore be justified in putting greater reliance on the major haemoglobins, when attempting to draw conclusions from comparisons with lampreys and other vertebrates.

Comparisons of Cyclostome Haemoglobin Sequences with those of other Vertebrates

Over the entire vertebrate series there are seven sites where no deviations have so far been recorded and a further nine positions where exceptions are limited to a single species. Of these invariant or near-invariant sites, the initial segment of the hagfish molecule shows identical residues at A3 and A12, where the sole exceptions are the human γ chain and the chick β-chain. In addition, the hagfish sequence, like that of the lamprey has a lysine at A5, where the only known substitutions are in myoglobin and the β-chain of the frog. On the limited evidence, it is probably safe to assume that most if not all, the other invariant residues will eventually turn out to be present in the hagfish. As a rough guide, we can estimate the probable percentage difference between the complete hagfish sequence and that of lampreys and other vertebrates by extrapolation from the known segment. Since this segment contains the unique agnathan sequence at its N terminus this will tend to exaggerate the differences in comparisons with other vertebrates. As a crude measure of divergence, the figures in Table 9.2 suggest that lamprey and hagfish sequences may differ to a rather similar extent from the fish α-chain, but a remarkable feature of the hagfish haemoglobin is that its first 30 residues shares no single amino acid in common with the homologous region of the bird α-chain. Consistent with the long period that has intervened since the divergence of the two cyclostome groups, the differences in the primary structure of their haemoglobins will almost certainly prove at least as great as those that separate the human and fish molecules, representative of the span of gnathostome evolution.

Table 17.2. Difference matrix for some vertebrate haemoglobins. Hagfish figures are based on the partial sequence.

	Human	*Carp*	*Lampreys*	
			L. fluviatilis	*P. marinus*
Carp	50	0		
L. fluviatilis	73	75	0	
P. marinus	72	75	5	0
Hagfish	78	75	61	61

Cyclostomes and the Functional Evolution of Haemoglobins

Based on the fact that the oxygenated lamprey monomer readily aggregates in the deoxygenated state to form dimers and temporary tetramers, it has been argued that the earliest vertebrate haemoglobins would have resembled those of lampreys, rather than the monomeric invertebrate globins. As in the lamprey, this primordial globin would have been capable of aggregating to form homotetramers, releasing oxygen more readily from its haem iron atoms. As seems to be the case in the lamprey, the first stages in the origins of cooperativity increased Bohr effects and decreased oxygen affinity may have been produced by the development of more constraining salt bridges in the deoxygenated state (leading to the aggregation of sub-units), but which break on oxygenation, with the liberation of protons and the separation of the subunits.

Throughout these evolutionary developments, selective forces might be expected to operate most intensely on those sites in the molecule most intimately involved in haem functions, interchain co-operativity and the salt bridges related to Bohr effects. Haem contact sites, which would be indispensable in any globin molecule, must already have been well-established at a very early stage in its evolution. Among the invariant residues, the histidine at F8 is a constant feature in both vertebrates and invertebrates, but at E7 the histidine of the vertebrates is absent in invertebrates, where it is replaced by the non-polar leucine or isoleucine. This has led to the suggestion that the appearance of this polar histidine for the first time in the cyclostomes, may have been a very significant step in the evolution of haem interactions, reduced oxygen affinity and the oxygenation-dependent aggregation — disaggregation equilibrium of the monomer.

Examination of the rates of amino acid substitution have shown that those sites that now function as haem contacts in the α- and β-chains of the higher vertebrates have been the most slowly evolving positions and were probably already established at about the time of the agnathan-gnathostome divergence, at about 500 million years. Some support for this view comes from comparisons of the residues listed as haem contact sites in human α- and β-chains with the corresponding positions in lamprey haemoglobin. This shows that nearly half these sites are occupied by the same residues, whereas the overall percentage of common residues for the molecule as a whole is only about 25%. Another crucial function is that concerned with interchain co-operativity, facilitating oxygen delivery, and here again there is evidence for selective constraint in the agnathan stage of vertebrate evolution. In mammalian haemoglobins, each α-chain makes contact with two β-chains forming $\alpha^1 \beta^1$ and $\alpha^1 \beta^2$ contact sites which favour haem interactions by placing these groups in close proximity. By comparing the homologous sequences in lamprey and mammalian haemoglobins, it has been claimed that it may be possible to distinguish which of the cyclostome residues are likely to be involved in the aggregation of sub-units during deoxygenation. A comparison of the $\alpha^1 \beta^2$ contact sites shows that about half are identical in the mammal and lamprey; a correspondence that again is much higher than that of the whole molecule. The fact that such a high proportion of these residues have been conserved, also suggests that the changes that occur in the conformation of lamprey sub-units during oxygenation may be very similar to those of mammalian haemoglobins. Riggs has also pointed out that at one of these sites (C6), glutamic acid replaces the arginine of mammalian haemoglobin and that an identical substitution also occurs in a human mutant haemoglobin, where it has the effect of reducing co-operativity.

The generally accepted interpretation of globin evolution postulates a series of gene duplications, leading first to the separation of myoglobins and α- and β-chains, followed by subsequent diversification within the mammals. Reconstructions of these events have been incorporated in phylogenetic trees, but these have differed widely in their time scales, evolutionary rates and in the order in which the significant events occur. For example, using a time scale derived from an average rate of globin evolution of 10.7 PAM 100 million years^{-1} calculated on the basis of sequence changes in frog β, carp α and mammalian chains, Dayhoff *et al.* (1972) put the myoglobin-haemoglobin divergence at 1070 million years followed by the β-α gene duplication at 500 million years. These authors believed that the lamprey sequence indicated a very ancient gene separation, which might have preceded the actual separation of the agnathan-gnathostome lineages. A different interpretation of these events has been advanced by Goodman *et al.* (1975). Their genealogical tree takes certain geological ages for important nodal points in globin evolution and, from these, rates of evolution have been estimated over defined periods in the history of the molecule. For example, the invertebrate-vertebrate dichotomy was placed at 680 million years and the agnathan-gnathostome (vertebrate-lamprey divergence) at 500 million years. Here, the gene duplications that produced the myoglobins and the vertebrate α- and β-chains are positioned after the separation of the lamprey haemoglobins at approximately 460 and 420 million years and would therefore be considerably more recent than the interpretation. This genealogical tree indicates an average rate of evolution over the period from 680 million years to the present, of 31 NR% when expressed in terms of the numbers of nucleotide replacements (100 codons/10^8 years), although the rate at different periods in the evolution of the globins has varied widely with the intensity of selection for functional improvement. For example, the evolutionary rate was apparently accelerated sharply in the early vertebrates at the time when gene duplications were opening new fields for molecular diversity, with the introduction of the myoglobin and α- and β-chains. Thus, in the period from 680 million years—300 million years, the average rate of evolution was 46 NR% falling to only 15 NR% in the subsequent 300 million years. Moreover, during a more restricted period which would include the α/β gene duplication, that is from 500-400 million years, the estimated rate was 109 NR%.

The sequence data may indicate that the age of a branch point in the phylogenetic tree is much older than the geological and palaeontological data would suggest, the solution may lie in a distinction between gene divergence and species divergence. It may be significant that statistical comparisons suggest that lamprey haemoglobin is closer to the mammaliah α-chain in its amino acid composition, while that of the hagfish, in all species and virtually all components, appears to bear a closer resemblance to the β-chain. In view of this, could it be that the gene duplication giving rise to α- and β-chains occurred at a much earlier stage in vertebrate evolution than has generally been supposed and that these genes were perhaps already present in the common ancestors of agnathans and gnathostomes? Subsequently, with the divergence of the lineages leading towards the myxinoids on the one hand, and the lampreys on the other, there could have been a loss of either the α type or β type genes in the two branches, resulting in divergent haemoglobins, which have retained their respective affinities, however distant to the alpha and beta chains of the higher vertebrates. Such a remote gene duplication, dating back to about 500 million years, has in fact been postulated by Dayhoff and Barker (1972), basing their estimate on sequence data and a constant rate of haemoglobin evolution.

Taken at their face value, the great variation in amino acid composition of hagfish haemoglobins might be taken as evidence for a very ancient separation of existing species. Moreover, this variability is in sharp contrast to the only minor differences that separate the haemoglobins of the two lamprey genera, *Lampetra* and *Petromyzon*. A much more likely explanation would be that this difference between myxinoid and lamprey species has arisen from the functional properties of their haemoglobins and that the greater degree of interchain cooperativity, increased Bohr effects and lower oxygen affinities of the lamprey molecules have involved much more intense selective pressures than those that have operated during the evolution of the hagfish haemoglobins. Showing little if any evidence of cooperativity or Bohr effects and with much higher oxygen affinities, hagfish haemoglobins, might have been able to tolerate a far higher rate of random mutation than could be accommodated by the more adaptable haemoglobins of the lamprey.

Cyclostome Insulins

Vertebrate insulin is a globular protein made up of two polypeptide chains α and β, linked by disulphide bonds. The biosynthesis of insulin involves the production of a large single-chained precursor molecule — proinsulin — containing about 86 amino acid residues. This is then enzymatically cleaved to produce the twin-chained hormonal molecule of insulin with 21 residues in the α-chain and 30 in the β-chain. The fragment remaining after cleavage of the proinsulin is known as the C-peptide. In the hagfish, the biosynthesis follows the same pattern, being preceded by proinsulin with a half life of about 12 hours for its conversion into insulin.

Amino acid Sequences of Hagfish and other Vertebrate Insulins

Comparisons of the amino acid sequences of hagfish insulin, extracted from the islet organ of *Myxine*, show that most of the features common to other vertebrate species are also preserved in the myxinoid molecule. For example, of the 24 residues that are invariant in other vertebrates all but one occur at the homologous sites in hagfish insulin. The exception is at position B18 where alanine is replaced for valine in the hagfish. In addition to these totally invariant sites, there are a further 5 positions where exceptions are so far known only from a single vertebrate species (A3, A5, B4, B22, B28) and at these sites also the hagfish insulin conforms to the normal vertebrate pattern. Conservative sites, that is positions where substitutions are usually restricted to amino acids with similar physico-chemical characteristics (acidic, aromatic, aliphatic, hydrophobic), have been identified at 16 positions in vertebrate insulin sequences. If hagfish insulin is compared with pig insulin, 9 of these positions are unchanged and a further three show amino acid replacements that retain the same characteristics. The remaining four positions represent radical changes in hagfish insulin. At 14 sites vertebrate insulins have been found to be highly variable. Of these, three are identical in pig and hagfish and three represent conservative substitutions. Residues which appear to be unique to the hagfish occur at no less than 16 sites, mainly in the β-chain and many of these represent radical substitutions.

As in the haemoglobin molecule, selective constraints appear to have varied with the position of residues in the three-dimensional molecule and with the part that they play in maintaining its essential structural and functional characteristics. Thus, in the insulin molecule the relatively small number of variable sites have been subjected to a rate of mutation about five times greater than that of the molecule as a whole. Because of this restriction, chance mutation and superimposed or back mutations will have been relatively more frequent. This will tend to obscure phylogenetic relationships and may account for the wide variations in sequence that have been observed even in species belonging to the same taxonomic group.

From a study of the three-dimensional structure of insulin it has been concluded that the side chains of the invariant or conservative amino acids are probably involved in stabilizing the structure of the monomer or favouring its aggregation to dimers or hexamers, while at variable sites, on the other hand, the residues mainly interact with the solvent at the surface. An interesting difference in hagfish insulin is that, although it may form dimers, it does not bind zinc ions and form tetramers. The explanation for this difference may lie in the replacement of histidine by aspartic acid at B10 and perhaps also to the substitution of phenylalanine and threonine for the hydrophobic residues normally present at B1 and B2. However, the fact that hagfish insulin retains most of the invariant or conservative residues suggests that its conformation may be similar to that of other vertebrate insulins and that the molecule is folded in a similar way. Furthermore, it has been pointed out that the 8 radical substitutions (compared to pig insulin), together with the 4 substitutions from the conservative group can probably occur without loss of biological activity. Three are in the a-chain disulphide loop on the surface of the molecule, two are at the beginning of the second helical region and three at each of the N and C terminals of the β-chain.

The biological effects of insulin result from its binding to the surface of the target cells and this binding region is thought to reside in relatively invariant sequences on the surface of the molecule. Because of this, the receptor binding affinities of the vertebrate insulins generally parallel their biological potency. In this respect, hagfish insulin is said to be unique. As measured by the lipogenic effect on rat fat cells its potency is about 5% of pig insulin, whereas its relative receptor binding affinity is about 23%. In contrast to these variable properties of the vertebrate insulins, the sites that are believed to be responsible for the negative co-operativity in insulin binding seem to have remained unchanged throughout the entire history of the vertebrates, since this same property has been observed in all the insulins tested, including that of the hagfish.

Amino acid Composition of Lamprey Insulin

Amino acid analyses of lamprey and hagfish insulins clearly show a general similarity, contrasting with the differences between the cyclostome molecule and that of various fish, bird and mammalian species. For 10 of the residues the proportions of individual amino acids are identical in both groups of cyclostomes and the SΔD value (40) indicates a relatively high degree of correspondence.

In the absence of sequence data for lamprey insulin it has been estimated that about 8 substitutions are likely to have occurred compared to the hagfish sequence. This is based on the assumption that most of the changes in sequence would have been confined to the highly variable sites. Even if this estimate should eventually prove to be rather low, it is unlikely that the general picture would be seriously affected. Bearing in mind the degree of correspondence in the amino acid composition of lamprey and hagfish insulin, it is clear that the evolutionary distance between the two groups as measured by the divergence in their insulin molecule is much smaller than that suggested by the differences in their haemoglobins. This of course, is in line with the much slower rate of evolution of the insulin molecule which has been put at an average rate of 4 PAM 100 residues^{-1} 100 million years^{-1}, compared with an overall average of 14 PAM 100 residues^{-1} 100 million years^{-1} for the globins. Not only is the insulin molecule much smaller than the haemoglobin chain, but it also contains a very high proportion of invariant sites; nearly 50% compared to less than 5% in vertebrate haemoglobin. Quite apart from these considerations, the matrices for insulin and haemoglobin suggest that other factors should be taken into account. For example, the probable percentage difference between lamprey and hagfish insulins (about 18%) is very much smaller than the difference between the hagfish and the mammal. On the other hand, for their haemoglobins the differences between the cyclostome and the higher vertebrates are not very much greater than the probable differences in the haemoglobins of hagfish and lamprey. The explanation for this disparity can probably be found in the different functions of the two molecules. Lampreys and hagfishes show many remarkable parallels in the morphology and cytology of their islet tissues and the physiological role of insulin must be very similar in both groups. This contrasts with the characteristics of their haemoglobins. That of the myxinoids still retains to a considerable degree the primitive features of a monomeric globin, with only slight haem interactions and Bohr effects. On the other hand, the haemoglobins oflampreys have become adapted to their much more active and migratory life and to the ability to move to and fro between varying marine and freshwater environments. This would have entailed more intense selection for functional improvements in the molecule, affecting co-operativity, haem interactions and Bohr effects. Since the metabolic demands on the insulin molecule are probably very similar in lampreys or hagfishes there would be less scope for this kind of divergent selection pressure.

CYTOCHROME C

As a basic element in the electron transport system of oxidative phosphorylation, cytochrome *c* occurs in all aerobic cells from the fungi to higher animals and plants. Like the globin molecule, it contains a haem group attached to the cysteine residues towards the amino terminal of the chain. Throughout the vertebrates, the number of residues varies only from 103-104. Because of its universality, information on the primary structure of the molecule is more complete and extensive than for any other protein, but sequence data for the hagfish is at present restricted to a short segment of 12 residues around the haem group. The evolution of cytochrome has been characterized by a high degree of conservatism and a correspondingly low rate of mutation. This conservatism is reflected in the large number of invariant or conservative sites and by a long sequence of 14 residues (75-88) where there have been only two substitutions in the entire range of living organisms. Estimates of the average rate of mutation have been put at only 3 PAM 100 residues^{-1} 100 million years^{-1} or 5 PAM 100 residues $^{-1}$ 100 million years^{-1} but, as in the case of haemoglobin, the rate of evolution is thought to have been accelerated in the earlier stages of vertebrate phylogeny.

Comparisons of the Primary Structure and Amino acid Compositions of Cyclostome and other Vertebrate Cytochromes

Within the short sequence of cyclostome cytochrome for which comparisons are possible there are the two completely invariant cysteines and the single histidine. Other invariant sites are the threonine at 18 and the lysine at 25. At positions 23 and 24 all gnathostomes have two glycines, but here the cyclostomesare unique; the lamprey alone having an alanine at 23, while at 24 the only organisms known to show substitutions are the hagfish and higher plants. On the rather slender evidence of this fragment, it has been suggested that hagfish cytochrome may show a closer resemblance to the invertebrates, whereas the lamprey is said to be closer to the higher vertebrates. Nevertheless, this point of view can be substantiated by evidence from the complete lamprey sequence as well as by statistical comparisons of amino acid analyses. For example, all tetrapods have asparagine at position 62, while in lampreys and other fish species this is replaced by serine. Similarly, all fish except the dogfish have, like the lamprey, a valine at 66 in place of threonine, which is near invariant throughout the tetrapods. The lamprey also shares with teleosts the same exceptional substitution at 68, while at 73 and 74 all the lower vertebrates differ from the tetrapods in showing amino acid replacement for the methionine and glutamic acid that are invariant throughout the higher vertebrates.

Comparisons of the amino acid analyses of cyclostomes and gnathostomes based on the coefficient SΔQ, show that the values for the hagfish are consistently very much higher than for the lamprey when comparisons are made with species representative of other vertebrate groups. In both cyclostome groups, the figures suggest a closer correspondence with the cytochromes of the fishes than with those of tetrapods and in the case of the lamprey, the agreement with the dogfish cytochrome is particularly striking. Significantly, the difference between the cytochrome of the hagfish and the lamprey is apparently greater than the disparities between lamprey cytochrome and those of other vertebrates.

Table 17.3. Values of the difference index S Δ Q for comparisons of the amino acid composition of cytochrome *c* from cyclostomes and other vertebrate species.

	S Δ D	
Group and species	*Hagfish*	*Lamprey*
Mammals (man, horse, dog, rabbit)	124	45
Birds (hen, duck, penguin)	105	34
Reptiles (turtle, rattlesnake)	111	44
Amphibia (frog)	100	42
Teleosts (bonito, tuna)	53	32
Elasmobranch (dogfish)	68	14
Lamprey	84	—

In its amino acid composition, hagfish cytochrome shows a number of marked variations from the usual vertebrate patterns. The most extraordinary example of this is the small number of lysine residues. This hydrophilic amino acid is regarded as one of the more conservative residues, showing rates of substitution considerably lower than those of the majority of amino acids. This indicates the vital part that it plays in maintaining the structure of the molecule and its functional properties. No less than 13 lysine residues are absolutely invariant over the entire spectrum of living organisms and at a further two sites, substitutions for lysine are only known in some invertebrate species. The lamprey conforms to the normal vertebrate pattern, showing 18 lysine residues, whereas the hagfish analysis shows only 12, in place of the 16-19 recorded for other vertebrates. The nearest approach to the hagfish condition is found in some Insects, but even here the number of lysine residues does not fall below 14 or 15. Other marked divergencies are the high number of alanine residues (10 in the hagfish compared to 5-7 in other vertebrates), a higher proportion of serine and lower numbers of threonine residues.

Cyclostome Skin Collagens

Collagen, the main component of the connective tissues is essentially a trimer, composed of three polypeptide chains forming left-handed helices and each containing about 1000 amino acid residues. These chains form in turn a three-stranded super helix. An important feature in maintaining the structure of the molecule is its content of proline and hydroxyproline (imino acids) and throughout the vertebrate series the trend towards increased stability has been correlated with an increase in the proportion of imino acids and a corresponding decrease in the proportion of serine and threonine residues.

Judging by the temperature required for denaturation, the secondary structure of the hagfish collagen is the least stable of any of the vertebrates and, in this respect, is in marked contrast to the much greater structural stability of lamprey collagen. These differences are reflected in the higher content of imino acids in the lamprey (173 proline + hydroxyprolines 1000 residues^{-1}) compared with only 154 in the hagfish. In this respect, the lamprey more closely resembles the higher vertebrates. This is also true of the lower number of combined serine and threonine residues. In the lamprey, these total 81 residues 1000 residues^{-1} compared with 98 in the hagfish. Invertebrate collagens are also characterized by higher proportions of serine and threonine varying in different species from about 55-135 1000 residues^{-1}.

Differences in the amino acid constitution of hagfish and lamprey collagens are emphasized in the matrix prepared by Pikkarainen (1968) which compares them with species representing the major vertebrate classes. In this case, the figures represent the simple sum of the differences in each residue, expressed as a proportion of 1000 residues. Except in comparisons with the ray or flounder, the differences between hagfish collagen and those of other vertebrate groups are very much larger than the corresponding comparisons with the lamprey. Moreover, the extent of the divergence between hagfish and lamprey collagens is similar to the difference between lamprey and dogfish and greater than the divergence between lamprey and frog.

Cyclostome Thyroglobulins

The biosynthesis of vertebrate thyroid hormones takes place within the protein thyroglobulin synthesized by the epithelial cells of the thyroid follicles. Characteristically, the thyroglobulin molecule has a sedimentation value of about 19 S and a molecular weight of about 680000, although other components have been recognized in higher vertebrates, including one sedimenting at 12 S.

Amino acid analyses of the thyroglobulins of the lamprey, dogfish and mammals indicate that the evolution of this protein has been extremely conservative. This is shown by the very small values for the difference indices between human, lamprey and dogfish thyroglobulins as well as by the cross-reactions that occur between lamprey thyroglobulin and antisera raised against mammalian thyroglobulin. On the other hand, there are significant differences in the carbohydrate composition of the iodoproteins and the total carbohydrate composition of the lamprey protein is only about half that of the dogfish molecule and approximately one third of the mammalian values. In addition, cyclostome thyroglobulins are remarkable for their very low concentrations of T_4.

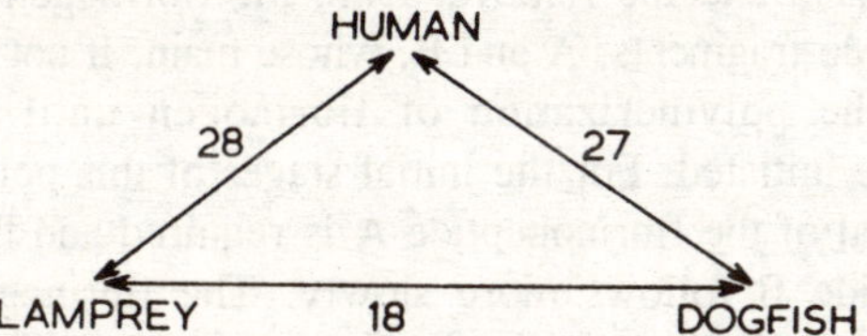

Fig. 17.1. Difference index S Δ Q for human, lamprey and dogfish thyroglobulins.

The ultracentrifugation pattern for the thyroglobulin of the hagfish, *E. stouti*, appears to be quite unusual in showing only a single peak at 3-8 S with no trace of a 19 S component. The molecular weight is estimated to be only 155000. In several species of lampreys a component sedimenting at 18-19S has been identified. The presence of a 12 S component was recognized in the iodoproteins of *L. fluviatilis* and a main 12 S fraction occurred in the purified iodoproteins of *L. tridentatus*. Investigations on the ammocoete of a nonparasitic lamprey, *L. reissneri*, suggested a parallel with conditions in the hagfish in so far as the main component sedimented at 3-5 S, with only a small proportion of 19 S iodoprotein whereas, after metamorphosis, the latter became the dominant form. This led to the suggestion that the presence of the lighter component in the larval endostyle and in the hagfish might be linked to the lack of any extracellular storage of iodoprotein and to the absence in the hagfish thyroid follicles of luminal colloid. This view assumes therefore that the 19 S fraction is the normal storage form of thyroglobulin and that the lighter components are intermediates associated with intracellular storage. The dominance of the lighter iodoproteins has not however been observed in the ammocoete of another non-parasitic species, *L. planeri*. Both in larvae and adults of this form the incorporation of ^{125}I- and (^{3}H)-labelled carbohydrates was followed over periods from 4-72 hours. Here the density gradient ultracentrifugation patterns consistently showed two main peaks at 18-19 S and 3-8 S with a shoulder representing the 12 S iodoprotein. With time, both these lighter components diminished, supporting the view that they are precursors of the 18-19 S thyroglobulin. The latter was synthesized by both ammocoete and adult, but in the endostyle its rate of elaboration proceeded more slowly, presumably in association with the purely intracellular storage of the thyroproteins.

Lamprey Fibrinogen and Fibrinopeptides

The process of blood clotting involves the thrombin-catalysed cleavage of the plasma glycoprotein, fibrinogen. This results in the formation of fibrin and the removal from the fibrinogen molecule of two fibrinopeptide fragments, A and B, whose main, if not sole function is to inhibit the polymerization of fibrinogen until the clotting mechanisms are initiated. For the initial stages of this polymerization, only the removal of the fibrinopeptide A is required and the release of the fibrinopeptide B follows more slowly. The fibrinogen molecule consists of three paired chains $\alpha_2\beta_2\gamma_2$ and in the process of clotting the fibrinopeptides A and B are released in sequence from the amino segments of the α- and β-chains; the cleavages taking place at arginyl-glycine bonds, leaving glycine at the amino terminus of both chains. During the formation of the clot, the fibrin is stabilized by the development of cross-linkages, involving the formation of γ-γ dimers and a slower cross linking of the α-chains.

Although the lamprey fibrinogen molecule seems to be significantly larger than that of other vertebrates so far investigated, it has the same $\alpha_2\beta_2\gamma_2$ sub-unit structure, the fibrinopeptides are cleaved at arginyl-glycine linkages and the fibrin molecule is stabilized by the formation of γ-γ dimers and multimers. However, the lamprey molecule displays a number of quite distinctive characteristics. For example, the α-chains are unusually large and may be almost twice the size of the γ-chains. On the other hand, the lamprey fibrinopeptide A, consisting of only six amino acid residues, is the smallest recorded in vertebrates and in the mammals the numbers range from a minimum of 13 in the guinea pig to a maximum of 19 in several species of artiodactyls. The β-chains are also comparatively large but, after the conversion to fibrin, they are greatly reduced by the splitting off of fibrinopeptide B. This is the longest fibrinopeptide chain on record, consisting of 36 residues with a covalently bound cluster of carbohydrates. Amongst the mammals, the smallest fibrinopeptide B chain is that of the rhesus monkey with only 9 residues and the largest which occur in the kangaroo and a number of species of artiodactyls have 21. The γ-chains, however, are of similar size in lampreys and mammals. There is still some uncertainty on the precise molecular weights of lamprey fibrinogen and its component chains, but Doolittle *et al.* (1976) have suggested that lamprey fibrinogen has a molecular weight of about 360000, with component α-, β-, and γ-chains of about 70000, 61000 and 47000.

In view of the evolutionary distance that separates the lampreys from the mammals, it is hardly surprising to find some species specificity in the operation of the thrombin-fibrinogen mechanisms from these two widely separated groups of animals. Thus, mammalian thombin is capable of removing only one of the two fibrinopeptides (B) from lamprey fibrinogen and lamprey thrombin releases only fibrinopeptide A from mammalian fibrinogen. In its ability to form a clot after the application of mammalian thrombin with the removal of only the fibrinopeptide B, the lamprey is unique amongst vertebrates, although in the normal course of events, lamprey thrombin will of course release both fibrinopeptides from lamprey fibrinogen.

Because of the absence of a specific physiological function for the fibrinopeptide fragments once they have been cleaved from the fibrinogen molecule, it is not surprising that these chains have been able to accept a high rate of mutation; in fact the rate of evolution of the fibrinopeptides is the most rapid so far observed and is put at about 90 PAM 100 million years^{-1}. Like the haemoglobin or immunoglobulin chains, it has often been assumed that the α-, β- and γ-chains of fibrinogen have been evolved from a common ancestral chain by a process of gene duplication. On this interpretation it might be expected that in a primitive vertebrate like the lamprey, there would be some evidence for this in a closer homology between these chains, when compared with the resemblances between these chains in the higher vertebrates. This has not turned out to be the case. Studies of the amino acid composition of lamprey fibrinogen have shown that the individual chains differ far more amongst themselves than do the mammalian chains. When the amino acid composition of the individual lamprey fibrinogen chains are compared with the corresponding human chains, it is clear that the differences are much more marked in the case of the α-chains and that of the lamprey is remarkable for its very high proportion of glycine, serine and threonine. These residues make up no less than 45% of the total composition, whereas in the human molecule they contribute only 30%. The composition of the lamprey β-chain is much less characteristic and to a somewhat lesser extent, this is true also of the γ-chain. In common with the immunological specificity of lower vertebrate fibrinogens, antibodies to mammalian fibrinogen do not cross-react with lamprey fibrinogen.

Table 17.4. Matrix for 'degrees of differentness' among α-, β- and γ-chains of lamprey and human fibrinogen.

	Lamprey α	*Lamprey β*	*Lamprey γ*	*Human α*	*Human β*
Lamprey β	47.2				
Lamprey γ	54.1	25.1			
Human α	35.3	25.5	34.5		
Human β	56.8	16.3	25.3	32.1	
Human γ	53.6	21.9	21.3	33.8	18.8

Studies on the amino acid composition and sequences of the lamprey fibrinopeptides have shown that although the sequences of the α- and β-chains on the fibrin side of the thrombin cleavage sites are almost the same as in the mammals, the fibrinopeptides of the lamprey are practically unrecognisable from those of the higher vertebrates. Comparing the two lamprey fibrinopeptides it may be seen that in spite of the vast disparity in their size, they nevertheless show some correspondence in the five N-terminal residues, although this homology is not a feature of any other vertebrate fibrinopeptides.

In view of the similarity in the terminal fibrin sequence of the α-chains of mammals and lampreys, it is surprising that mammalian thrombin should be unable to liberate the lamprey fibrinopeptide A. This may, as Cottrell and Doolittle (1976) suggest be due to the substitution in the lamprey fibrinopeptide of leucine for the valine in the penultimate position and this possibility is enhanced by the fact that out of 99 fibrinopeptides for which data is available, none has a leucine at this position. In fact, out of 52 fibrinopeptide A chains, valine occupies this site in all but three.

In keeping with the well-established interspecific variability of the fibrinopeptides, the lamprey chains show little resemblance to those of other vertebrates except that they exhibit a negative charge. However, fibrinopeptide B does share a tripeptide sequence Asp-Tyr-Asp with the B chain of mammalian species and, as in the latter, the tyrosine is sulphated.

The picture that emerges from a study of these lamprey peptides and the parent fibrinogen chains is one that bears convincing evidence of the common ancestry of the molecule in the shape of its basic structure and physiological function, but which nevertheless exhibits some features which as far as we know at the present time, are unique to the cyclostome. These include the ability of lamprey fibrinogen to form a clot after the removal of only the fibrinopeptide B, as well as the unusual sizes of the α- and β-chains. In the absence of a wider range of comparative data from other lower vertebrates, and especially from the myxinoids, it is quite impossible to say whether the special characteristics of lamprey fibrinogen or fibrinopeptides are to be regarded as primitive, or whether, as has been postulated by Doolittle *et al.* (1976), they may be attributed to large-scale deletions or duplications.

Immunoglobulins and Immune Responses

In the past there was a tendency to regard specific adaptive immunity as a vertebrate innovation, whose first manifestations were to be sought, albeit in rudimentary form, in the cyclostomes. The criteria normally used in identifying these defensive mechanisms include the ability to reject foreign tissue from an unrelated member of the same species (homograft rejection), the development of delayed hypersensitivity (allergy), immunological memory as exhibited by second set homograft rejection or secondary responses to antigens (anamnestic responses), the proliferation of immunocompetent cells after antigenic stimulation and the production of circulating antibodies (immunoglobulins). Although there is now considerable evidence for cellular immunity amongst various invertebrate groups, the cyclostomes are still important for comparative immunologists since they are the first organisms in the phylogenetic series to show concurrent cellular and humoral immune responses and the first to exhibit circulating antibodies of defined specificity and of the immunoglobulin type.

In general, the range and intensity of the immune responses are correlated with the degree of development of the lymphoid tissues. In cyclostomes, these tissues are relatively undifferentiated and respond weakly and only slowly to antigenic stimulation. Although lymphocyte-like cells occur in cyclostomes, plasma cells — the source of antibodies in higher vertebrates — do not appear to be present. Lymphohaemopoetic tissues comparable with the spleen and bone marrow are present in lampreys and in the pronephros and submucosa of the hagfish intestine, but it is doubtful whether either group possesses the equivalent of the vertebrate thymus.

Cell-mediated and Humoral Responses

Lampreys

Although restricted in their scope, lampreys show the usual repertoire of vertebrate adaptive immunity; producing antibodies of the immunoglobulin type and showing cell-mediated immunity and skin graft rejection. The cells responsible for antibody production have not yet been identified, but proliferation in the lympho-haemopoetic tissues occurs after antigenic stimulation by killed bacterial cells. Agglutinating antibodies have been produced to certain bacterial cells and to the H antigen of human red cells but, on the other hand, lampreys have been unresponsive to a wide range of soluble or particulate antigens. The response to killed *Brucella abortus* demonstrates the lamprey's capacity for immunological memory; antibody being produced after a sub-threshold priming dose, followed by a second injection of the same concentration. As in the anamnestic responses of higher vertebrates, repeated administration of antigen leads to higher titres of antibody. Delayed allergic reactions are also displayed after injections of old tuberculin, following previous sensitisation with Freunds adjuvant and *Mycobacterium tuberculosis*. Cellular responses in the form of skin allograft rejection began at three weeks and took about 6 weeks to complete. Second set rejection occurred more rapidly than the first accompanied by inflammatory infiltration and cellular destruction.

Hagfishes

Earlier work had suggested that hagfishes were immunologically unresponsive, but with the application of improved handling and using temperatures above those to which the animals would normally be exposed in their normal environment, it has now been possible to demonstrate both cellmediated and humoral immune mechanisms. Skin allografts are rejected accompanied by the usual features of cellmediated immunity — lymphocyte infiltration, haemorrhagic inflammation and the destruction of pigment cells. For first set allografts the median survival time at 18-19°C was 72 days and 28 days for second set grafts. Chronic allograft rejection, although more extreme in the cyclostomes, is nevertheless characteristic of elasmobranchs, chondrosteans, apodans and urodelans and is in sharp contrast to the acute and rapid rejection that generally occurs in teleosts and anuran amphibians. The more primitive and chronic type of rejection has been related to a poorly developed system of histocompatability genes. These determine the degree of compatability between host and donor tissues and the development of general immunocompetence in the

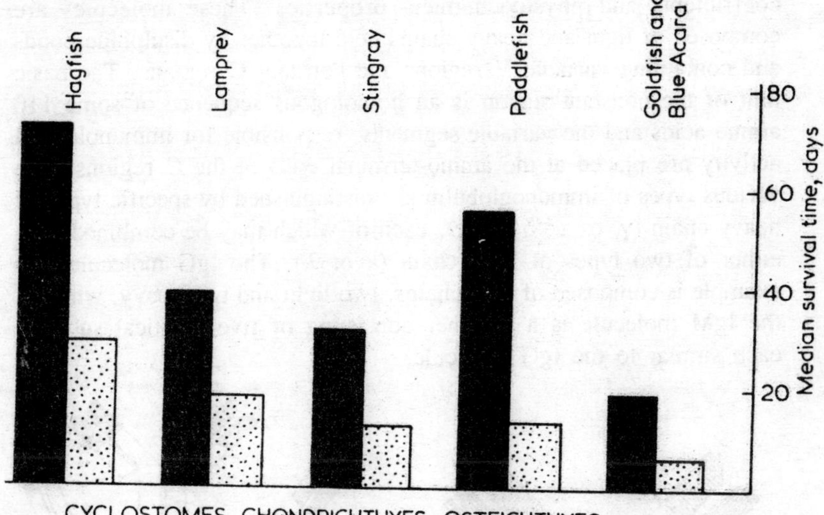

Fig. 17.2. Comparisons of skin allograft survival times in various species of fish and cyclostomes.

vertebrates may have been dependent on the increasing differentiation of tissue antigens.

Normal hagfish serum and mucus contains heat labile agglutinins to mammalian erythrocytes, in addition to specific induced agglutinins that are produced by repeated injections of sheep red cells, and which result in increased titires. Unlike the naturally occurring agglutinins, these induced antibodies are heat stable and highly specific. Hagfish antibodies have also been induced to keyhole limpet haemocyanin. Bactericidal responses of the hagfish in response to injections of *Salmonella typhi* and a gram-negative bacterium from the gut of the spiny lobster have been studied by Acton *et al.*, (1969, 1971).

Cyclostome Immunoglobulins

Vertebrate antibodies are associated with a diverse group of serum proteins the immunoglobulins — normally composed of multiple polypeptide chains and representing a wide range of molecular weights and physico-chemical properties. Since the terminology of the these immunoglobulins was derived from conditions in the mammals, there are some difficulties in applying it to the antibodies of lower vertebrates. The various classes of mammalian immunoglobulins are distinguished as IgA, IgD, IgE, IgG and IgM representing a range of molecular weights from 150 000-950 000, with characteristic sedimentation coefficients and physicochemical properties. These molecules are composed of light and heavy chains held together by disulphide bonds and containing variable V regions and constant C regions. The basic unit of the constant region is an homologous sequence of some 110 amino acids and the variable segments, responsible for immunological activity are placed at the amino-terminal ends of the C regions. The various types of immunoglobulin are distinguished by specific types of heavy chain (γ, α, μ, δ and ε), each of which may be combined with either of two types of light chain (k or λ). The IgG molecule, for example is composed of four chains, two light and two heavy, whereas the IgM molecule is a polymer consisting of five identical subunits each similar to the IgG molecule.

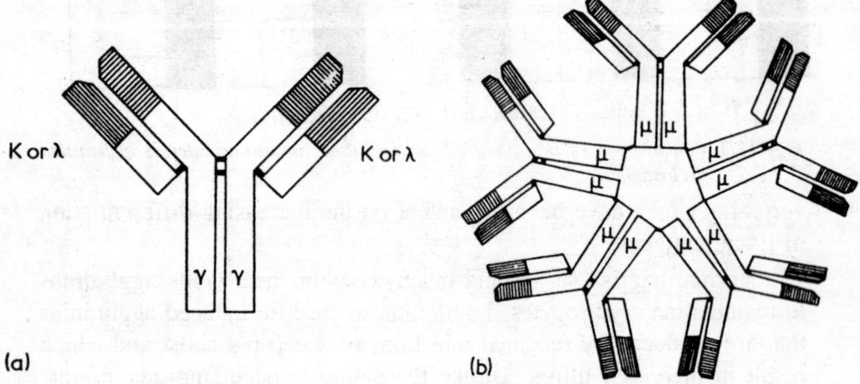

Fig. 17.3. Molecular structure of IgG (a) and IgM (b) immunoglobulins.

In the extent of its repertoire of immune response, the hagfish may be somewhat restricted and this has been linked to its intestinal peritrophic membranes, which may serve as a protection against pathogens. The inducible bactericidal antibodies to *Salmonella typhosa* 'H' antigen and to a bacterium isolated from the spiny lobster were heat-labile substances, showing no increase in titre with a second immunization, although the response was accelerated. Associated with a macroglobulin fraction these antibodies were of only limited specificity, reacting to other gram-negative, but not to gram-positive organisms. The active serum fractions showed a major 20 S component after ultracentrifugation, with minor components sedimenting at 9 S and 13 S. After reduction with thiols there was no evidence of separation into light and heavy chains. The inducible antibodies to sheep erythrocytes or to the haemocyanin of the keyhole limpet were heat-stable and there was a marked increase in titres with repeated immunization. The anti-keyhole limpet haemocyanin antibody is described as IgM-like, occurring in the macroglobulin fraction of the serum and having a sedimentation coefficient of about 28 S.

Lampreys have produced antibodies to a number of antigens, including human 'O' cells, killed *Brucella abortus* cells, *Salmonella* antigens and the F^2 bacteriophage. In the estimation of molecular size by ultracentrifugation, mammalian antibodies show two main peaks at

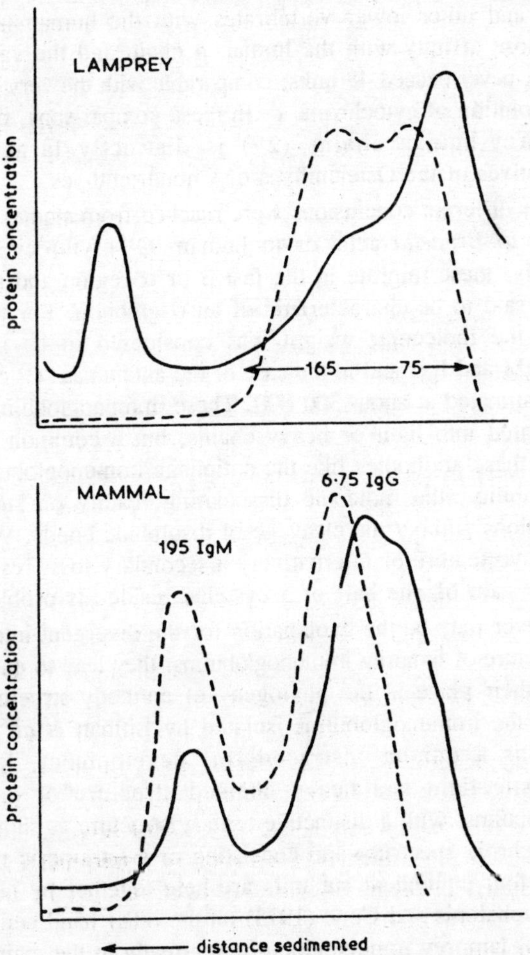

Fig. 17.4. Analysis of lamprey and mammalian antisera by ultracentrifugation on linear sucrose gradients.

19 S and 6.7 S, corresponding to IgM and IgG immunoglobulins. Lamprey F[2] antiphage antibody activity however, differs in being spread over a wide range from 7-16 S, but purification yielded two antigenically identical antibodies with sedimentation coefficients of 6.6 Sand 14 S; the latter regarded as an aggregate of the lighter subunits. From gel electrophoresis it was concluded that the 6.6 S component consisted in its turn of light and heavy chains similar to those of higher vertebrates.

Comparisons of the amino acid composition of the heavy chains of lampreys and other lower vertebrates with the human heavy chains show a close affinity with the human μ chain and the values for the SΔQ index never exceed 48 units; comparable with the very conservative rate of evolution of cytochrome *c*. In these comparisons, the value for the lamprey-human chains (27) is distinctly larger than for representatives of the Osteichthyes or Chondrichthyes.

Rather different conclusions were reached from studies on lamprey antibodies to *Brucella* cells or to human 'O' erythrocytes. Electrophoretically, these migrate in the fast β or α region and their charge density is said to be characteristic of an α-globulin. For the *Brucella* antibody, the molecular weight was considered to be intermediate between IgM and IgG and in the case of the antihuman 'O' cell antibody this was estimated at about 300 000. These immunoglobulins could not be dissociated into light or heavy chains, but a common and unique feature of these antibodies like the antiphage immunoglobulin, is their extreme lability; the molecule dissociating readily on storage or in weak solutions without the cleavage of disulphide bonds. Whether this is a primitive feature or alternatively a secondary loss, resulting from the loss or gain of one half of a cysteine residue is problematical.

Whatever may be the explanation for the divergent interpretations of the structure of lamprey immunoglobulins, they lead to quite different views on their place in the phylogeny of antibody structure. On the one hand, the immunoglobulins isolated by Litman *et al.* (1970) are regarded as a unique 'ostracoderm' development, lacking the characteristic light and heavy chained structure of gnathostome immunoglobulins, with a distinctive tertiary structure as shown by their circular dichroic spectrum and consisting of a tetrameric molecule in which the four equivalent subunits are held together by non-covalent bonds. Marchalonis and Cone (1973) on the other hand believe that an IgM type of lamprey immunoglobulin is already in the main stream of vertebrate immunoglobulin evolution and that genetic mechanisms coding for light and heavy chains had already been evolved by the agnathan stage of vertebrate evolution. With the realization that immunoglobulins are universal throughout the vertebrates, it was natural that a search for clues to the origin of these molecules should be extended to the invertebrates, where naturally occuring bactericidins and agglutinins are widespread. Since similar bactericidins and haemagglutinins are known in the cyclostomes, could not these molecules be regarded as stepping stones in the plylogeny of the immunoglobulins?

In unimmunized lampreys, natural haemagglutinins are large molecules with a sedimentation coefficient of 48 S and their subunits can only be separated by reagents that cleave disulphide bonds. Like the agglutinins of invertebrates, they are potentiated by divalent ions and their amino acid composition is said to resemble that of the horseshoe crab (*Limulus*) and to an ever greater extent that of the starfish, *Asterias*. This correspondence and the especially low S Δ Q values for starfish agglutinin when compared to the heavy chains of the lamprey or the eel, led Carton (1974) to suggest that the *Asterias* protein might well represent an immunoglobulin precursor, particularly in view of the supposed chordate affinities of the echinoderms. From what is known of the structure of the *Limulus* agglutinin, Marchalonis (1977), on the other hand, has argued that the invertebrate proteins are unlikely to have been directly ancestral to the immunoglobulins. Composed of 18 identical sub-units with molecular weights of 22,500, the molecule is described as a cylindrical toroid, bearing no resemblance to an immunoglobulin. However, as Marchalonis points out the question of molecular homology can only be resolved when complete sequence data are available for these invertebrate proteins.

The poorly developed lymphoid tissues of cyclostomes and what has been regarded as their relatively primitive immunological systems, have led to a belief that this would be reflected in a low incidence or absence of neoplasms. In fact, in an examination of some 25,000 *Myxine* from the West coast of Sweden, about 10% showed tumours in various organs, although three years later the incidence was less than 1%. For liver carcinoma, there was a clear correlation with length (and therefore age) and a similar relationship has been noted in the lamprey *L. fluviatilis*, where islet cell tumours tended to be more frequent towards the close of their life cycle.

Other Biochemical Characteristics of the Cyclostomes

Acid Mucopolysaccharides of Skin, Notochord and Cartilage

The mucopolysaccharides of the connective tissues consist of carbohydrate chains in the form of linear polymers, made up of repetitive units of disaccharide, linked covalently to protein or polypeptide. In the skin, the usual vertebrate components include hyaluronic acid, dermatan sulphate and various dermatan polysulphates. The skin of the lamprey conforms to this pattern, in which the main components are hyaluronate and dermatan sulphate. The skin of the hagfish, on the other hand, is unique in possessing a dermatan sulphate containing trisulphated disaccharide residues. Lamprey cartilage contains a normal vertebrate 6-sulphate (chondroitin sulphate C), but the hagfish possesses a chondroitin 4-sulphate (chondroitin sulphate E), otherwise found only in the cartilage of invertebrates.

The mucopolysaccharide constitution of the cyclostome notochord is equally interesting. In the lamprey, this contains a chondroitin sulphate A, which also occurs in gnathostomes, but in the hagfish a novel component, chondroitin sulphate H has been identified. In all these respects therefore, the myxinoids are separated from the rest of the vertebrates, whereas lamprey tissues contain a similar range of mucopolysaccharides to those of gnathostomes. Although the myxinoid pattern has been described as primitive, it should be noted that the mucopolysaccharide content of the skin of *Amphioxus* is said to have a vertebrate rather than an invertebrate character and the constitution of its notochord is quite different from that of the cyclostomes.

Bile Salts

In his review of vertebrate bile salts, Haslewood (1968) remarked that their structure does not support the view that the cyclostomes are a 'monophyletic' group, or that the divergence of lampreys and hagfishes is of comparatively recent date. The unique bile salt of the hagfishes, *Myxine* and *Eptatretus*, myxinol disulphate (3α, 7α, 16α, 26(or 27) tetrahydroxy-5 cholestane, 3, 26 (or 27) disulphate) was described as the most primitive form known, in so far as it still contains the complete carbon skeleton of cholesterol and contains 27 carbon atoms with only four OH groups. All other bile alcohol C 27 sulphates have at least five OH groups, of which three are in the 3, 7 and 12 positions, and with only one sulphate ester.

Myxinol is said to be a rather ineffective emulsifying agent and would presumably be required in rather large volumes during the digestion of lipids. This has been advanced as a possible explanation for the relatively high levels of plasma cholesterol in the hagfish, acting as a precursor in the synthesis of the bile alcohols.

The main component in the lamprey, *P. marinus* is an entirely different form, showing changes at five positions compared to myxinol.

Petromyzonol, as it has been named (3α, 7α, 13α, 24-tetrahydroxy-5α cholane) has 24 carbon atoms, only one sulphate ester group and three of its OH groups are in the 3α, 7α, and 12α positions.

Composition and Crystalline Structure of Cyclostome Otoliths

The calcareous bodies (otoliths) of the vertebrate labyrinth occur either as single large statoliths or as masses of small particles varying from 1-50 μm—statoconia — which may be united in an organic gel. In the lamprey labyrinth (*L. fluviatilis*) the otoliths consist mainly of statoconia from 2-25 μm in diameter, but in some instances two or three may be fused to form larger bodies up to 250 μm. These are composed mainly of apatite — calcium phosphate — with small amounts of carbonate. In the hagfish, *M. glutinosa*, only statoconia were present, also consisting of apatite and with apparently even less carbonate. In these respects the cyclostomes are said to differ from all other vertebrate groups, where the otoliths (statoconia and statoliths) are entirely composed of calcium carbonate in the form of either aragonite, calcite or vatarite.

Nitrogen Metabolism

In many of the aquatic vertebrates, including the actinopterygian fishes, nitrogen is mainly excreted as ammonia and the small quantities of urea that they produce probably originate in the hydrolysis of arginine, or by the degradation of uric acid via the purine pathway. Such forms lack at least the complete series of enzymes that are involved in the ornithine-urea cycle, although arginase is usually retained.

Urea is present in only small amounts in the blood of lampreys and hagfishes and most of their nitrogen is excreted as ammonia. Neither in hagfish nor lamprey have the enzymes of the purine pathway been detected. The hagfish liver is apparently unable to synthesize urea, since of the five enzymes involved in the ornithine-urea cycle only arginase was identified by Read (1975) and the absence of carbamoyl synthetase and ornithine transcarbamylase has been cited by Fänge *et al.* (1973). This cycle is also said to be inoperative in the lamprey and Read (1968) was able to find evidence for only arginase and carbamoyl phosphatase; the latter perhaps involved in pyrimidine synthesis.

The fact that these enzyme deficiencies are to be found in both groups of cyclostomes must throw some doubt on the widely held view that their absence in other aquatic vertebrates has been a secondary loss, attributable either to gene repression or deletion and it now seems quite possible that a complete series of enzymes of the purine pathway or the ornithine-urea cycle may not have been developed in the common ancestors of agnathans and gnathostomes.

Lens Proteins

Soluble lens proteins are classified according to their electrophoretic mobilities and molecular weights, into α, β and γ crystallines. The highest concentrations of lens proteins transferred during evolution from one vertebrate class to another (organ-specific) are to be found in the more negatively charged β crystallines, while the proteins that tend to be restricted to one particular class (species specific), are most abundant in the less negatively charged γ crystallines. In the α crystallines, organ-specific proteins are more numerous than the species-specific forms.

In mammals the α component accounts for about 30% of the soluble lens protein and has a molecular weight of about 500 000 consisting of two main polypeptide chains. Since α- and β-chains are believed to be represented in lamprey crystallines, these are thought to have arisen as a result of gene duplication in the common ancestor of agnathans and gnathostomes. From the degree of divergence between the amino acid sequences of the α- and β-chain of the cow, it has been estimated that this duplication may have occurred at about 400 million years.

The extremely low rate of evolutionary change in the lens proteins of vertebrates (estimated as about 2.5 PAM 100 million years^{-1}) is reflected in the fact that vertebrates from different classes share a number of lens antigens in common. This has been demonstrated in the elegant studies of Manski (1967a, b), in which antisera raised in the rabbit against several aquatic vertebrates (including the lamprey) were analysed by immuno-electrophoretic techniques against lens extracts of species representing the entire vertebrate series. As might be expected, these studies showed a complete absence of common lens antigens in cephalopod and lamprey lens proteins, but the lamprey pineal was found to possess some antigens in common with lamprey paired eyes, although these were not shared by the paired eyes of other vertebrates. Six agnathan antigens were identified in the lamprey, of which 5 components were recognized in the Dipnoi and six in members of the Actinopterygii. Elasmobranchs only retained four of the lamprey antigens and the same was found to be true of mammals.

Lactate Dehydrogenase Isozymes

The enzyme lactate hydrogenase concerned in the interconversion of lactate and pyruvate in the glycolytic pathway, exists in a number of forms or isozymes differing in their molecular structure. These have been intensively studied as a model for gene evolution or as a tool for the genetic analysis of animal populations. In the vertebrates these isozymes are tetramers with a molecular weight of about 140 000; the sub-units having masses of about 35 000-37 000. These sub-units, which are encoded by two structural genes A and B (believed to have arisen by gene duplication), may associate, either as homotetramers A_4 or B_4 or as heterotetramers A_3B_1, A_2B_2 and A_1B_3, thus producing the five isozymes characteristic of the birds and mammals. Within vertebrates there are characteristic differences in the distribution of the A and B sub-units; the A chain predominating in the white muscle, the B chain in heart muscle or brain. This suggests that the different isozymes vary in their metabolic roles. Where, as in the lower vertebrates, individual isozymes may be missing, this is usually explained by the failure to polymerize as heterotetrameric forms.

In an examination of the tissues of large numbers of the Pacific hagfish, *E. stouti* no fewer than 9 LDH phenotypes were identified electrophoretically by Ohno *et al.* (1967), suggesting the presence of two alleles at each of the two A and B gene loci. From the apparent absence of the heterotetrameric isozymes it was concluded that the hagfish was unique amongst vertebrates in having a monomeric LDH molecule, like that of its haemoglobin. However, this conclusion was not substantiated by Arnheim *et al.* (1967) who, in the same species, estimated the molecular weight as about 140 000, similar to that of other vertebrate LDH molecules.

In the tissues of the Atlantic hagfish, *M. glutinosa*, two LDH isozymes have been distinguished electrophoretically; a slower moving cathodal (LDH-1) type in the parietal and visceral lingual muscles and a faster moving anodal (LDH-5) type in the continuously active red fibres of the craniovelar and heart muscle.

In view of the evidence that the hagfish conforms to the usual vertebrate pattern, with two gene loci coding for A and B chains it is surprising to find that only one LDH isozyme has been identified in the lamprey although, like that of other vertebrates, this is a tetramer, showing homology with the A sub-units of other fishes when tested by immunological methods. In this respect, therefore, we are driven to the conclusion either that the lamprey is more primitive than the hagfish and had not undergone the initial gene duplication that produced the A and B genes of other vertebrates including the myxinoids or, alternatively, that the B gene has been lost in the course of petromyzonid phylogeny.

Cyclostome Chromosomes and the Size of the Genome

The chromosomes of the lampreys of the Northern Hemisphere (family Petromyzonidae) are characterized by their small size and exceptionally high diploid numbers. For seven species representing four genera (*Lampetra*, *Petromyzon*, *Okkelbergia* and *Ichthyomyzon*) these numbers have varied only within a narrow range from 164-168 and are greater than in any other vertebrate species. Preliminary studies on the karyotype of the Southern Hemisphere genus, *Geotria*, suggest that this will turn out to have diploid numbers even higher than those of the Petromyzonidae, but this is in sharp contrast to the other Southern genus, *Mordacia*, where there are only 76 diploid chromosomes. A further distinction between the Mordaciidae and the Petromyzonidae is that, whereas in *Mordacia* species the centromere is usually located at or near the centre of the chromosome (metacentric), in the Northern lampreys it is usually positioned terminally or sub-terminally (acrocentric).

Although lamprey chromosomes are generally very small (in most cases about 0.5-1.5 μm in length) their total area is greater than in mammals and exceeds that of other amniotes. This is all the more surprising in view of the fact that the nuclear DNA content of lampreys is generally only 40% of human values. A further interesting feature of the genome size is that, in spite of the similarity in chromosome numbers within the Petromyzonidae, one species, *P. marinus*, has a much higher DNA value (60%) than the other Northern species. Furthermore, the members of the genus *Mordacia*, whose chromosome numbers and chromosomal areas are much smaller than those of the Petromyzonidae, nevertheless show very similar DNA values.

Counts on several hagfish species suggest that this group of cyclostomes is probably characterized by chromosome numbers of 46-52 combined with very high DNA values representing 78% of human values. Most of these chromosomes are very large in comparison with those of lampreys and the chromosomal area also approaches mammalian figures. A majority of the chromosomes appear to be acrocentric. For a local Swedish population of *Myxine glutinosa* there has been a report of a remarkable range in chromosome numbers, varying from 17-116 and this has been related by Fernholm (1972) to their characteristic morphological and histological variability.

How are the differences between lamprey and hagfish karyotypes to be interpreted and what light do these differences shed on their relationships? Although it has been suggested that the high chromosome numbers of the Petromyzomidae may have arisen through the division of metacentric chromosomes, a more widely favoured alternative is that they have been produced by polyploidy. However, if this view were to be accepted, how are we to explain the divergent karyotypes of the Mordaciidae and the Petromyzonidae? One possibility might be that these low numbers in *Mordacia* have been produced by centric fusions, in which the centromeres of acrocentric chromosomes fuse to form metacentrics. This mechanism would be consistent with the presence in *Mordacia* of a majority of metacentric chromosomes. A common ancestry of the Northern lampreys and the Mordaciidae at a remote geological period and the subsequent isolation of the latter forms within the Southern Hemisphere would be in harmony with the morphological evidence, which suggests that this group is the most specialized of all lampreys. However, by itself, this view of the origins of the *Mordacia* kayotype does not entirely explain the low numbers, which are less than half those of the Northern forms. Either a further loss of chromosomes (or a subsequent increase in those of the Petromyzonidae) must be a possibility.

Attempts to trace evolutionary trends in the karyotype cannot be divorced from a consideration of changes in genome size, as measured by nuclear DNA content. Throughout the vertebrate series there has been a general tendency towards increased amounts of DNA and the values in the mammals, for example are far greater than those of the protochordates and some fish species. This was interpreted by Ohno (1970, 1973) as indication that tetraploidisation may have occurred twice in the mammalian lineage; the first during the evolution of the ancestral vertebrate stock and the second at a later and possibly reptilian stage. Of the two principal means of increasing the gene loci, polyploidy or tandem gene duplication, Ohno believes the first to have been a more important factor in the gene duplications that have occurred throughout the history of the vertebrates and of which many examples have been given in previous chapters. Because the functional gene loci form only a small fraction of the total genome, tandem gene duplication would be expected to result most frequently in the accumulation of non-functional or trivial DNA, rather than leading to a duplication of new structural genes. Tetraploidisation on the other hand, will by its very nature, duplicate every functional locus.

For Ohno, a chromosome number of 48 had a special significance and was regarded as the chromosomal complement characteristic of the ancestral vertebrate, combined with a DNA value of about 20%

of the mammalian genome. For example, about half of the mammalian species have chromosome numbers from 40-56, with a peak frequency at 48, and the same numbers occurs widely in many teleost orders. In his detailed interpretation of the evolution of the vertebrate karyotype, Ohno was influenced by an earlier report of a diploid chromosome number of 94-96 in the lamprey, *L. reissneri* (Nogusa, 1960), which suggested that tetraploidisation had occurred in the line from the ancestral vertebrate to the agnathans. In this scheme, the hagfishes would retain the ancestral complement of 48, while vastly increasing their DNA content, presumably by tandem gene duplication.

With the more complete information that is now available on lamprey karyotypes, it is possible to modify Ohno's scheme, retaining an ancestral form with 48 chromosomes, but with a DNA value of only 10%. Here, tandem gene duplication is invoked to give a basal teleost DNA content of 20% and polyploidy is held to be responsible for the higher chromosome numbers of the genera, *Lampetra*, *Geotria* and *Ichthyomyzon*. It would also explain satisfactorily their 40% DNA values. In the scheme proposed by Ohno, the history of the vertebrate genome was traced back through the cephalochordates to the urochordates, where the chromosome number of 24 and the DNA value of 5-6% could plausibly be related by tetraploidisation to the ancestral vertebrate genome. However, it now seems unlikely that *Amphioxus* can be regarded as on the main line of vertebrate evolution and the relationships of the protochordates as a whole to vertebrate ancestry remain obscure.

INDEX